The Continuing Story of The International Space Station

Springer
London
Berlin
Heidelberg
New York
Barcelona
Hong Kong
Milan
Paris
Santa Clara
Singapore
Tokyo

Peter Bond

The Continuing Story of The International Space Station

Springer

Published in association with
Praxis Publishing
Chichester, UK

Peter Bond
Press Officer for the Royal Astronomical Society
Consultant for the European Space Agency
Cranleigh
Surrey
UK

SPRINGER–PRAXIS BOOKS IN ASTRONOMY AND SPACE SCIENCES
SUBJECT *ADVISORY EDITOR*: John Mason B.Sc., Ph.D.

ISBN 1-85233-567-X Springer-Verlag Berlin Heidelberg New York

British Library Cataloguing-in-Publication Data
Bond, Peter, 1948–
The continuing story of the International Space Station. –
(Springer-Praxis books in astronomy and space sciences)
1. Space stations – International cooperation 2. Astronautics
– International cooperation
I. Title
629.4′42

ISBN 1-85233-567-X

Library of Congress Cataloging-in-Publication Data
Bond, Peter, 1948–
The continuing story of the International Space Station/Peter Bond.
p. cm.
Includes bibliographical references and index.
ISBN 1-85233-567-X (alk. paper)
1. International Space Station. 2. Space stations. I. Title.

TL797.B66 2002
629.44′dc21 2002021064

Printed by MPG Books Ltd, Bodmin, Cornwall, UK

Project Copy Editor: Alex Whyte
Cover design: Jim Wilkie
Typesetting: BookEns Ltd, Royston, Herts., UK

Printed on acid-free paper supplied by Precision Publishing Papers Ltd, UK

To

Josephine Mary Calvert

Table of contents

List of Illustrations. i
List of Tables. ix
Acknowledgements. xv
Introduction xvii

1 BUILDING A GIANT. 1

2 FROM DREAM TO REALITY 21

3 HANDSHAKE IN SPACE 64

4 METAMORPHOSIS 100

5 COMING AND GOING 129

6 CONSTRUCTION SITE IN SPACE 164

7 LIFE ON THE SPACE STATION 227

8 VALUE FOR MONEY? 273

9 FUTURE UNCERTAIN 318

Appendix 1 ISS Assembly November 1998 to January 2002 339
Appendix 2 Future Assembly Schedule 340
Appendix 3 ISS Crew Biographies 343
Selected Web Sites 376
Selected Reading List 380
Index 382

List of illustrations

A selection of coloured illustrations can be seen between pages 206 and 207.

The completed International Space Station 4
The Canadian-built Mobile Servicing System 5
One of the US solar panels during testing 7
The European-built cupola 10
The Soyuz TM-33 spacecraft docked to the Pirs module 12
The Transhab module proposed for use on the ISS 15
The AERCam Sprint satellite during a Shuttle test flight. 17
Edward Everett Hale's Brick Moon 22
A US Army V2 rocket at White Sands, New Mexico 26
Wernher von Braun's wheel-shaped space station 27
Pickering, Van Allen and von Braun with a model of Explorer 1 29
Soviet Chief Designer Sergei Korolev with Yuri Gagarin 30
Salyut 1, the world's first space station 33
The doomed crew of Soyuz 11 34
The USAF Manned Orbiting Laboratory 37
Skylab in orbit 40
Evolution of USSR space stations 42
Vladimir Remek, the first foreign cosmonaut, on board Salyut 6 46
The Salyut 7–Cosmos 1686 complex 51
Artist's impression of the Mir core module 53
The Mir core module, Kvant 1 and Soyuz TM-3 54
Progress M-18 undocking from Mir 57
Sergei Krikalev and Klaus-Dietrich Flade inside Mir 59

Krikalev's homecoming after 312 days in space 60
Valeri Poliakov admires the view from a window on Mir 62
Artist's impression of Apollo–Soyuz 65
Artist's impression of the Shuttle docked to Mir 66
Norman Thagard sleeping in one of Mir's crew cabins 69
Shuttle Atlantis undocking from Mir's Kristall module 70
Thomas Reiter during the first spacewalk by a European 72
IMAX image of Shuttle Atlantis undocking from Mir 73
Shannon Lucid reading a book in the Spektr module 75
The completed Mir station 76
STS-79 astronaut Terence Wilcutt in the Shuttle docking module 77
Jerry Linenger dons his Sokol pressure suit 79
The six-man crew shortly before a near-disastrous fire on Mir 80
Michael Foale exercising on the Mir treadmill 83
A close-up of the damaged solar panel on Spektr 84
Valeri Korzun inside the transfer node on Mir 86
Vladimir Titov's spacewalk on 1 October 1997 87
Solovyov, Thomas and Eyharts enjoy a meal on Mir 89
Valeri Ryumin greeting Talgat Musabayev 91
Mir photographed from Shuttle Endeavour in January 1998 93
The penultimate crew to visit Mir 96
Mir burning up over the Pacific on 23 March 2001 99
Spacelab in the Shuttle payload bay 102
A 1981 space station design 106
1984 concept for a triangular space station 109
The "power tower" design of July 1984 110
The "dual keel" space station design 111
Spacewalk to test space station construction techniques in November 1985 112
The 1987 modified design of the Freedom space station 113
The scaled down version of Freedom unveiled in 1991 115
The Mir 2 space station 119
Artist's impression of the Option C space station design 120
The International Space Station at completion of Phase Two 124
ESA's Hermes spaceplane docked to Freedom 125
Aerial View of Kennedy Space Center in Florida 131
Space Shuttle Discovery arrives at Launch Pad 39B 132
Map of the main launch facilities at Baikonur 134
Launch of Space Shuttle Discovery 138
Proton launch of the Zarya module, 20 November 1998 141
Launch of the Expedition One crew, 31 October 2000 142

Launch of the third Ariane 5 from Kourou 144
The Soyuz TM 148
Cutaway of the Progress M spacecraft 149
A European ATV approaches the ISS 151
HTV docking with ISS 152
Artist's impression of HOPE X spaceplane 154
The X-38 lands on Rogers Dry Lake in California 156
Mikhail Tyurin with the Pirs docking probe 158
View of the ISS through the crew optical alignment system 159
The ISS Mission Control Center in Houston 162
Size comparison of Mir and the completed ISS 164
Zarya seen from Shuttle Endeavour on 6 December 1998 167
IMAX view of Zarya and Unity 168
STS-88 astronauts inside a Pressurised Mating Adapter 170
Jeffrey Williams during spacewalk on 21–22 May 1999 175
Susan Helms carrying a treadmill into the ISS 176
Artist's impression of Zvezda deploying its solar arrays 177
Shuttle Atlantis docked to the front end of the ISS 179
Koichi Wakata in the supply-laden Zarya module 180
Front view of the ISS showing the newly installed Z1 truss 182
The Expedition One crew on board the ISS 183
Yuri Gidzenko carrying a ventilation fan 185
Carlos Noriega during an EVA to install solar arrays 187
Artist's impression of the ISS with the P6 solar array assembly 188
The Destiny laboratory being moved by the Shuttle's robotic arm 191
The STS-98 crew join ISS commander Shepherd inside Unity 193
The crews of Expedition One, Expedition Two and STS-102 195
Susan Helms carries out battery maintenance in Zarya 197
James Voss and Susan Helms operate the Canadarm 2 201
The ISS and its robotic arm as seen from Shuttle Endeavour 203
The first space tourist, millionaire businessman Dennis Tito 204
Canadarm 2 manoeuvres the Quest airlock 208
James Reilly during the first spacewalk from the Quest airlock 210
Dan Barry and Patrick Forrester on the ISS main truss 213
The crews of Expedition Two, Expedition Three and STS 105 214
A plume of smoke over New York on 11 September 2001 216
Pirs approaches the ISS 217
Claudie Haigneré in the Destiny module 220
The ISS Expedition One crew during winter survival training 230
The Andromède crew in a Soyuz simulator at Star City 232
Michael Lopez-Alegria inside a packed Zarya module 237

Yuri Usachev testing the Vozdukh air purification system 240
Claudie Haigneré climbs into an Orlan pressure suit 245
Three crews sharing a meal in the ISS wardroom/galley area 251
Yuri Usachev with a Russian water container 255
Yuri Gidzenko inside the Leonardo Multi-Purpose Logistics Module 257
James Voss conducts maintenance work in Zvezda 260
Yuri Usachev works out on the cycle ergometer 263
Yuri Lonchakov and Yuri Usachev on the ham radio 266
Yuri Usachev in one of the "private" rooms in Zvezda 267
A digital photo of glaciers in southern Chile 277
An EXPRESS rack 280
The Advanced Protein Crystallisation Facility 282
Yuri Usachev with a plant experiment 283
Artist's impression of the XEUS X-ray observatory 285
STS-98 astronauts move a rack into position in Destiny 290
Artist's impression of the European Columbus module 292
Artist's impression of the Japanese Experiment Module 294
Yuri Usachev with a weightless pizza 299
The ISS Payload Operations Center in Huntsville, Alabama 303
Yuri Gidzenko takes a photo on Zvezda 305
The Phantom Torso radiation experiment 307
Claudie Haigneré using the COGNI apparatus 313
The Reconfigured Russian Segment 323
Artist's impression of Mini Station 1 333
The Shenzhou 2 spacecraft 335
Portrait of the Expedition One crew 344
Portrait of the Expedition Two crew 348
Portrait of the Expedition Three crew 352
Portrait of the Expedition Four crew 356
The Expedition Five crew 360
The Expedition Six crew 364
The Expedition Seven crew 368
The Expedition Eight crew 371

List of tables

Table 1.1	ISS final configuration: facts and figures	2
Table 5.1	Shuttle launch profile	136
Table 5.2	Space Shuttle vital statistics	137
Table 5.3	Proton launch profile for Zvezda module	140
Table 5.4	Soyuz ascent timeline	142
Table 5.5	Ariane 5V statistics	143
Table 5.6	Ariane 5V/ATV launch timeline	145
Table 5.7	H-IIA vital statistics (212/HTV version)	145
Table 5.8	Soyuz TM vital statistics	147
Table 5.9	Progress M and M1 comparison	150
Table 5.10	ATV vital statistics	151
Table 5.11	HTV vital statistics	153
Table 5.12	HOPE-X vital statistics	154
Table 7.1	Expedition Three Timeline Review	234
Table 7.2	ISS Extravehicular Activities	248
Table 7.3	International Space Station Food List	252
Table 8.1	NASA ISS Commercial Prices	301
Table 8.2	Price limits for ISS Russian Segment Resources	302

Acknowledgements

This book is the product of many years of research. Trying to follow the numerous changes that have taken place in the Freedom, Alpha and International Space Station programmes over the years has been a major challenge, and it would not have been possible to keep up with the latest developments without the help of many people and organisations across the world.

First and foremost, I must pay tribute to the public information officers and other employees of the major space agencies and companies, who helped me with my many requests for information and interviews, particularly Debra Herrin, Kyle Herring and Kay Grinter of NASA; Clare Mattock, Elena Griffoni and Andrei Krasnov of ESA; Mireille Campan of NASDA; and Linda Billings of Spacehab. Most of the images used in the book have been obtained through the kind assistance of Jody Russell and Debbie Dodds at Johnson Space Center, and Stephane Corvaja of ESA.

Many other sources have been consulted, including the vast range of Internet sites and magazines devoted to space exploration, notably the British Interplanetary Society's monthly publication *Spaceflight, Space News* and *Aviation Week & Space Technology*. I have also been most grateful for access to a draft text entitled *The International Space Station Handbook,* written by Professor Michael Rycroft and Alice Houston.

The rapidly expanding library of books devoted to space exploration that has been published by Springer–Praxis and Springer–Verlag has also been an important source of information. My thanks go to publisher Clive Horwood and his editorial adviser Stuart Clark for their support and confidence and for giving me the opportunity to add to that impressive body of literature.

Last, but not least, I must thank Linda Arrowsmith, whose secretarial skills saved me so much time, and my wife Edna, whose countless cups of coffee, patient forbearance and constant support encouraged me to bring this project to fruition.

Peter Bond
Cranleigh, Surrey
January 2002

Introduction

The history of the International Space Station is one of the longest running sagas of modern times – a story that covers more than three decades of political intrigue, financial bungling and duplicity, technological wizardry, human courage and dreams of a brighter future.

From the imaginary flights of fancy of 19th-century fiction writers and the visions of more practical men of science, people began to envisage a time when humanity would leave its terrestrial cradle and reach for the stars. Hastened by the development of rocket-powered missiles as weapons of mass destruction, the space age dawned on 4 October 1957 with the launch of the world's first artificial satellite. Fourteen years later the first steps along the road towards humanity's permanent occupation of space were taken when the Soviet Union launched a space station named Salyut 1.

Now, in the year 2002, the night sky has gained a new star – a man-made creation that circles the planet once every 90 minutes and puts all of its predecessors to shame in terms of scale and ambition. In an unprecedented demonstration of international collaboration, 16 nations are working together to assemble, piece by piece, the largest, most expensive, most complex structure ever to orbit the Earth. Like the exotic creation of some mad scientist, a gigantic collection of cylinders and metal beams, bristling like a porcupine with solar arrays, antennas, and robotic arms, is gradually taking shape 400 km above our heads.

Stanley Kubrick's 1968 movie classic *2001: A Space Odyssey* portrayed a time when passengers would be able to take a scheduled Pan Am flight to colonies on the Moon via a Hilton space station. Regrettably, this celluloid vision of the early 21st century has not materialised. Instead of the stream of commuters travelling to and from a space hotel, only one fare-paying private citizen – an American multimillionaire – has succeeded in completing a brief commercial flight to a space station. Despite early promises that an Earth-orbiting station would open the door to further exploration of the Solar System, colonies on the Moon remain the stuff of science fiction, and the world's lone space station shows little promise of becoming a hub for interplanetary travel.

And yet ... the International Space Station cannot just be condemned as a multibillion dollar white elephant. By bringing together East and West, communist and capitalist, rich and not-so-rich, the project has transformed the ways in which space exploration and exploitation are conducted. The first era of spaceflight, characterised by "the space race" and Cold War competition, can now be consigned to the dusty pages of history. In its place is a new era of international collaboration – an age of partnership and open dialogue that has the potential to revolutionise future methods of scientific research and the commercialisation of space.

Scientists, engineers, politicians and bureaucrats from the United States, Canada, Japan, Russia, Brazil and 11 European countries are now intimately involved in a 15-year programme to construct and operate a permanently occupied human outpost in space. Apart from the unique opportunities offered to science and technology, the ISS provides an unprecedented opportunity for former adversaries to work together as partners, colleagues and friends, sharing in a joint endeavour and relying upon each other's expertise and resources.

NASA Space Station Programme Manager Tommy Holloway summed it up like this:

> When the world looks back on the International Space Station, they will see one huge team accomplishing an incredible mission. And through integrity, trust and respecting people, no obstacle, whether technical or cultural, is preventing this world space flight team from achieving our goals.

Chapter One

Building a Giant

The Sun sets and the orange glow fades into the blackness of night, allowing the stars to appear as if by magic against the velvet background. The tension builds as you scan the western sky for signs of movement. Suddenly, there it is! A bright, starlike object climbing higher in the sky and becoming more luminous with every second.

As it passes 400 km (245 miles) overhead, only the Moon, Venus and Jupiter outshine this man-made intruder, a metallic mirror that so admirably reflects the hidden Sun's rays. Following an unseen arc across the heavens, the wandering star continues its inexorable journey towards the eastern horizon. Within minutes it has gone. All that remains is the awe and wonder associated with this technological marvel – the International Space Station. After four decades of human space travel, there is still something special about staring upwards at a distant speck of light that is occupied by three people, fellow citizens of planet Earth.

THE INTERNATIONAL SPACE STATION

The starlike image of the International Space Station (ISS) could hardly provide a more misleading impression of its size or shape (Table 1.1). By the end of 2001, the enormous structure already weighed more than 150 tonnes, making it the largest man-made object ever to orbit the Earth. By the time it is completed, sometime after 2006, the space station will comprise more than 100 separate components and probably weigh three times as much – around 450 tonnes (one million pounds). It will then measure 109 m (356 ft) across and 88 m (290 ft) in length, large enough to fill a football stadium.

Table 1.1 ISS final configuration: facts and figures

Width (truss):	108.5 m (356 ft)
Length (modules):	88.4 m (290 ft)
Mass (weight):	453,592 kg (about 1,000,000 lb)
Operating altitude:	385 km average (240 miles)
Orbital inclination:	51.6 degrees to the equator
Orbital velocity:	8 km/s (5 miles/s)
Atmospheric pressure inside:	1,013 mbars (14.7 lb/in^2) – same as Earth
Pressurised volume:	1,218 m^3 (43,000 ft^3) including six laboratories
Crew size:	3, increasing to 7 by completion in 2006

Although its physical dimensions will be dominated by more than 3,200 m^2 (almost an acre) of solar panels, the heart of the station will be the pressurised modules that are permanently occupied by a crew of six or seven. With half a dozen scientific laboratories, perhaps two habitation modules and various other compartments, the multinational crews will be able to float around a home with an internal volume equivalent to that of a Jumbo Jet. More than 900 researchers worldwide are working on experiments that will take advantage of the unique microgravity environment offered by these labs and associated external platforms.

In order to bring this enormous project to fruition, a consortium of 16 nations has come together to construct and launch the various elements. More than 40 launches of the Space Shuttle and Russian expendable rockets will be required to deliver the different pieces, with at least 850 hours of spacewalks to assemble and maintain the station during its construction phase alone. With a targeted lifetime of at least 10 years, the entire programme will have cost in the region of $100 billion by the time it comes to an end.

As if these vital statistics are not sufficiently impressive, the ISS programme will also make its mark in the annals of space exploration through a remarkable series of space "firsts": the first time a partnership of nations has owned and operated a space station; the first time western agencies have been given access to previously secret Russian facilities; the first time training and control centres in many countries have been linked; the first time traffic involving multiple spacecraft built by many countries has been coordinated; the first attempts at large-scale commercial ventures, including space tourism; the deployment of the largest solar arrays ever placed in orbit; the first time a robot has handed over to another robot in human space flight; and the list keeps on growing.

None of this would have been possible without the introduction of new forms of intergovernmental agreements and Memorandums of Understanding between five space agencies in the United States, Canada, Europe, Japan and

Russia, tying together one huge international team from Belgium, Brazil, Canada, Denmark, France, Germany, Italy, Japan, the Netherlands, Norway, Russia, Spain, Sweden, Switzerland, the United Kingdom and the United States. These legal documents carefully outlined each partner's contributions, and share of common operating responsibilities, utilisation opportunities and crew time. Also unique to the ISS programme are the barter arrangements that allow partners to trade their shares in these responsibilities across hardware, operations, people and science.

THE STATION'S CORE

The design of the International Space Station has inevitably been determined by the size and lift capability of the available launch vehicles, particularly the US Space Shuttle and the Russian Proton rocket. This has resulted in a multimodular layout that comprises a number of cylindrical sections, the largest of which measure approximately 13 m (43 ft) by 4.2 m (14 ft) and weigh around 20 tonnes. In addition, the dominance of the Americans and Russians is reflected in a division that sees one half of the station under the control of each major partner.

At the heart of the complex are the four modules around which the rest of the station will be assembled. The core section is an American-financed, Russian-built module known as Zarya, which provided the first propulsion and power for the fledgling ISS. Its 16 fuel tanks can hold more than 6 tonnes of propellant, and orbital boost or changes in attitude could be performed with the aid of two main engines, 24 large thrusters and 12 smaller thrusters.

To its rear is another piece of Russian hardware, the Zvezda service module, where the resident crews spend most of their free time. This section houses the main living quarters and life support systems, the galley, the bedrooms and the sanitary facilities. Both of these modules are equipped with multiple docking ports for use by visiting manned and unmanned Russian spacecraft, and for the eventual expansion of the Russian part of the station.

By the end of 2001, the only Russian module to be added to this duo was a small docking compartment named Pirs, which enables pressure-suited cosmonauts to exit the station in order to handle experiments and carry out assembly and maintenance tasks.

Docked to the front end of Zarya is the first American-built contribution to the ISS. The 5.5-m (18-ft) long Unity node acts as a vestibule between various rooms on the station. It has six docking berths, one on each side, to which other modules can be attached with the aid of cone-shaped pressurised mating adapters. Unity was launched with two of these – one to link up with the front

Artist's impression of the completed International Space Station. In the foreground, attached like spokes of a wheel to Node 2, are the Columbus (left), Kibo (right) and Centrifuge modules. Perpendicular to the pressurised modules is the huge truss with its solar arrays and Mobile Servicing System. To the rear is the Russian segment with a fully equipped Science Power Platform (now simplified). A European Space Agency Automated Transfer Vehicle is coming in to dock at the far end of Zvezda. (ESA/David Ducros image)

end of Zarya, and the other for docking with the Shuttle. This latter adapter was removed in January 2001 to allow the addition of the US Destiny module, and then reattached to the front of the newly arrived lab.

Destiny is the first – and so far the only – science laboratory delivered to the ISS. The laboratory is 9.2 m (30 ft) long – including its Common Berthing Mechanism – and weighs 14.5 tonnes (32,000 lb). Its main purpose is to provide a pressurised, shirt-sleeve environment where astronauts can carry out a variety of scientific experiments. These will mostly be located in 12 drawer-like racks. The computers in Destiny control power, thermal and vacuum conditions for the experiments, and also monitor the health and status of each payload. The module is also equipped with a number of life support systems, including cooling, air revitalisation, temperature and humidity control. Two computers in Destiny keep the space station in its proper orientation (attitude).

The other American-built pressurised section delivered by the end of 2001 was a 6-tonne joint airlock, known as Quest, which opened the way for space station occupants to take a walk in space using either Russian or US pressure suits and without the presence of a Space Shuttle. The airlock was attached to the starboard side of the Unity module with the aid of the station's Canadarm 2 robotic arm.

Manual carrying and lifting of bulky, massive pieces of hardware is not possible for pressure-suited astronauts, even in the weightless environment of space, so robotic assistance is essential. The 17.6-m (57.7-ft) long Canadian-built arm, which was delivered to the ISS in April 2001, is able to lift an object

Artist's impression of astronauts working with the Canadian-built Mobile Servicing System on the ISS truss. (Canadian Space Agency)

weighing 120 tonnes on Earth. With seven motorised joints and a hand at each end, it can "inchworm" (move end over end) between anchor points on the outside of the station. Its mobility will be enhanced when a Mobile Base System is added. This small "truck" will enable the Canadarm 2 to move along rails on the station's main truss structure, carrying modules and truss segments to their desired positions.

Other cranes and robotic arms will be used for routine tasks. Two manually operated Russian Strela cranes, which proved so useful on Mir, have been placed outside Zarya and the Pirs compartment. This proximity allows cosmonauts to use them in tandem to manoeuvre equipment during spacewalks.

Additional automated appendages will be available later in the ISS programme. They include a European Robotic Arm, for use on the Russian segment, and a similar crane to move payloads around outside the Japanese Kibo science laboratory.

POWER TO THE PEOPLE

The largest, most noticeable features of the completed space station will be its beam-like truss and huge solar panels. The truss is the backbone of the ISS. When it is completed, it will be the length of a football field, aligned perpendicular to the station's main axis. Laboratories, living quarters, payloads and systems equipment will be directly or indirectly connected to it. Also attached will be eight US solar arrays supplying enough power to light a small town.

The support section, the 9-tonne Z1 truss, was delivered to the space station by Shuttle Discovery in October 2000 and attached to the Unity node by means of the Shuttle's robotic arm. It houses gyroscopes for attitude control and communications equipment that provides enhanced voice and television capability.

The remainder of the aluminium truss will be assembled, piece by piece, over a period of about two years, starting with the central S0 section in April 2002. Eventually, there will be four truss segments on each side of the S0 carrying four sets of photovoltaic modules, each comprising two arrays and a thermal radiator required to cool them. Electricity will then be sent around the station through some 13 km (8 miles) of cable.

The first pair of power-generating solar arrays arrived on the station in December 2000, when they were temporarily attached to the Z1 truss. This P6 module will eventually be located at the far end of the main truss on the port side.

One of the 34-m (112-ft) long US solar panels during testing. (Lockheed Martin photo 98-03)

The photovoltaic arrays, which are the largest ever deployed in orbit, provide the essential power that enables the crews to live comfortably, safely operate the station and perform complex scientific experiments. Each module comprises two panels measuring 34 m (112 ft) long by 12 m (39 ft) wide. The 33,000 silicon cells that cover a single panel create up to 30 kW of electricity when in sunlight. To ensure that they receive as much radiation as possible, the arrays have motors that enable them to rotate and track the Sun.

Each photovoltaic module has its own thermal control system. Heat from the electronic boxes is removed by liquid ammonia coolant which is pumped through tubes and into large, flat radiator panels that radiate the heat into space.

Like a city power plant, the modules generate primary electricity at 130 to 180 volts d.c., too high for consumer use. The power is stabilised at 160 volts, then shunted to batteries for storage and to switching units that route it to local transformers. The transformers convert or "step down" the electricity to 124 volts d.c., then distribute it to laboratories and living quarters on the ISS.

Even though the station spends about one-third of every orbit in Earth's shadow, the electrical system will continuously provide at least 78 kW to ISS systems and users. In order to provide power when the space station is in eclipse, energy from the solar arrays is stored in rechargeable nickel–hydrogen batteries. In an emergency, these can provide enough energy to keep the station operating at reduced consumption for one complete orbit.

When the entire set of solar panels is fitted, the eight arrays will theoretically be capable of producing almost 250 kW of electricity – enough to supply a community of some 200 homes. However, after 15 years in orbit, their output will decline to around 185 kW, of which 30 kW will be made available for scientific payloads.

A different system is used in the Russian segment. Both Zarya and Zvezda have a pair of silicon solar "wings" that generate up to 10 kW of power and supply 28 volts d.c. to the Russian modules. This energy is usually consumed in the modules themselves, although some power from Zarya was made available to Unity in the early stages of assembly. In order to make power sharing possible, special step-up and step-down units are installed to convert the electricity from one voltage to the other. Surplus energy for use during eclipses is stored in six nickel–cadmium batteries in Zarya, with another eight in Zvezda.

At present, the Russian part of the station does not have sufficient power resources for major expansion and large-scale scientific research. A Science Power Platform (SPP) that would supply about 50 kW of electrical power for the Russian segment of the ISS has been under development for some years. Originally designed for the Mir 2 project, the SPP was intended to comprise a pressurised module that would contain all of its subsystems, a European Robotic Arm, and a special telescopic truss with eight solar panels, radiators and scientific payloads.

This now seems to have been scaled down and simplified by eliminating the pressurised section and attaching only four solar panels. Its launch aboard the Space Shuttle, previously planned for 2002, has now been put back at least two years. The SPP will eventually be located on the Pirs module after the docking compartment has been moved to the upper port on Zvezda's transfer section. The attachment will take place with the aid of the Shuttle's robotic arm.

INTERNATIONAL INVOLVEMENT

To date, most of the hardware delivered to the space station has been made in either the United States or Russia. The most significant exceptions have been the Canadian robotic arm, which was required to lift the Quest airlock into position, and the data management system provided for the Russian segment by the European Space Agency (ESA).

The station's Mobile Servicing System (MSS) is Canada's main contribution to the ISS programme. Intended to reduce the time astronauts spend on spacewalks in the hostile environment of space, the MSS will support assembly and maintenance of the complex structure and manoeuvre external payloads into position.

Apart from Canadarm 2, the system will include a work platform and a Special Purpose Dexterous Manipulator, a highly advanced robot that has two arms and sophisticated feedback mechanisms that allow it to touch and feel much like a human hand. Equipped with lights, a video camera and four tool holders, the Manipulator will perform sophisticated operations such as installing batteries, power supplies and computers. NASA is contributing a Mobile Transporter, that will enable the complete MSS to travel the length of the ISS truss, and robotic work stations that will enable the crew to control the Transporter from inside Destiny.

Computers play an essential role in the operation of the International Space Station. The task of interconnecting and testing 13 major software systems (including life support, guidance, navigation, propulsion and communications) tops three million lines of computer code - an enormous endeavour never before accomplished.

In the case of the European-made data management system installed in Zvezda, this takes the form of two fault-tolerant computers connected to Zvezda's central computer, and two control posts with laptop computers that enable the crew to command experiments and the European Robotic Arm. Apart from providing control and command functions for the Russian part of the station, the system handles guidance, navigation and control for the entire ISS. These computers will be reused for two of ESA's other main contributions to the programme, the Columbus laboratory and the Automated Transfer Vehicle.

If all goes according to the original plan, six scientific laboratories will eventually be attached to the ISS. Like its American counterpart (Destiny), the 6.7-m (22-ft) long Columbus will be equipped with standard payload and system racks located around each of the four "walls". It will accommodate several specialised facilities devoted to experiments on materials science, biology, fluids and human physiology. External platforms will be available to carry out technology and astrobiology experiments, as well as observation of the Earth and the stars.

On the opposite side of Node 2 (at the forward-facing end of the station) will be the larger Japanese Kibo laboratory. Apart from the main research area, Kibo has been provided with an upper airlock, which allows equipment to be transferred to and from the vacuum of space without having to depressurise the entire module.

Of particular interest to scientists is the Exposed Facility, the largest external platform of its kind on the station, which will be able to cater for up to 10 payloads weighing between 500 and 1,500 kg (1,100 and 3,300 lb). Experiments already identified include an X-ray astronomy observatory, a prototype of a laser communication system, an ozone-monitoring instrument and equipment to measure space particles and debris. The Japanese module also has its own small robotic arm for payload operations on the exposed platform.

The other Japanese-built unit (largely financed by NASA) is the Centrifuge Accommodation Module, which will be attached to the top of Node 2. Already six years behind schedule and plagued by technical problems, the centrifuge is regarded by scientists as an essential addition to the station. Its main purpose will be to provide varying levels of artificial gravity – from one-hundredth the force of gravity found on Earth to twice normal gravity. This will enable scientists to make comparisons on how these different levels of gravity affect organisms housed under otherwise identical conditions.

Earth observation will be one of the most time-consuming occupations for crew members, and ESA has offered to provide one or more cupolas – pressurised observation bays fitted with six lateral windows and one circular top window. Apart from use for Earth photography and spare time recreational use, the observation posts will be of value during external assembly and maintenance tasks.

The European-built cupola will offer a pressurised observation post. (ESA/David Ducros)

Less spectacular, but just as important to the scientific viability of the station, are the contributions from Brazil. These include the Express pallet – an unpressurised platform which is designed to supply power and data links for up to six external payloads – a smaller payload support fixture known as the Technological Experiment Facility, a rack that houses Earth observation equipment and an unpressurised cargo container.

Little is known about the two proposed Russian–Ukrainian laboratories, which are likely to suffer lengthy delays due to shortage of funds, and whose research role will be limited unless a way can be found to provide more power from the simplified Science Power Platform.

ACCESS

Scientific laboratories would be of little consequence without a means of delivering large amounts of experimental hardware and raw materials, then delivering the results safely to Earth for subsequent analysis. In addition, simply maintaining a three-person crew (which may eventually increase to six or seven) requires regular supplies of food, water, oxygen and other consumables.

Such routine tasks currently involve three types of spacecraft – the US Shuttle, the Russian Soyuz ferry and an unmanned Progress cargo ship. In the early stages of construction, a Spacehab module installed in the Shuttle's payload bay was used to carry cargo to the space station, but since March 2001 Shuttle deliveries have been made with the aid of three Italian-built Multi-Purpose Logistics Modules. These purpose-built cylinders can be crammed with up to 5 tonnes of supplies and equipment. A further 2 tonnes is delivered every three months by a Progress craft, while small amounts of equipment can be flown up or down at six-monthly intervals during Soyuz "taxi" missions.

However, as the station expands, with the possibility that the crew complement may double, additional traffic will be necessary. Eventually, this will entail no less than five distinct vehicles (US Shuttle, Russian Soyuz and Progress, European Automated Transfer Vehicle and Japanese H-II Transfer Vehicle) delivered by four different launch systems based in four launch centres scattered around the globe. Most of these will also have the capability to reboost the station's orbit before it decays and to dispose of unwanted waste by burning up during re-entry into the atmosphere.

This unprecedented flow of spacecraft to and from the space station will require a complex system of coordination and communication. At the same time, thousands of engineers and scientists will be demanding continuous links between mission control and payload control centres in Houston, Huntsville,

The Soyuz TM-33 spacecraft docked to the Pirs module of the International Space Station. (ESA/CNES photo)

Moscow, Montreal, Oberpfaffenhoffen, Toulouse and Tsukuba. The logistical tasks are complicated even more by the need to train crews and ground operators in multiple locations on three continents and to orchestrate some 1,000 hours of spacewalks while performing on orbit assembly of 47 major elements (each one unique and different) and more than 100 smaller elements.

In order to enable such interwoven networks to function efficiently, fast, reliable communications are essential. Various systems using different frequencies are employed for different purposes.

S-Band steerable horn and omnidirectional antennas, pre-installed on various truss segments, handle voice communications together with the telemetry, tracking and control data. However, the primary high-data-rate communications link for video and payload data, as well as two-way file transfer, uses the Ku-Band frequency. Its 1.9-m (74-in) diameter steerable antenna is capable of automatically tracking NASA's Tracking and Data Relay Satellite System (TDRSS) communication satellites while transmitting at data rates of up to 75 Mbps.

However, despite these sophisticated systems, communications between the station and the ground are not as reliable as might be expected. Signal blockage caused by the ISS itself means that coverage per orbit is typically no more than 50% when using two TDRSS satellites. Although Russian communication coverage is nearly continuous while using the network of Russian ground stations, these antennas are only available for part of an orbit. A Luch data relay satellite is also available, but coverage only lasts for approximately 45 minutes or 50% of each orbit.

Also available during extravehicular activity (EVA) and Shuttle rendezvous and docking operations is a UHF system. Four antennas mounted on Destiny and the ISS truss are designed to receive signals from the Shuttle over a distance of 7 km (4.4 miles). For EVA, communication availability provided by these antennas is nearly 100% (with all four antennas functional). Pressure-suited astronauts can communicate with each other inside the Quest module, using an antenna in the airlock. If the Shuttle is present, its UHF subsystem can also communicate with people working outside the station.

As the monolithic space station grows in size and crew members are often widely scattered, it is obviously essential that they have some way of keeping in contact with each other. An internal audio system allows the crew to speak to each other through an "intercom" in each module, but it can also be configured for EVA, Shuttle and air-to-ground use by tying into the UHF or S-Band systems.

WHAT NOW?

Until early 2001, the main doubts over the future configuration and assembly schedule of the space station seemed to involve the Russian contributions. Since then, escalating costs, financial constraints and technical difficulties have cast a shadow over the entire programme.

After a highly critical assessment of NASA's ISS management by an independent task force, several pieces of U.S. hardware have been put on hold and threatened with cancellation, including the Habitation Module – which was to contain the galley, toilet, shower, sleep stations and medical facilities – the Node 3 passageway with its advanced life support system, and the Crew Return Vehicle. If implemented, the changes will inevitably prevent the expansion of the resident crew size from the current figure of three to the originally planned six or seven.

Since it is generally accepted that two and a half crew members are required simply to maintain the space station, the suggestion to limit the crew size to three for the foreseeable future implies a stagnation of the entire scientific programme. Not surprisingly, these proposals have upset the international partners, who see their costs soaring and the value of their investment in underused state-of-the-art facilities shrinking.

While NASA and Congress contemplate the consequences of the enforced belt-tightening exercise, other ISS components are suffering from lengthy delays. One of the most important from a scientific point of view is a facility known as the Centrifuge Accommodation Module, which has dropped six years behind schedule after technical problems forced a major redesign.

Meanwhile, lack of government funds has forced a rethink of how the Russian segment will be extended in years to come. Unable to meet the costs of developing indigenous ISS components, leading Russian space enterprises have been forced to seek western partners to develop and market new modules that will be offered for commercial use. At the same time, the main power source for the Russian segment, the Science Power Platform, has been severely delayed and downgraded, while the two proposed science labs exist only on paper.

So is it all doom and gloom? At present, despite diplomatic pressure from the international partners, it seems that the US Congress will agree to place NASA and its new boss Sean O'Keefe under probation for the next two years. Only then will a decision be made on whether to give the agency the go-ahead to add its so-called "enhancements" and complete the assembly of the station.

In these circumstances, the programme seems to offer little opportunity for innovation and initiative in the fields of technology or science. However, despite these unpromising conditions, engineers and scientists around the globe have been striving to develop new, imaginative hardware that may one day see the light of day on the space station.

TRANSHAB

One of the earliest examples of this approach was a NASA project to develop an inflatable ISS module called Transhab (transportable habitat), a concept derived

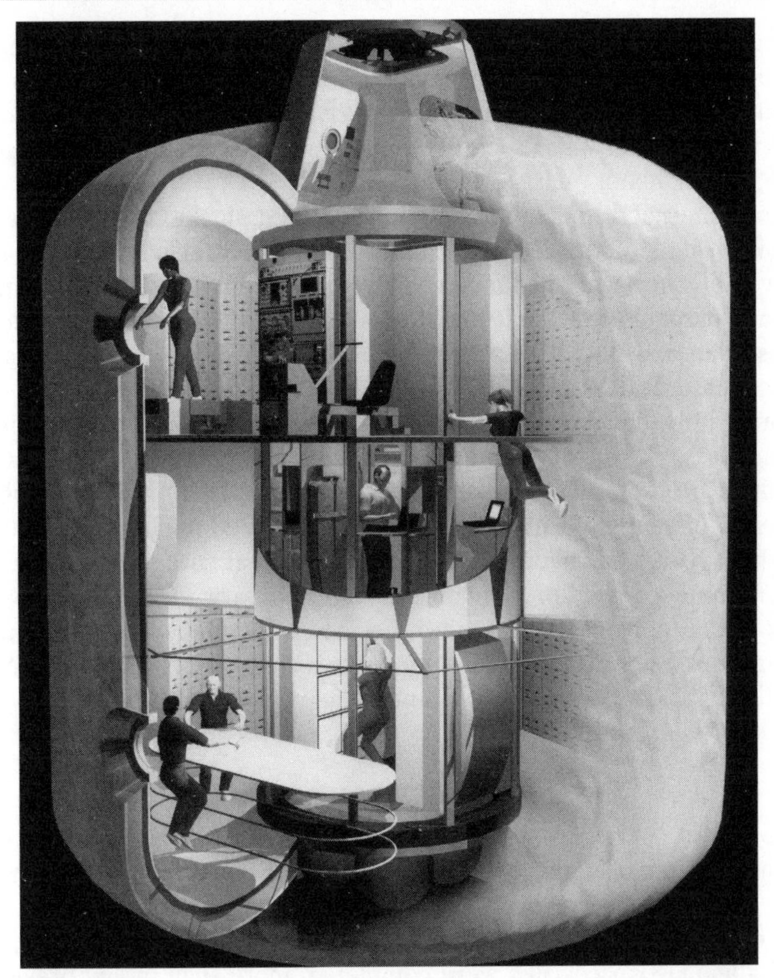

Artist's impression of the Transhab module proposed for use on the ISS. The lower level contains the wardroom and galley, the middle level is the crew quarters, and the top floor is used for crew hygiene, exercise and health monitoring. (NASA photo S99-05363)

from a preliminary design for a shelter on Mars. Envisaged as a possible substitute for the US Habitation Module, which was put on hold in 1997, Transhab can be stowed inside a Space Shuttle cargo bay and inflated to full size in orbit.

Transhab would contain the same amenities – toilet, shower, galley, sleeping areas – as the Habitation Module, but instead of a metal casing, the flexible module would be made of layers of puncture-resistant fabric, foam rubber and Kevlar (the major component of most bullet-proof vests), so it would actually provide better protection from micrometeorites than a metallic design. A shell of carbon composite ribs maintains its shape in orbit.

One advantage of Transhab is that it offers 30% more room than the aluminium-skinned Habitation Module, so it could accommodate up to six astronauts compared with four for its rival. Also, unlike the traditional cylindrical design of the Boeing-designed module, it can be divided into "floors", rather like a multistorey house.

Several internal layouts have been suggested, including one where the module is divided into three levels connected by a central ladder/access tunnel. This revolutionary layout has received favourable comments from a number of astronauts, particularly since it would open the way towards more people living on the station and might eventually help NASA to develop a crew transport vehicle for future Mars missions.

Unfortunately, development progress has been limited by funding shortfalls in recent years. Since completing vacuum and micrometeorite strike tests on a $1.2 million prototype at Johnson Space Center in 1999, the Transhab team has run into difficulties. Unwilling to provide the $200 million needed to build the module, NASA has been seeking to strike a deal with private interests to build an improved version. Meanwhile, in February 2001, George Abbey, director of the Johnson Space Center, decided to curtail a number of projects, including Transhab, in an effort to rein in the soaring cost overruns on NASA's International Space Station programme.

FLYING ROBOTS

Since the beginning of the International Space Station's construction, more than 30 spacewalks have been performed. This represents more than one-quarter of all American spacewalks carried out in NASA's four-decade history. In 2001 alone, 18 spacewalks were completed, more than in any previous 12-month period in the history of human spaceflight. This record was expected to be broken again in 2002, when 22 spacewalks were planned from the Shuttle and station.

Apart from its cost and complexity, such extravehicular activity is time-consuming and potentially life-threatening, so it is important to try to minimise the number of spacewalks and the need for human intervention in sometimes routine tasks. At present, various mechanical arms and cranes attached around the station's exterior are available as EVA aids, but engineers in America and Europe are looking at more revolutionary robotic methods of helping ISS crews.

Perhaps the most advanced of these is NASA's Robonaut, a "humanoid" robot built to resemble a legless astronaut in a space suit, with two arms, two five-fingered hands, a head and a torso. Its most important feature is its pair of arms and end effectors based on the human hand. With the aid of more than

The AERCam Sprint satellite during a test flight on 3 December 1997. (NASA photo STS087-752-035)

150 sensors in each arm, Robonaut will be capable of handling detailed and complex tasks. The robot will be operated under telepresence, i.e. by an astronaut using a virtual reality headset and gloves who will see and control what is happening to his or her mechanical counterpart. It will be able to interface with the station's existing Mobile Servicing System, serving as a spacewalker's assistant or stand-in for tasks too dangerous for humans.

One of the most innovative inventions is a small, spherical "personal satellite assistant" that bears a close resemblance to the small, floating "droid" that sparred with Luke Skywalker in the movie *Star Wars*. The first space trial of such a system took place in the Shuttle cargo bay in late 1997, when the beachball-sized AERCam (Autonomous Extravehicular Activity Robotic Camera) Sprint was released during a spacewalk. The free-flying camera platform, which was covered in a soft, cushioning material to prevent damage in the event of a collision, moved slowly around the payload area using a cold gas propulsion system. On this occasion, Sprint was controlled from the Shuttle flight deck by pilot Steve Lindsey with the aid of a joystick, but future versions were expected to have infrared distance sensors and autonomous capability.

More advanced "satellite assistants" under development at NASA's Ames Research Center in California could be fitted with a variety of sensors for use as early warning systems. For example, one could be sent into a sealed module to search for hot spots, leaks or noxious gas concentrations before lives were risked by sending in astronauts to investigate the situation. A similar onboard satellite could also be used as a "floating, talking clipboard", with the "intelligence" to respond and give helpful instructions to an astronaut undertaking a complex task. Prototypes have already been constructed and a test flight on a KC-135 aircraft is planned for 2002.

Another system suitable for performing surveys of the ISS exterior is being developed in Germany. In its original form, the hexagonal Inspektor satellite was space-tested during a three-month flight that began in December 1997. Despite an early navigational problem, the small spacecraft successfully returned video images of the Mir space station and the Earth.

Following the success of this "space first", Daimler-Chrysler Aerospace (now Astrium) began to develop a second-generation Inspektor satellite for the ISS. Initially intended as a contender for NASA's Advanced Engineering and Technology Development Programme, the project had to be reconfigured when the US space agency ran into funding difficulties. The current model, known as Micros, resembles the soft, spherical "personal assistant" being designed by NASA–Ames, with a weight of only 8 kg (18 lb) and a diameter of about 30 cm (12 in).

Equipped with 16 nitrogen gas thrusters, three microcameras and a "sniffer" to detect leaks form the space station, the space ball would be "launched" and recovered by spacewalking astronauts. Although its flying time would be limited to no more than 12 hours, it could also survive for much longer outside the station by using an electrostatic gripper system to attach itself to a module and wait for retrieval. Once recovered, it can be checked and refuelled in the Quest airlock and reused. There is even a possibility that a future generation could be equipped with a robotic arm for ISS maintenance.

SOFT LANDING

In the world of space transportation, change takes place slowly. The US Shuttle fleet began operation 20 years ago with the launch of Columbia – the only one of the four Shuttles not modified to dock with the ISS – while many of the expendable launch systems now in use are based on military missiles developed in the late 1950s and 60s.

Although the construction and maintenance of the ISS would not be possible without the heavy lift capability and versatility of the Space Shuttles, the reusable orbiter is extremely complex and costly to operate. With each launch costing in the region of $500 million, approximately one-quarter of NASA's annual budget is spent on Shuttle operations alone. In an effort to reduce this phenomenal bill, NASA has handed over the servicing and maintenance of the fleet to a privately owned group called the US Space Alliance and initial discussions have begun to find a way of completely privatising the Shuttle launch system.

Part of the reason for the Shuttle's ability to soak up cash is the requirement to deliver a human crew to orbit and return them safely to Earth. Life support systems and safety backups inevitably increase the cost of development and maintenance. Furthermore, the total reliance on the Shuttle for US manned space activities means that a repeat of the 1986 Challenger launch disaster would seriously disrupt – or even terminate – International Space Station construction and operation. However, even expendable rockets are extremely expensive and they tend to be much less reliable. With a customer typically expected to pay around $10,000 for each kilogram of payload launched into low Earth orbit, expectations that commercial space activities will develop in the near future seem highly optimistic.

In the absence of the Shuttle, ways to deliver experiments and hardware safely to Earth are extremely limited. Although the Russians have experimented with a return capsule attached to their unmanned Progress supply ships, these have currently been discontinued, and no other spacecraft designed to visit the International Space Station have a capability of soft-landing with more than a few kilograms of material.

In an effort to change this situation, Europe's Astrium company has been working with Russia and the European Space Agency to develop a novel method of safely returning samples to Earth – a lightweight, inflatable heat shield that can protect sensitive payloads during the fiery re-entry into the atmosphere.

The first test flight of this Inflatable Re-entry and Descent Technology (IRDT) took place in February 2000. After completing five orbits and several engine firings, the Fregat rocket stage and its IRDT dummy payload separated and began to re-enter the upper atmosphere. Both the small heat shield on the demonstrator and the large shield on the Fregat inflated at an altitude of 50 km (31 miles), then functioned as parachutes to deliver their cargo safely back to Earth. About 17 minutes after the final Fregat burn, the upper stage and the demonstrator hit the flat Russian steppes at a velocity of 13 m/s.

The inflatable heat shield has a number of advantages over existing designs. Not only can it be folded into a very small package, but its low weight and low cost offer the potential for future cut-price transportation of samples and cargoes from the International Space Station back to Earth.

It is appropriate, perhaps, that the reinvention of one of the most basic space technologies should be associated with the International Space Station – a pioneering programme that will one day be seen as a staging post between the long-gone glory days of Gagarin and Glenn, and a future that may eventually lead us back to the Moon, to the red sands of Mars, and beyond.

Chapter Two

From Dream to Reality

As the massive bulk of the International Space Station materialises some 400 km (250 miles) above our heads, it is worth remembering that the concept of an artificial orbital structure has existed in the minds of science fiction writers, scientists and engineers for more than 130 years.

Fantasies about travelling into space go back to the second century, when the Greek writer Lucian described an imaginary voyage to the Moon. However, it was not until 1,500 years later that Isaac Newton described the scientific principles involved in placing an object in orbit. In his book *Philosophiae Naturalis Principia Mathematica*, which was printed in 1687, Newton explained how the balance of gravity and centripetal forces would enable a fast-moving projectile "to revolve in an orbit and go round the whole earth".

The first fictional account of a space station did not appear until 1869, when Edward Hale, an American clergyman, published a story called "The Brick Moon" in the *Atlantic Monthly* magazine. However, Hale's hollow, white-painted satellite, which was launched by an enormous flywheel assembly driven by a waterfall, bore little resemblance to the space station of today. Its main function was to act as a navigational aid for ocean-going ships.

TSIOLKOVSKY

The first real breakthrough came in 1911, through the intellect and imagination of Konstantin Tsiolkovsky, a Russian teacher of mathematics. Tsiolkovsky, however, was no ordinary schoolteacher. Overcoming the

"The Brick Moon" described in Edward Everett Hale's 1969 fictional story. (NASA photo)

handicap of deafness from the age of 9, he became fascinated by all aspects of space travel, and by applying Sir Isaac Newton's Third Law of Thermodynamics – for every action there is an equal and opposite reaction – to the problem of leaving Earth and travelling through the vacuum of space, Tsiolkovsky was able to envisage multistage rockets fuelled by liquid oxygen and liquid hydrogen.

In his first magazine article, published in 1903, the visionary wrote of a future space infrastructure involving rockets and satellites. Once he became aware of the potential of rocket power, it only required a small leap of his imagination to consider the deployment of larger, human-occupied structures in low-Earth orbit, and in 1911 he began to think about space stations. To Tsiolkovsky, it was patently clear that humanity's destiny lay far beyond Earth's cradle, in the vast expanse of the Universe.

In his 1933 work, *Album of Space Travels*, he presented his design for a large habitation in orbit. Among its most notable features were artificial gravity created by rotating the structure on its longitudinal axis, and a park planted with vegetables and trees as part of a bioregenerative life support system.

For many years, however, these visions came no closer to reality. In the turmoil of the Soviet Union during the years of the First World War and the Communist Revolution, Tsiolkovsky's work was largely ignored, but, as the century progressed, others began to take an interest in space travel.

One of the early pioneers inspired by Tsiolkovsky was German scientist Hermann Oberth, who envisaged space stations circling 1,000 km (620 miles) above the Earth that could be used for military surveillance, astronomical and weather observations, and even as refuelling stations for interplanetary flights. However, it was an American, Robert Goddard, who pioneered the first practical step towards space travel. Seen as an eccentric by many of his contemporaries, Goddard nevertheless persuaded the Smithsonian Institution to give him a grant of $5,000 to pursue his golden grail – the development of a liquid-fuelled rocket.

Finally, on 16 March 1926, the moment of truth arrived. Goddard, his wife and two assistants posed for photographs alongside a small, spindly rocket that was about to make history. Moments later, this unimpressive creation soared to a height of 56 m (184 ft) and reached a maximum speed of 97 km/h (60 miles/h). The liquid-fuelled rocket had been born.

Despite this and further successes, there was still a long road to tread before a satellite could be lifted beyond the grasp of Earth's gravity and inserted into orbit. The United States government generally ignored Goddard's work, and in Europe and the Soviet Union a similar fate met the small but enthusiastic groups of amateurs who struggled to find funds to develop the space vehicles of the future.

Among these visionaries was a young Slovene named Herman Potocnik. In 1928, the year before his premature death at the age of 37, he published a book called *The Problem of Space Travel* under the pseudonym of "Noordung". Eventually translated into many languages, it was the first book to be devoted largely to space station design and had a considerable influence on the work of technicians and researchers, as well as science fiction writers. It also inspired many space station designs during the 1950s, including the famous wheel in the movie *2001: A Space Odyssey*.

Noordung's "Inhabitable Wheel" had an outer diameter of 50 m (164 ft), and rotated about its axis in order to create artificial gravity in the inhabitable outer ring. This contained cabins, laboratories, workshops, kitchen and bathroom. There was also a circular viewing gallery with portholes for observing the Earth and the stars. A lift and two staircases led to the hub, which had a rotating airlock and docking adapter.

The station would be powered by two large concave mirrors that focused solar radiation onto pipes containing a liquid that would vaporise, then operate turbines to produce a continuous electrical current. The vapour would cool and condense in other pipes that were shaded from the Sun.

The first person to envisage a space station in geostationary orbit, so that it "hovered" above the same spot on the Earth's surface, Noordung considered various uses including weather and military observations, warnings of icebergs for shipping and mapping the Earth.

Although his work was ahead of its time and unlikely to become reality for decades, concrete progress with rocket propulsion was being made by small groups of enthusiasts in the Soviet Union and Germany. They, too, were underfunded and poorly supported, but they pooled their resources to design and build pocket-sized rockets.

VON BRAUN

Only with the remilitarisation of Germany in the 1930s was rocket research elevated to a national priority. Backed by the German Army, a brilliant young rocket designer named Wernher von Braun moved into the spotlight. Appointed technical director of a top secret Army Experimental Station on the Baltic island of Peenemünde, von Braun set to work to build on the inspirational work of Goddard and Oberth, who had gained considerable publicity with his design for a giant two-stage booster.

The determined young man grasped the opportunity to design and construct the first, devastating, ballistic missile while, at the same time, furthering his ambition to develop a rocket capable of reaching the edge of space. Despite the lack of enthusiasm shown by the Führer – no doubt fuelled by numerous failures – von Braun's team persevered until, in October 1942, they were rewarded with a resounding triumph. Their dart-shaped A-4 rocket, later to be known around the world as the infamous V2 or "revenge weapon 2", soared to an altitude of more than 80 km (50 miles) before plunging into the sea. The age of the missile had been born, and the Space Age was just around the corner.

Fortunately, the V2 was still far from failure-proof, and mass production of the new superweapon by imported slave labour took some time to organise. By the time the supersonic slaying machine began to descend in large numbers on the long-suffering inhabitants of London, the war had already swung in favour of the anti-Nazi alliance. Of around 3,000 V2s that were built, just over half were fired in anger, bringing death and devastation in their wake.

As the war turned inexorably against Germany, von Braun's team began to consider their increasingly insecure futures. Unwilling to be captured by the

Soviets, who were closing in from the east, von Braun and his leading engineers surrendered to the advancing American Army. However, with the war drawing to a close, the Western democratic powers had to draw back and consider the future of their uneasy alliance with the revitalised Communist dictatorship, and the decision was made to ship von Braun's team of experts and some 100 V2 rockets to the safety of Huntsville, Alabama, rather than leave this formidable arsenal in the custody of the Soviets. This hoard formed the basis of a fledgling rocket development programme, but just as important were the blueprints for more advanced rockets that were smuggled across the Atlantic without the knowledge of the Soviet authorities.

When the Soviet experts arrived at the V2 underground assembly plant in Nordhausen, the nightmarish tunnels had apparently been stripped of everything of value. Nevertheless, a thorough sweep of the complex produced valuable documents missed by the Americans during their haste to evacuate the scene, and some 3,500 lesser lights in the rocket programme were scooped up and transported thousands of kilometres to new homes in the Soviet Union.

In the Cold War that followed, the relatively harmonious relations between East and West evaporated, only to be replaced by a titanic struggle for hegemony. One of the keystones of the ideological struggle was the struggle for nuclear superiority, which involved not only the development of more powerful atomic and hydrogen bombs, but also the creation of long-range missiles capable of delivering their deadly payloads to the doorsteps of the enemy.

As a by-product of this military rivalry, both superpowers took advantage of the V2 technology and expertise that they had acquired to build vehicles that might also be utilised for a more peaceful, constructive purpose – the conquest of space.

On the American side, von Braun's German contingent had settled into the military arsenal at Huntsville, where they built a series of impressive launch vehicles that seemed more suited to space exploration than rapid response to a Soviet nuclear attack. In 1956, the Huntsville-developed Jupiter C entered the record books when it flew 5,280 km (3,300 miles) at a speed of 25,600 km/h (16,000 miles/h). It seemed only a matter of time before von Braun's team gained the distinction of placing an American satellite into orbit.

As his fame spread, the dashing, blond-haired aristocrat became a national celebrity, first as the author of a series of articles on space exploration in *Collier's* magazine, and then as a TV pundit. The Technical Director of the Army Ordnance Guided Missiles Development Group was perceived not only as an expert on his subject but, curiously for a man who devoted much of his life to creating weapons of mass destruction, as a powerful advocate for the peaceful exploration and exploitation of space.

A US Army V2 rocket is prepared for launch from White Sands, New Mexico. (US Army photo)

In order to realise his dreams, von Braun realised that he must catch the eye of the politicians and loosen the public purse. His campaign to mobilise support began in early 1952, with the first of his articles in *Collier's*:

> Within the next 10 or 15 years, the Earth will have a new companion in the skies, a man-made satellite that could be either the greatest force for peace ever devised, or one of the most terrible weapons of war – depending on who makes and controls it. ... Development of the space station is as inevitable as the rising of the Sun; man has already poked his nose into space and he is not likely to pull it back.
>
> Scientists and engineers know how to build a station in space that would circle the Earth 1,075 miles (1,720 km) up. ... The job would take 10 years, and cost twice as much as the atom bomb. If we do it, we can preserve the peace and take a long step towards uniting mankind.

He went on to describe a majestic 80-m (250-ft) wide wheel made of reinforced nylon that would later become the inspiration for the gleaming white station depicted in the 1968 movie *2001: A Space Odyssey*. Although the wheel concept was based on Herman Potocnik's 1928 design, it was von Braun who put the elegant solution into a modern form.

Von Braun envisaged his space station rotating to provide artificial gravity for the comfort of its occupants. The graceful, spinning wheel would be multifunctional, operating as a navigational aid for ships and aircraft, a meteorological station, a military platform and a staging post on the road to the stars.

The interior of the wheel's circular tube would be divided into three decks. Apart from living space for the crew, these would contain rooms for communications equipment, Earth observatories, military control centres, weather forecasting centres and navigational equipment. Power would be provided by turbines driven by mercury vapour. "From this platform, a trip to the Moon itself will be just a step, as scientists reckon distance in space," he wrote.

Von Braun's futuristic concepts, brilliantly illustrated by artists such as Chesley Bonestell, captured the public imagination. Four million copies of the first *Collier's* magazine were printed, and six more articles were printed between 1952 and 1954. Further exposure came with a book in which he described his plans for a large space station.

The next step was to spread the message even wider. With the blessing of movie mogul Walt Disney, von Braun's dreams of reaching into space were disseminated via three TV shows to an adoring public who eagerly accepted the cartoon images of Moon rockets and space stations as the inevitable future of American technological supremacy.

In many ways, von Braun was decades ahead of his time. Only now, at the beginning of the 21st century, are nations coming together to unite in a

The classical wheel-shaped space station envisaged by Wernher von Braun and later made famous in the movie 2001: A Space Odyssey. *(Digital photo)*

construction of a huge space structure – the International Space Station. His vision was flawed, however, because it failed to take into account the political and economic realities of the time. In the 1950s and 1960s, national prestige was at stake and superpower competition, not cooperation, was the maxim.

FROM SPUTNIK TO THE MOON

The stakes were raised as the International Geophysical Year (1957–58) approached. The United States had publicly announced its intention to launch an artificial satellite. Less publicised was the similar declaration by their Soviet rivals. An overconfident President Eisenhower inadvertently handed the honour of launching the first artificial Earth satellite to the Soviet Union when, to the chagrin of von Braun's team, he asked the US Navy to take the lead role with its Vanguard rocket.

The rest is history. On 4 October 1957, a small, metal sphere named Sputnik delivered a body blow to American self-esteem when a secret rocket built by an unknown chief designer delivered the world's first satellite into orbit. Only after Sputnik's incessant beeping signals alerted the slumbering US giant to the ambitions and capabilities of the Soviet Union did Eisenhower realise his error and turn to von Braun to restore American pride. Within a matter of months his team duly obliged by launching the grapefruit-sized Explorer 1 into orbit atop a modified Jupiter C rocket. But it was too late. The damage had been done.

Worse was to follow as subsequent Soviet launches confirmed that its rockets could lift much heavier payloads than their American counterparts. The 1958 launch of a dog named Laika indicated that it would not be long before a Soviet citizen would be following in her footsteps. The world watched and waited to see how a wounded America would respond.

The first step was to reorganise the disparate US effort so that it could compete with the secretive Soviet space programme. In October 1958, a National Aeronautics and Space Agency (NASA) was created from the existing National Advisory Committee for Aeronautics (NACA). Among the personnel transferred to the new agency were von Braun and his 4,200-member group at Huntsville, along with all of their facilities to develop, test, build and launch space vehicles. Within a few months, the American public was treated to the sight of seven "hotshot" pilots posing for the cameras and proclaiming their eagerness to become the first humans in space.

Once again, the promise was not fulfilled. To the chagrin of all Americans, the silver-suited Mercury astronauts were beaten into space by an inexperienced 27-year-old Russian Air Force lieutenant. Yuri Gagarin went

Dr James Pickering (left), Dr James Van Allen and Dr Wernher von Braun (right) hold up a model of Explorer 1, the first American satellite, shortly after its launch on 31 January 1958. Von Braun strongly believed in the importance of orbital stations, while Van Allen later became an outspoken critic of the Freedom and International Space Stations. (MSFC photo 00285)

into the history books by becoming the first human to leave the planet and circle the Earth.

Once again, the world's most advanced technological nation was shocked into action. Hardly had the first Mercury astronaut set foot on dry land after a 15-minute lob into the Atlantic than President John F. Kennedy set America a new goal. Seeking a competition that the US could win, the young President aroused nationalistic passions by declaring the intention "before this decade is out, of landing a man on the Moon and returning him safely to Earth".

Six years after his death, Kennedy's Moon Race was won as Neil Armstrong and Edwin "Buzz" Aldrin set foot on the lunar Sea of Tranquillity. In order to achieve this historic achievement, a mobilisation of American industry that was unprecedented in peacetime was required.

Among those leading the way in the $25 billion programme was von Braun, now the first director of NASA's Marshall Spaceflight Center in Huntsville. Once again, his inspirational leadership weaved its magic as the Huntsville team developed the monumental Saturn Moon rockets, the largest and most successful family of rockets the world has ever seen.

With Moon fever gripping the country, von Braun's dream of building a permanent space station was pushed further down the list of NASA priorities. He knew that he would have to wait until the Moon Race was won before the opportunity would arise to present the agency, the politicians and the public with his vision of a permanent orbital laboratory and habitation.

THE BATTLE OF THE GIANTS

Thousands of kilometres away, on the eastern side of the Iron Curtain, an equally charismatic, but anonymous figure was also driven by an obsession with the conquest of space. He, too, was employed by his masters to design missiles that could carry nuclear devastation to the other side of the world. But, like his former Nazi rival, Sergei Korolev is remembered more for his remarkable contributions to the human exploration of space rather than for the development and enlargement of the world's arsenal of intercontinental ballistic missiles.

Until his death in 1967, Korolev remained an unknown genius, the anonymous mastermind behind the series of tremendous Soviet successes that staggered the richest nation in the world. The first artificial satellite; the first living creature to fly in space; the first human space traveller; the first spacecraft to photograph the far side of the Moon; the first woman to orbit the Earth; the first three-man crew; the first walk in space. The list of Korolev's triumphs resounds through the annals of spaceflight.

One more space "milestone" was first conceived during his reign as head of the OKB-1 design bureau, although he never lived to see it come to fruition – the creation of a space station.

On 10 March 1962, 10 months after President Kennedy threw down the gauntlet to the Soviets by declaring a finishing date for a race to the Moon, Korolev's team produced a report called "Complex for the Assembly of Space Vehicles in Artificial Earth Satellite Orbit".

Soviet Chief Designer Sergei Korolev (right) with the world's first human space traveller, Yuri Gagarin. (Sotheby's)

Since the Moon Race was now in progress, the main purpose of this document was to discuss a possible scheme for a manned circumlunar mission. The report largely focused upon the in-orbit construction of a four-module spacecraft that could carry three men around the Moon.

However, the same principles and techniques could be used to build a space station. Indeed, the same report included a reference to a small space station that would be assembled from independently launched modules. It also mentioned a ferry spacecraft called Siber (North) that would deliver the station's three-man crew.

The station would include a habitation module and a "science-package" module, each weighing about 6 tonnes at launch. The station modules and Siber ferry would be delivered into low-Earth orbit by Korolev's Vostok rocket, the vehicle that had already made its mark by launching Gagarin and his colleagues on their pioneering flights.

The following year, this concept was modified to a less complicated conglomeration that comprised three modules. This time, the proposed circumlunar complex comprised a manned spacecraft (Soyuz A), an unmanned propulsion module (Soyuz B) and an unmanned fuel tanker (Soyuz C).

However, the preoccupation of Korolev's OKB-1 with the Soviet Moon programme left a window of opportunity for the rival OKB-52, led by Vladimir Chelomei, who was entrusted with the development of the first space station. On 12 October 1964, his bureau began to design a military station with a design life of one to two years that would be occupied by a crew of two or three cosmonauts.

With photographic reconnaissance (spying) on NATO ships and land forces as its main objective, there was no need to aim for a costly, multimodule design. Instead, Chelomei's group conceived a single section station, which they dubbed Almaz (Diamond), and a crewed supply ferry known as the Transport Logistics Spacecraft (Russian acronym TKS).

Each of these stations would have a launch weight approaching 19 tonnes, almost three times heavier than Korolev's Soyuz modules. This would be possible through use of Chelomei's new three-stage Proton rocket, which was already under development for the circumlunar Moon programme.

As ambitions and competition between Soviet design bureaux rose, Korolev tried to regain ascendancy by proposing a 90-tonne space station that could be placed into Earth orbit by a single launch of the new N-1 Moon rocket that was being designed by his team. The monolithic structure was to be equipped with a docking module with ports for four Soyuz transport craft.

Not surprisingly, Soviet officials recognised that Korolev's proposal was more of a spoiling ploy than a serious proposal. His bureau was already swamped by other work related to the Moon programme. In 1967, an interdepartmental commission gave the official go-ahead to OKB-52 and its Branch No. 1 for construction of the Almaz station.

By that time, Chelomei's remarkable rival had been dead for a year – a casualty of a botched operation for peritonitis – and the way seemed clear for Almaz to blaze a trail through the heavens as the world's first space station. But, although development of the main module was proceeding well, lengthy delays built up with its subsystems. With the race to the Moon already won by the Americans, the Soviet authorities were becoming impatient for a return on their sizeable investment in space.

In a move that Korolev would have greatly appreciated, the Soviet Ministry of General Machine Building decided in February 1970 to transfer overall control of the Almaz programme to OKB-1. The bureau, now under the leadership of Korolev's former deputy, Vasili Mishin, was expected to inject new impetus to the programme, with the particular objective of beating the US Skylab into orbit.

By combining subsystems from the Soyuz manned spacecraft of OKB-1 with the major structural components developed by OKB-52 Branch No. 1, the space station (officially known by its Russian acronym DOS-1, which meant Long Duration Station 1) was completed in just 12 months. Another addition was about to be made to the roll call of glorious Soviet space triumphs.

Not that all preparations went according to plan. A record-breaking 18-day marathon, undertaken by Andrian Nikolayev and Vitali Sevestyanov in Soyuz 9, was intended to pave the way for long-term occupation of the new space station. However, despite a regime of exercise and medication, the crew did not cope well with their prolonged spell of weightlessness in cramped conditions, and had to be carried from their capsule on stretchers.

THE SOVIET SALUTE

The countdown for the world's first space station was marked by a minor crisis that typifies what can happen in a top-heavy bureaucratic system. It was decreed that the DOS-1 station should be known to the world as Zarya (Dawn), and this was the name actually painted on the side of the first space station to reach orbit. However, shortly before lift-off, it was realised that Zarya was also the call-sign of the Soviet mission control centre. The station was hurriedly renamed Salyut (Salute) in honour of the deceased Soviet cosmonaut and space hero Yuri Gagarin. (The name "Zarya" was later revived for the first element of the International Space Station, which was launched into orbit on 20 November 1998.)

Salyut 1 lifted off from Baikonur Cosmodrome in Kazakhstan on 19 April

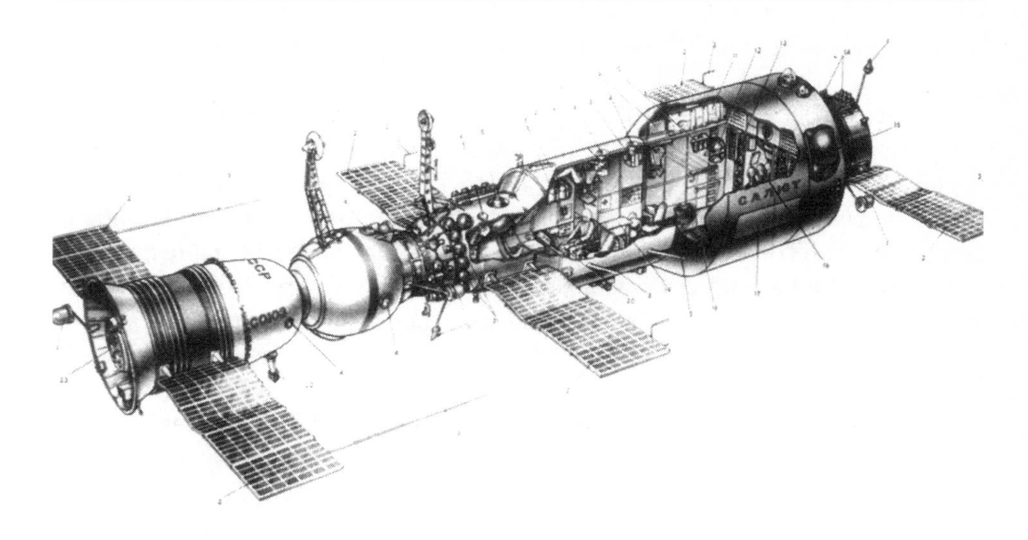

Cutaway diagram of the world's first space station, Salyut 1. (Peter Bond/Soviet Academy of Sciences photo)

1971, two full years before the American Skylab. The cylindrical station comprised two pressurised compartments and an unpressurised service module which contained the main propulsion system. Electricity was provided by two pairs of silicon solar arrays.

The forward transfer compartment, which measured 2 m (6.6 ft) wide and 3 m (10 ft) long, contained the docking port and a hatch that could be used for spacewalks.

The rear work compartment was divided into a small wardroom, 2.9 m (9.6 ft) wide and 3.8 m (12.5 ft) long and a research section 4.15 m (13.7 ft) wide and 4.1 m (13.5 ft) long. Removable panels covered the station's walls. Each wall was painted a different colour (light and dark grey, apple green and light yellow) to aid orientation in weightless conditions. Most of the cosmonauts' spare time was spent in the wardroom, which was fitted with a dining/work table, cassette player, sketch pad and a small library of books. The large-diameter work module contained the main control panel, a sanitation-toilet unit, two refrigerators for food and sample storage, and space for sleeping bags. Scientific equipment in the larger section included the Orion 1 telescope, the Oasis greenhouse and other biological experiments, including colonies of tadpoles and fruit flies.

Only four days after the station reached its operational orbit at an altitude of 200 × 220 km (125 × 139 miles), a three-man crew blasted off in Soyuz 10. The expected docking was completed within 24 hours, but, to general amazement, the craft remained linked for only 5 hours. Without entering the station, the disappointed crew hastily returned to Earth, appropriately landing in darkness on the morning of 25 April. Years later, the Soviets admitted that the Soyuz had been unable to complete its pressure seal with the Salyut.

Unsure of the reason for the mysterious failure, the Soviets prepared a second crew to go up and investigate. However, the experienced primary crew had to be replaced at short notice when an X-ray showed a spot on the lung of cosmonaut Valeri Kubasov. This was the reason that Soyuz 11, launched on 6 June, carried a back-up crew which included rookies Georgi Dobrovolsky and Viktor Patsayev. This time, the docking went perfectly. Commander Patsayev opened the hatch and floated through the tunnel into his new home, closely followed by his two companions.

Over the next four weeks, the crew became media celebrities as they frequently starred on TV broadcasts. The trio were kept busy testing the station's systems, monitoring Earth's surface and atmosphere, observing distant stars and conducting biomedical experiments. One of their favourite occupations was tending the first "space garden".

The doomed crew of Soyuz 11, Georgi Dobrovolsky, Viktor Patsayev and Vladislav Volkov, during training. After a 23-day stay on Salyut 1, the men perished during their return to Earth. (Peter Bond photo)

Not all went according to plan. A fire caused by a short in an electrical cable created panic and caused the crew to demand an immediate return to Earth, but Chief Designer Mishin was able to calm them down. A few days later, Patsayev toasted his 38th birthday in fruit juice. The much-valued presents from his colleagues were an extra onion and lemon from their fresh food rations.

Once again, "serious troubles in the space station's systems" arose, prompting the authorities to curtail the mission and bring the crew home a week early. On 30 June, the men loaded their flight log, Earth observation film and experimental results into the Soyuz for the return trip. Wearing only their woollen flight suits, they settled back in their individual body-contoured couches and undocked from the station.

As their craft braked to re-enter the upper atmosphere, explosive bolts fired to separate their descent module from the other sections of the Soyuz. Communication with the cosmonauts ceased, but no one on the ground was concerned – radio blackout was a normal consequence as the spacecraft followed its fiery dive through the atmosphere. The parachutes and soft-landing system worked perfectly, but still the hovering helicopter crews were unable to establish a voice link with the returning heroes.

Then came the shock. As the recovery team opened the Soyuz hatch and peered inside, there was no sign of movement or excited greeting from any of the cosmonauts. Instead, they lay motionless, silent, in their seats. All three were dead. Despite desperate efforts to resuscitate them, there was to be no happy ending. The official cause of death was depressurisation of the Soyuz cabin. Without the protection of pressure suits, the trio had quickly lost consciousness and expired.

The subsequent inquiry showed that a ball joint in an air pressure equalisation valve had been dislodged by the explosive jolt created during descent module separation. Patsayev tried to stop the leak with his finger, but to no avail.

Stunned by the disastrous finale of what should have been a triumphant display of the supremacy of communist technology, the Soviets set to work to redesign the flawed Soyuz. Meanwhile, Salyut 1 continued to orbit the planet unoccupied. It was eventually deorbited on 11 October, after 175 days aloft. It had been occupied for just 23 days.

Worse was to follow. With Skylab looming on the horizon, the Soviets were eager to regain the initiative. On 29 July 1972, the second DOS-type station was destroyed in an unpublicised Proton launch failure.

The next attempt for global recognition, before the advent of Skylab, occurred on 3 April 1973 when the first Almaz station was launched under the misleading name Salyut 2. Before it could be occupied, the new station lost orbital stability and began tumbling. In the usual Soviet manner, Moscow issued only vague details of the station's status and objectives. When Salyut 2 broke up on 14 April, the official statement simply declared that its mission had been completed. Years later, it was revealed that an electrical fire had caused the Almaz to depressurise and eventually break into more than 20 pieces.

The third DOS station reached orbit on 11 May 1973, just three days before Skylab. However, the Soviet obsession with hiding failures once again surfaced. Designated as Cosmos 557, a certain sign that something had gone wrong, the giant spacecraft remained empty throughout its 11-day lifetime. Once again, speculation in the West was rife, but the truth was only revealed much later. Shortly after reaching orbit, a malfunction in the ion sensors that controlled the station's attitude caused the spacecraft to exhaust most of the fuel used by its orientation system. One account states that a command to raise the station's orbit was sent while the craft was pointing the wrong way, causing it to re-enter the atmosphere.

Although the damage to Soviet prestige from this series of embarrassing débâcles was minimised to some extent by the veil of secrecy thrown around their programme, all opportunities to counteract the impact of Skylab had been wasted. However, space exploration has always been an unpredictable business, and the Americans almost gift-wrapped a negative publicity coup to their rivals.

SKYLAB: AMERICA'S APOLLO LEGACY

Throughout the 1960s, two American space station programmes were competing for government funds. The US Air Force wanted to fly a Manned Orbiting Laboratory – a modified two-man Gemini spacecraft enlarged by a partially pressurised workshop. In time it became increasingly obvious that this concept had passed its sell-by date, and in June 1969, one month before the first Apollo Moon landing, the Nixon Administration cancelled the programme.

However, the civilian options offered by NASA's Apollo Applications Programme were also dividing the agency. When it became clear that the almost unlimited funds available to NASA during the lunar race would be drastically reduced, ambitious proposals for multiple orbital workshops and manned ferry craft were quietly dropped. Instead, von Braun's team at Marshall Spaceflight Center backed the launch of an empty Saturn IV-B rocket stage that could be outfitted in orbit. The Johnson Space Center in Houston favoured the easier, cheaper alternative of launching a fully equipped laboratory on a single Saturn V booster.

The Texans won the day, and in July 1969 Project Skylab was born. Four

The USAF Manned Orbiting Laboratory was based upon the two-man Gemini spacecraft. The programme was eventually cancelled in June 1969. (USAF artist's impression)

years later, the 85-tonne giant sat on the launch pad at Cape Canaveral, cocooned inside an enormous protective canister. The plan was to launch the station on a two-stage Saturn V rocket, then send a three-man crew to make Skylab operational.

At first, all seemed to go well as the huge Saturn V slowly rose from the pad and headed eastward over the Atlantic Ocean. However, unknown to mission control and the watching multitudes, the mission was already in trouble less than one minute after lift-off. Air seeping under the protective shroud ripped a thin aluminium meteorite shield and one of the two main solar panels from the lab.

Only after the disabled Skylab reached orbit 430 km (270 miles) above the Earth did ground controllers realise that their $250 million creation was in serious difficulties. Although the x-shaped array of solar panels on the Apollo telescope mount deployed correctly, there was no glimmer of power from the two main solar panels. There was no choice but to postpone the crewed Skylab 2 flight.

Meanwhile, the station's internal temperature began to soar. While they tried to discover how the station had arrived in such dire straits, controllers turned Skylab in order to minimise its exposure to the Sun, even though this reduced the electricity supply from the available solar panels.

An emergency repair kit was hastily assembled and the Skylab 2 crew were given a crash training course which, it was hoped, would familiarise them with the most likely problems they were likely to face and how to rectify them.

Eleven days after the disastrous debut of Skylab, the crew of Charles "Pete" Conrad, Paul Weitz and physician Joe Kerwin set off for the unknown. Their worst fears were confirmed when the huge station drifted into view. One solar panel was entirely missing, while the other was snagged by an aluminium strap. The spacecraft completed a soft dock at the station's forward docking port while the crew discussed how best to proceed.

Confident that they could dispose of the tangled debris, they plunged into their makeshift tool kit and pulled out a "pruning hook". Hindered by a lack of footholds, Weitz had to anchor himself in the Apollo hatch by Kerwin grimly holding on to his legs. Despite his best efforts, the aluminium strap would not budge. Repeated attempts to dock once more with Skylab led to further frustration, and a torrent of colourful language from commander Conrad, before they finally pulled in at the docking port.

The next day, the crew ventured inside the ailing laboratory. Although the temperature was quite high, their gas masks proved to be unnecessary. Given the go-ahead to occupy the workshop permanently, the crew wasted little time in assembling a makeshift parasol and deploying it through a small airlock. The station's temperature began to drop almost immediately, and by 4 June, it was down to a respectable 24 °C.

Top priority was now to boost the power levels. Without power, Skylab would continue to resemble a giant whale floundering on a beach. On 7 June, Conrad and Kerwin ventured outside with the 8-m (25-ft) long pole and bolt cutter. In the absence of a decent foothold, they expended a great deal of energy in simply positioning the cutter on the errant strap.

Once Conrad had edged along the pole to attach a tether to the solar panel, Kerwin began the struggle to sever the metal strip. Suddenly, it parted, but to their chagrin the stuck panel refused to budge. Only after a joint tug-of-war with the rope did the panel break free, sending both men tumbling at the end of their tethers. Their remarkable efforts of endurance and courage had saved the station.

With 3 kW of power now surging through Skylab's systems, the crew was able to spend the last two weeks of its 28-day assignment concentrating on the scientific programme and enjoying the most luxurious facilities ever provided for space travellers.

Their home certainly bore no resemblance to the serenely spinning wheel envisaged 20 years earlier by von Braun. Like the Salyuts, Skylab comprised a

series of cylindrical sections, but the American station dwarfed its Soviet counterparts. Based on the third stage of the Saturn V rocket, it provided 278 m^3 (10,000 ft^3) of work space. The main workshop module was divided into two "rooms" by a metal grid.
Each astronaut had his own sleeping compartment. Entertainment came in the form of a dartboard with Velcro-tipped darts and magnetised playing cards that stuck to the work/dining table.

Since there was no way to bring fresh supplies to the station, all water, food and other consumables, including clothes and towels, were stored on board the cavernous station. All trash – including soiled clothes and used urine bags – was pushed into an airlock and dumped in a huge waste disposal tank. Skylab also carried the first zero gravity toilet to be flown on an American spacecraft.

Daily meals for each member of the crew had been selected in advance from a menu that included 72 food items. Heaters were built into their individual food trays, while a freezer and refrigerator were also available.

Bone loss and weakening of muscles was combated with the aid of exercise machines and a negative pressure device that simulated gravity by pulling blood into the men's legs. A collapsible shower was provided, but the long, complicated procedures required to set it up, use it and then dismantle it without spraying the surroundings with water meant that it was rarely used.

Despite numerous minor gripes with the design of certain pieces of equipment, the first crew returned home on 22 June 1973, delighted with the successes of their rescue mission and the subsequent scientific bonanza of Earth and solar X-ray images.

Six days later, the second crew set off for Skylab. After a miserable start, when they all succumbed to space sickness, the super-efficient trio enthusiastically entered into their research programme. By mid-August they thought nothing of working for 12 to 14 hours a day, and even put in a request for extra activities. Three spacewalks were also carried out, including a record-breaking 6½-hour EVA to deploy a second sunshade over the deteriorating parasol. Their adventure came to a successful conclusion on 25 September when the Apollo capsule splashed into the Pacific after 59 days aloft.

The third and final crew – all newcomers to spaceflight – found this a hard act to follow. They got off to the worst possible start when Chief Astronaut Alan Shepard reprimanded mission commander Gerald Carr for trying to hide the severe space sickness of crew member William Pogue.

Morale hit rock bottom as they struggled to find all of the equipment scattered around the station by their predecessors, while at the same time attempting to meet the impatient demands of mission managers down on the ground. Eventually, an agreement was reached to adjust work schedules and increase the time allotted for leisure activities.

Skylab, the only American space station, photographed by the third crew as they headed for home. Note the rectangular sunshield that was deployed by the second resident crew. (NASA SL4-143-4706)

However, as they entered the second half of their extended stay, the crew gained second wind and things began to improve. By the end of their record-breaking 84-day stay, they had travelled 1,260 times around the planet and covered 55 million km (34.5 million miles) – equivalent to a flight to Mars. They had also conducted four spacewalks, most notably a seven-hour spell outside the station on 26 December 1973.

Despite their early difficulties, the crew had gathered some 20,000 pictures of Earth, 30 km (19 miles) of data on magnetic tape and 75,000 images of the Sun, as well some exquisite images of Comet Kohoutek. Apart from the new information they provided on the way the human body adapted (or otherwise) to zero gravity, they produced improved crystals and metal alloys in a small furnace, and completed a series of trials of a revolutionary astronaut manoeuvring unit.

Commander Carr described the main lesson of their record-breaking odyssey: "That man can live a normal existence in space and that he can accomplish a great many things that can't be done on the ground."

The final crew left behind a "time capsule" of food and other items for future visitors, but no one ever set foot inside Skylab again. The assumption that the station's orbit could be raised to a safe altitude by the new Space Shuttle proved flawed, partly because of prolonged delays in completing the revolutionary reusable spacecraft, and partly because enhanced solar activity swelled Earth's atmosphere. In July 1979, the increased atmospheric friction caused the gigantic ghost ship to plunge out of control into Earth's protective blanket of air, scattering debris over the Australian outback.

SUCCESS AND FAILURE

In the Soviet Union, the desire for secrecy continued to feed speculation in the West about how its space programme would rebound from the Soyuz 11 tragedy and the subsequent miserable failures of Salyut 2 and Cosmos 557.

Two solo flights of the modified Soyuz ferry verified the validity of the design changes and gave an opportunity to try out some of the experiments to be flown on future space stations. However, the most notable change was the need for crews to wear pressure suits, with the result that only two cosmonauts could fit inside the revamped Soyuz.

Finally, on 24 June 1974, Salyut 3 was successfully placed into a 219 by 270 km (136 by 168 mile) orbit. The suspicions of Western observers were soon aroused by this low altitude, by the choice of an Air Force crew, and by the use of radio frequencies that were normally allocated for military use.

Salyut 3 was, indeed, intended primarily for military reconnaissance. Details released in recent years reveal that it was, in fact, the second of the Chelomei-designed Almaz stations. Although it was similar in size to the civilian Salyuts, the Almaz was quite different in appearance. It comprised three cylindrical sections – an airlock chamber with four access points to the vacuum of space, a 4.15-m (13-ft) diameter work module and a slightly narrower living module.

Sleeping quarters for the two-man crew in the living section comprised one permanent bunk and one foldaway bunk. The station was also equipped with a shower and a prototype water regeneration system that condensed moisture from the cabin atmosphere.

Since spying was at the top of the crew's agenda, the most important piece of equipment was a high-resolution camera. Film could be developed on board and scanned by a TV imaging system for broadcast to Earth, or, as the end of the mission approached, it could be loaded into a small capsule that was sent Earthwards through a chamber in the airlock.

Like its predecessors, Salyut 3 failed to live up to its potential. Pavel Popovich and Yuri Artyukhin successfully checked out the station's systems during a two-week sojourn on the station, but their would-be successors had to make a rapid return to Earth after a problem with their Soyuz ferry.

By the time Salyut 3 was deorbited on 24 January 1975, its civilian counterpart was already in orbit. Salyut 4 closely resembled the earlier OKB-1 designs, the main difference being the provision of three steerable solar panels

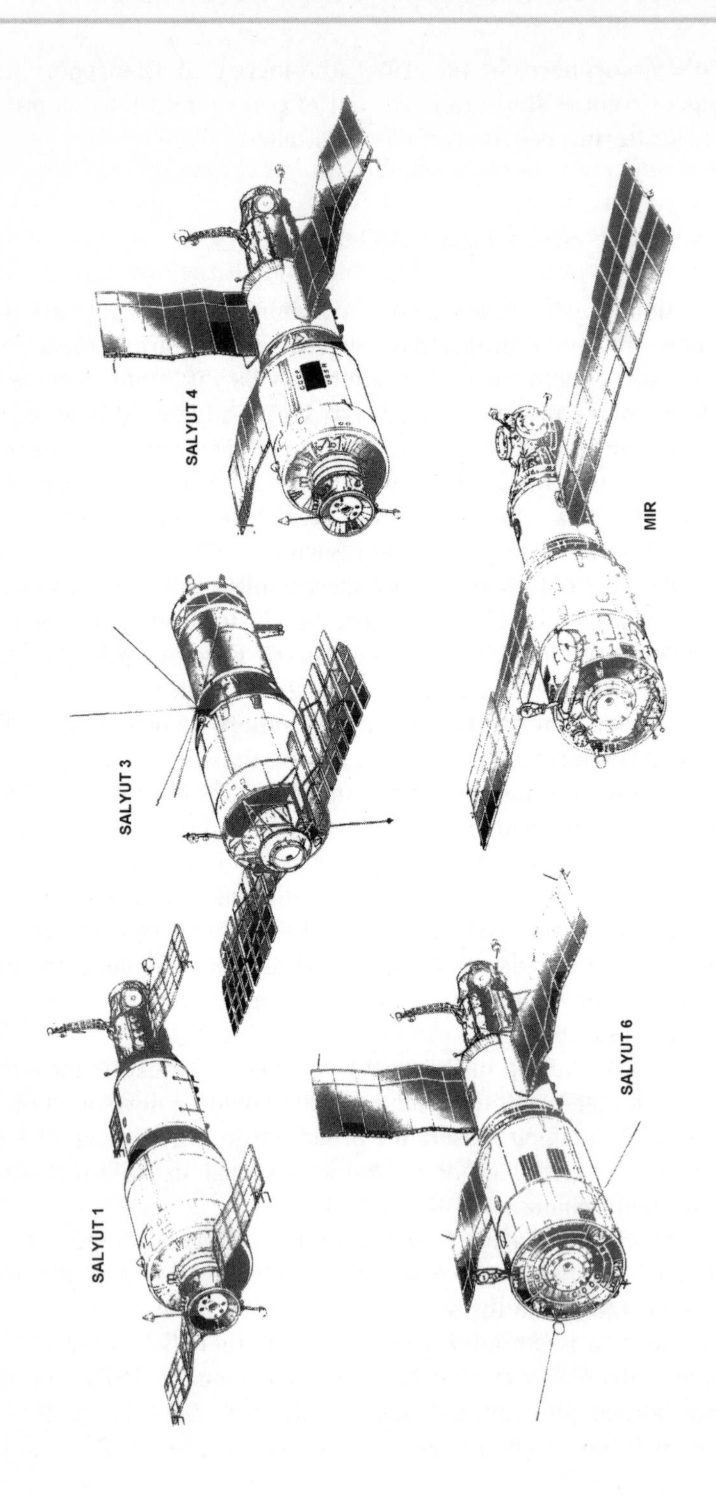

Evolution of USSR Space Stations: Artist's Impression. (Teledyne Brown Engineering)

in the shape of a T. Another innovation was a bicycle ergometer which generated electricity when in use.

From its relatively high 350-km (220-mile) circular orbit, the station's occupants were able to devote a considerable amount of time to astrophysical observations, as well as plant cultivation. A major problem soon arose when a failure of the solar telescope's pointing system caused its main mirror to be damaged. In a remarkable display of on-orbit dexterity, Alexei Gubarev and Georgi Grechko resurfaced the mirror and worked out a way of aiming the telescope using a stethoscope, stopwatch, and sounds generated by the mirror's motion.

Once again, the Soviets struggled to capitalise on a promising start. A second crew was almost killed in April 1975 when their launcher veered off course shortly after lift-off. Seven weeks later, the Soyuz 18 back-up crew succeeded in entering the station. During the next two months, they continued the scientific programme, watched their Soyuz 19 comrades lift off from Baikonur en route to an historic rendezvous with a US Apollo spacecraft, and managed to survive a breakdown in the Salyut's air conditioning system. Although they were weakened by their lengthy spell aloft, a strict exercise regime paid off as the cosmonauts insisted on walking from their return capsule to greet the media.

The most noteworthy event during the final months of Salyut 4's life was the first automatic rendezvous and docking by an unmanned ship, a dress rehearsal for future Progress cargo craft.

By the time Salyut 4 broke up harmlessly over the Pacific on 3 February 1977, its Almaz successor had been in place for more than seven months. The military Salyut 5 closely resembled Salyut 3 and was mainly utilised for the same purpose – orbital reconnaissance – although searches for mineral deposits and observations of storms and forest fires were also of benefit to the national economy. The first use of various furnaces for materials processing also took place on board Salyut 5.

Once again, a bright start was spoiled by subsequent difficulties. The first crew returned to Earth in a hurry after 48 days. It was later revealed that "acrid odours" in the cabin and the illness of a cosmonaut Vitali Zholobov were to blame. A replacement crew failed to dock with the station and almost drowned in a frozen lake upon landing. The third crew entered Salyut 5 wearing breathing masks, then proceeded to completely flush out and replace the air in the station from storage tanks in their Soyuz 24 craft. Once again, the all-important spy film was returned in a retrievable capsule the day after the crew bade the station farewell.

A fourth Almaz was scheduled for 1979, but the entire programme was cancelled by Defence Minister Dmitri Ustinov. As a result, the only long-term legacy of Chelomei's manned programme was the TKS transport system, which was revived for use in the next generation of civilian space stations.

SALYUT'S SECOND GENERATION

Until the advent of Salyut 6, the flexibility and orbital duration of space station missions was severely restricted by the inability to bring up supplies and fresh crews or return to Earth the results of research. This all changed with the second generation of Soviet space stations which made its debut on 29 September 1977.

Outwardly, Salyut 6 closely resembled its predecessors, with three main cylindrical modules and three solar arrays – two lateral and one upright like a fin. The main alteration was the addition of a second docking port at the rear, which allowed two spacecraft to link up simultaneously with the station.

This single change revolutionised space travel by opening the way for long-duration occupations. However, yet another docking failure had to be overcome before the theory could be put into practice. Despite four attempts to link up with the station's front port, the two-man crew of Soyuz 25 could not achieve a hard dock and had to make a hasty return to Earth. A giant question mark lingered over the future of the station: was Salyut's forward port damaged or did the problem lie with the Soyuz system?

Two months later, Soyuz 26 pulled in at the rear docking port on a mission to commission the new station and find out what had gone wrong during the previous flight. The provision of a second entrance was already paying off.

On 20 December 1977, Georgi Grechko and Yuri Romanenko made the first EVA from a Soviet space station – the first such excursion by cosmonauts since 1969. Grechko's inspection of the forward docking port showed that it was in pristine condition and confirmed that the Soyuz 25 must have been to blame. This good news was rapidly followed up by the launch of Soyuz 27 carrying Vladimir Dzhanibekov and Oleg Makarov, the first crew ever to visit an occupied space station in orbit.

Although the resident crew retired to their Soyuz in case a hasty emergency evacuation was required, the docking of the second ferry at the forward port went perfectly. After the televised bear hugs and toasts in fruit juice, the four men settled down to enjoy the special delivery of letters, books, newspapers and research equipment.

Following five days of combined activities, Dzhanibekov and Makarov prepared to head homewards, following a sequence of events that would be repeated many times in the years ahead. After transferring their personally contoured couches from one Soyuz to the other, they loaded the old Soyuz 26 with film and research results and pulled away from the rear port, leaving it free for a fresh replacement.

More space history was made on 22 January 1978 when an unmanned Progress craft arrived at the recently vacated back door of the Salyut. The new visitor delivered a tonne of fuel together with a similar weight of compressed air, food (including fresh fruit and bread), water, film and other cargo. It took the crew more than a week to shift all of these supplies into the station, leaving room for rubbish to take their place.

Another revolutionary breakthrough followed on 2 February when fuel and oxidiser began to pump through the pipes linking the two craft. Having replenished Salyut's fuel supply, the Progress further prolonged the station's operational life by acting as a tug and boosting it into a higher orbit. Then, its useful life over, the robotic cargo ship was ordered to self-destruct over the Pacific Ocean.

One more landmark event occurred on 2 March with the arrival of a "guest" cosmonaut, Czech Army Air Force captain Vladimir Remek, the first space traveller not to be an American or Soviet citizen. Remek's week-long stay on the station was the forerunner of many such visits by representatives of Soviet satellite countries in the years ahead.

No sooner had the international crew departed than the first long-term expedition started preparing for a heroes' homecoming. The space station was finally evacuated on 16 March. Their 96-day occupation had smashed the space endurance record established by the Skylab 3 astronauts.

Three months later, the second principal expedition arrived at Salyut 6, ready to push the endurance record even further. During their 140-day stay, Vladimir Kovalyonok and Alexander Ivanchenkov played host to three Progress supply ships and two international crews that included an East German and a Pole. However, the most notable "first" was the relocation of a Soyuz spacecraft from the aft docking port to the forward port, leaving the rear entrance free for an automated rendezvous and docking by Progress 4. The crew pulled away to a distance of 200 m (660 ft) and waited while ground control ordered the unmanned station to swivel through 180 degrees, leaving the way clear to return to their orbital home.

As 1979 dawned, the Soviet space programme seemed to be on the crest of a wave, having passed the cumulative space time recorded by their American rivals, while the NASA manned programme floundered in the doldrums, beset by delays with the new Space Shuttle.

However, storms were on the horizon. The third Salyut 6 long-term crew arrived on 26 February 1979, equipped with a special tool kit and a commission to prevent hazardous fuel leaks from contaminating the entire propulsion system. In an unprecedented work programme, Vladimir Lyakhov and Valeri Ryumin had to spend almost two weeks emptying a damaged fuel tank, venting any fuel dregs into space, filling it with nitrogen gas and then repeating the entire process until the system could be safely sealed off from the remaining two fuel tanks.

The first foreign guest cosmonaut, Czech Vladimir Remek (left), on board Salyut 6. Also at the dining table are Alexei Gubarev, Georgi Grechko and Yuri Romanenko. (Peter Bond photo)

The duo's 175-day marathon turned into an endurance exercise that tested the mettle of both occupants. Much of their time was spent on repairs, with remedial work required for the treadmill and exercise cycle, the communications and TV systems, the air conditioning and life support systems. Despite these distractions, the men succeeded in taking some 9,000 photographs of the Earth, carried out numerous astronomical observations with the station's infrared, radio and gamma ray telescopes, completed more than 50 furnace smeltings and undertook various biological experiments.

Hardest to bear, however, was the unprecedented long-term isolation. After a main engine failure on the incoming Soyuz 33 on 11 April, the Soviet authorities insisted upon a redesign of the craft's propulsion system before they would risk further lives. However, this, in turn, created another safety problem – how could the ageing Soyuz 32 "lifeboat" be replaced before its systems seriously deteriorated?

The answer was to send up an unmanned version that would dock automatically at the vacant second port. On 8 June, Soyuz 34 delivered 200 kg (440 lb) of cargo and enabled the crew to return experiment results and broken equipment in the unoccupied Soyuz 32, secure in the knowledge that a fresh "lifeboat" was now available.

Further supplies arrived in May and June on board two Progress ships, but there was to be no human company to break the solitude. However, the monotonous routine was interrupted with the arrival of an experimental radio telescope that unfurled in the aft port as Progress 7 pulled away. After three weeks of astronomical and Earth observations, it was time to jettison the telescope's 10-m (33-ft) diameter antenna, but to the crew's dismay, the giant dish snagged on the station's aft docking target.

In order to free the antenna and clear the way for future spacecraft, it was essential for the tired, unprepared crew to carry out an emergency spacewalk. On 15 August, Ryumin donned a pressure suit and struggled along the full length of the station to complete the hazardous task of cutting the metal dish free. The perspiring cosmonaut gratefully sent it on its way with the aid of his boot.

Salyut 6 was eventually mothballed and abandoned on 19 August, but the Soviets had not yet finished with the sturdy station. An improved version of the manned ferry, known as Soyuz T-1, docked automatically with the station on 19 December, where it remained for 95 days. No sooner had the modified ferry departed than Progress 8 arrived with fresh supplies for future human occupants.

Within 12 days, Soyuz 35 pulled alongside with the station's next residents. Flight engineer on the two-man crew that entered the station on 10 April 1980 was none other than Valeri Ryumin, who had completed a remarkably rapid period of recuperation since his return from the station eight months earlier. To his surprise, Ryumin had been called upon for a rapid reacquaintance with Salyut when prime crew member Valentin Lebedev had badly damaged a knee during a trampoline accident.

Ryumin's second spell aboard the redoubtable station proved to be much less stressful than the first. The lengthy list of visitors included a brief rendezvous with the Soviet crew of Soyuz T-2, week-long tours by crews that included guests from Hungary, Vietnam and Cuba, and four of the now-routine Progress resupply missions. Another landmark in space station history was passed when water was piped for the first time from a Progress ship to the station's tanks.

By the time the much-travelled cosmonaut touched down on the steppes of Kazakhstan on 11 October, he had spent almost a year in orbit, long enough to fly all the way to Mars, with no long-lasting side effects.

Two more major expeditions were sent to Salyut 6 before it was retired from duty. In November, the first three-man Soviet crew to fly since the Soyuz 11 tragedy nine years earlier arrived for an 11-day stay on the station. The final long-term crew's 75-day sojourn was marked by brief visits from Intercosmos crews that included a Mongolian and a Romanian. The last occupants of Salyut 6 departed on 26 May 1981 – six weeks after the maiden flight of the US Space Shuttle – but the empty station still had a role to play.

An experimental TKS spacecraft, originally developed by Chelomei's OKB-52 bureau as a supply ship for the Almaz space stations, pulled in at the forward docking port on 19 June after jettisoning its (unmanned) Merkur crew return capsule. Cosmos 1267, which was almost as large as the space station itself, was used several times to raise the orbit of Salyut 6 before both craft were deliberately destroyed during re-entry on 29 July 1982.

The significance of this impressive newcomer became clear when the Soviets announced that it was "designed to test systems and elements of the design of future spacecraft and for training in the methods of assembly of orbital complexes of a big size and weight." In other words, ships such as Cosmos 1267 would soon become the building blocks upon which a new generation of modular space stations would eventually be founded.

Even before the demise of Salyut 6, its successor was already in orbit and awaiting its first occupants. It was immediately clear that, despite some minor modifications, Salyut 7 closely resembled its highly successful predecessor and was intended largely to consolidate the Soviet lead in long-endurance spaceflight experience.

First on board the pristine station were Anatoli Berezovoi, who was making his first trip into space after 12 years as a cosmonaut, and Valentin Lebedev, now fully recovered from the injury to his knee. With remarkable frankness, Lebedev related in his diary the trials and torments, as well as the moments of triumph and pleasure, involved with life on board a space station.

"The most difficult thing in flight is not to lose temper in communication with the Earth and within the crew, because accumulating fatigue makes for frequent slips in contacts and in the work with the Earth," he confided. "In the crew, too, these explosive moments happen, but no eruption must be allowed, for a crack, if it appears, will grow wider."

The Soviet propaganda machine went into overdrive for the week-long visits of Frenchman Jean-Loup Chrétien and Svetlana Savitskaya, the second woman to fly in space. The daily routine was also broken by a 2½-hour spacewalk and the opportunity to launch two small Iskra satellites from the "trash airlock".

As the 211-day marathon neared the finishing line, Lebedev's longing for home was countered by a warm familiarity with his isolated dwelling.

"Everything in it is so near and dear to me now," he wrote. "We've touched every square millimetre and object in here. We know exactly where every piece of equipment is mounted, not from documentation but from memory. Many little details, such as photographs on the panels, children's drawings, flowers, and green plants in the garden, turn this high-tech complex into our warm and comfortable, if a little bit unusual, home."

The replacement crew was a long time in coming. On 10 March 1983, Cosmos 1443, another in the series of large TKS craft, arrived at the station's front port. However, after a failed docking attempt by Soyuz T-8, it was left to the back-up crew of Vladimir Lyakhov and Alexander Alexandrov to eventually float through the rear hatch of Salyut 7–Cosmos 1443 on 28 June.

Further misfortunes continued to plague the programme. In early September, the main oxidiser line of the Salyut 7 propulsion system ruptured during refuelling. As one Western analyst commented, this left Salyut 7 "dead in the water" and totally reliant on Progress craft for attitude and orbital manoeuvring.

Then the Soyuz T-8 crew, who had been trained to install two additional solar panels on the station, suffered a further near mishap on 26 September when the rocket carrying their Soyuz T-10 craft exploded on the launch pad. Cosmonauts Titov and Strekalov barely escaped with their lives.

In the absence of the trained maintenance team, it was left to Lyakhov and Alexandrov to boost the station's power supply. During two tricky spacewalks on 1 and 3 November, the duo successfully achieved their objective of boosting the station's power supply by 50%, by attaching the extra panels to the "top" solar array.

Despite concerns about its condition after almost five months in space, the Soyuz T-9 ferry operated perfectly to bring the hard-driven duo home on 23 November.

As 1984 dawned, the Soviets continued to pour scorn on Western media reports that Salyut 7 was in serious trouble. To prove their point, a three-man crew blasted off on 8 February to reoccupy the station. Their craft was labelled Soyuz T-10, as if the previous aborted mission had never existed.

In an unusual departure from normal practice, the third main expedition to Salyut 7 contained physician Oleg Atkov – a clear indication that they expected to be setting new endurance records.

After the temporary distraction of a visit by an Indian "guest" cosmonaut, Leonid Kizim and Vladimir Solovyov were able to concentrate on their primary task, the repair of the station's main propulsion system. During a remarkable series of four spacewalks in 10 days, the duo cut through the station's external insulation and outer hull before inserting two bypass conduits.

A fifth EVA on 18 May saw a second set of additional solar arrays attached to one of the main lateral panels, further boosting electricity supplies. The overhaul of the propulsion system was completed during spacewalk number six on 8 August, when the ruptured fuel line was finally sealed. However, despite this impressive demonstration of how intervention by humans can overcome disaster in space, the cosmonauts' efforts were largely irrelevant in influencing the future of the station – Salyut 7's main propulsion system was never again used to boost its orbit.

Once Salyut 7 was mothballed on 2 October 1984, the heroic trio returned to Earth after a record-breaking 237 days in orbit. As the months passed, it seemed that the Soviets had decided to bring the station's career to a premature end. Then in February 1985, communications between the station and ground control abruptly ceased. It seemed clear that a major malfunction had left the space station wallowing helplessly in a decaying orbit.

Contrary to all expectations, the Soviets decided to mount a hazardous, almost reckless, rescue operation. Unsure of what they would find, the would-be salvage crew of four-time space traveller Vladimir Dzhanibekov and Salyut expert Viktor Savinykh successfully completed a manual docking with the lifeless hulk on 8 June 1985.

Donning their gas masks, the pair ventured inside. As their torches illuminated the dark, cavernous cabin, their attention was drawn to a packet of rusks and some salt tablets that had been left as a traditional welcoming gift by the previous occupants. However, as they continued to study their bleak surroundings, the task of resurrecting the station seemed to become ever more hopeless. In the absence of electricity or sunlight – the metal window shades were drawn – the internal temperature had plummeted to below zero, resulting in a frosted coating on walls and windows. Even the drinking water was frozen.

The only hope was to find a way of restoring some power. With the aid of their Soyuz ferry, the cosmonauts turned the station so that its solar panels captured the maximum amount of sunlight. One by one, each of the chemical batteries was fully charged. By 13 June, they were able to reactivate the station's orientation system, clearing the way for a supply delivery by a Progress craft.

Meanwhile, despite their fur boots, hats and gloves, the sub-zero temperature remained a major handicap. Not until 10 June was there sufficient power to turn on the cabin air heaters and begin the slow process of raising the station's temperature. The crew returned to the relative comfort of the Soyuz to snatch some sleep, and managed to eat some hot food by improvising a stove from a metal container and a photographic lamp. Until they could defrost the water tanks, drinking water remained in short supply, so the men rationed every drop, including water from the EVA suits and condensation from various pipes and hoses. The working conditions were complicated still further by the total reliance on the Soyuz life support systems. The poor ventilation led to an accumulation of exhaled carbon dioxide, resulting in tiredness and headaches.

The remarkable renaissance continued after Progress 24 docked on 23 June.

On board were much needed water and fresh food, together with replacement parts for the reviving Salyut. By the end of June, the station's water tanks were ice-free, although another month passed before atmospheric humidity returned to normal. After another Progress (dubbed Cosmos 1669) brought further supplies on 21 July, the repair men *par excellence* enjoyed the thrill of a walk in space to attach a pair of solar array extensions to the third of Salyut 7's main panels.

The dramatic 112-day rescue mission had cleared the way for further manned flights to the station. In the first ever hand-over of a primary crew, Viktor Savinykh stayed on board to join the incoming pair of Vladimir Vasyutin and Alexander Volkov, the station's fifth group of long-term occupants. The mission began well with the arrival of Cosmos 1686, a modified TKS, on 2 October. However, the flight was suddenly terminated after Commander Vasyutin became ill, forcing an emergency return to Earth on 21 November.

With the launch of the new Mir space station on 20 February 1986, Salyut 7 once again seemed to have reached the end of the line. But it had one more task to perform. On 5 May, in another first for the Soviet manned programme, cosmonauts Vladimir Solovyov and Leonid Kizim transferred from one space station to another. In order to catch up with Salyut 7–Cosmos 1686, which lay some 3,520 km (2,200 miles) ahead of Mir, the crew manoeuvred their Soyuz T-15 spacecraft into a lower, faster orbit. Docking with the elderly station took place the following day.

For the next seven weeks, the pair carried out the series of experiments originally intended for their predecessors. Of particular significance were the first live televised spacewalks in Soviet space history. On 28 May, viewers saw the men practise future space construction techniques by unfolding and then

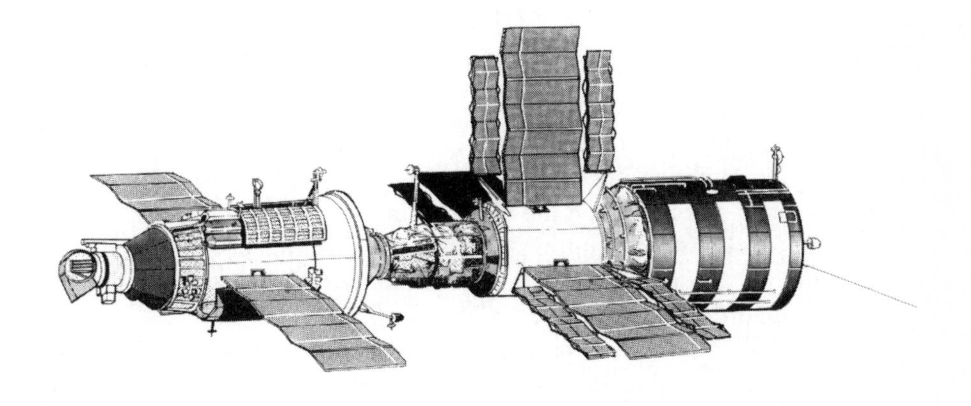

The Salyut 7–Cosmos 1686 complex. By this stage, additional solar panels had been added to each of the main solar arrays on the Salyut 7 space station. (Teledyne Brown Engineering Diagram)

dismantling a 12-m (40-ft) girder. More records were shattered as the duo chalked up their eighth spacewalk on 31 May, bringing their total EVA experience to 31 hours 40 minutes. This time, they practised welding several joints after deployment of the girder.

Salyut 7's brief return to the limelight ended on 26 June, when Solovyov and Kizim transferred back to Mir, taking with them some 400 kg (880 lb) of equipment. In mid-August, the engines of Cosmos 1686 were fired to boost the orbit of the abandoned station to a record altitude of 475 km (300 miles). However, even this was not high enough to overcome the increased frictional drag caused by expansion of the Earth's atmosphere during a period of maximum solar activity. Short of fuel and suffering from major systems failures, the 40-tonne Salyut 7–Cosmos 1686 combination crashed to Earth on 7 February 1991, scattering debris across the sparsely inhabited mountainous regions of Chile and Argentina.

MULTIMODULAR MIR

The arrival of Mir (Peace) signalled the beginning of a new era in the history of manned spaceflight. Although the newcomer was similar in size and design to the civilian Salyuts, there were a number of importance differences. The most significant of these was the presence of six docking ports, five in a spherical node at the front end and another at the rear. It was clear that the Soviets were intending to use these ports to enlarge the station by permanently attaching additional modules to the base section.

The other main external difference was the provision of only two solar panels, although this was partly compensated by the use of more powerful gallium arsenide photovoltaic cells.

Internally, Mir's cylindrical work compartment was subdivided into two separate areas in the traditional way, but the absence of specialist experimental equipment meant that it was much more spacious and comfortable. Its facilities included two small, individual sleeping cabins, exercise machines and improved life support and ventilation systems. Many of the station's operations were automated and each cosmonaut also had his own personal computer. Videos and music were available for leisure use and during the lengthy, monotonous exercise sessions.

Solovyov and Kizim, the first crew to enjoy the gleaming new station, spent two periods on board, punctuated by their jaunt to Salyut 7. During their absence, an upgraded, unmanned ferry craft known as Soyuz TM successfully completed its maiden flight to Mir. By the time of their departure for Kazakhstan on 16 July, Kizim had made history by becoming the first human to spend an entire year in space.

Artist's impression of the Mir core module, which was launched on 20 February 1986. A Soyuz spacecraft is shown attached to one of the five docking ports in the station's spherical transfer compartment. Mir went on to become the largest, longest-lived space station in history. (CNES/David Ducros)

Although the expansion of Mir was not long in coming, the operation was far from smooth. Despite some initial computer problems and the loss of a vital data relay-communication satellite – harbingers of things to come – the second prime crew of Yuri Romanenko and Alexander Laveikin seemed well established by the time the Kvant (Quantum) module was launched on 31 March 1987.

Five days later, after a series of major orbital manoeuvres, the crew retreated to their Soyuz craft while the Kvant attempted an automatic docking. To their dismay, the Igla rendezvous system lost its lock as the 20-tonne spacecraft began the final leg of its approach. The horrified duo watched and prayed as the Kvant, like some giant albatross, glided past, only 10 m (33 ft) from the station.

A nerve-racking second attempt on 9 April seemed to go according to plan, but the new module refused to complete a hard dock with Mir's aft port. Romanenko and Laveikin were sent outside to find out what had gone awry. To everyone's surprise, they found a rubbish bag lodged in the docking unit. Ground control commanded the Kvant to extend its probe, allowing the cosmonauts to retrieve the waste item. At long last, the way was clear for Kvant to seal its connection with Mir, then, on 12 April, the newcomer's propulsion section was separated from the space station module and jettisoned.

Apart from increasing Mir's pressurised living space, Kvant significantly improved the station's capabilities. An Elektron unit was available to generate oxygen from water, while other equipment removed carbon dioxide and other harmful trace gases from the station's atmosphere. Also on board were six gyrodynes, which permitted extremely accurate pointing of the station and allowed considerable fuel saving. Science also benefited from the delivery of an international astrophysical observatory, which comprised an X-ray telescope and an ultraviolet telescope. The timing proved to be perfect as the nearest supernova in several centuries exploded soon afterwards.

The Mir core module (left) after the arrival of Kvant. Soyuz TM-3 is docked at the rear port. (NPO Energia)

A supplementary solar panel delivered by Kvant was installed on Mir's exterior during two spacewalks in June, so increasing the station's power output to 11.4 kW. Then came the announcement that Laveikin was suffering from an irregular heartbeat. Deciding to play safe, the Soviet authorities ordered him home with the next visiting crew, Alexander Viktorenko and Syrian Mohammed Faris. The third man on board Soyuz TM-3, Alexander Alexandrov, joined Romanenko for the remainder of his record-breaking stay.

The pair's occupation passed fairly peacefully, the main excitement being a redocking trial with the unmanned Progress 32. However, as Romanenko began to noticeably tire towards the end of his marathon, the daily work schedule was reduced to only 4½ hours. Even as the exhausted cosmonaut headed back to Earth for a well-earned rest after 326 days aloft, an even more ambitious expedition was already under way.

Vladimir Titov and Musa Manarov extended the non-stop spaceflight duration record to almost 366 days – an entire year away from home, family and normal gravity. During their extended stay, they enjoyed the diversion of two visiting crews, each of which brought its own excitement.

At the end of August 1988, Soyuz TM-6 delivered physician Valeri Poliakov to the station, providing some valuable medical expertise for the final few months of the prime crew's occupation. However, his companions, Vladimir Lyakhov and Afghan Abdul Mohmand, almost perished when the main engine of their aged Soyuz TM-5 malfunctioned, almost stranding the men in orbit. Only after two manual firing attempts did they succeed in completing their mission.

The second visit came as Titov and Manarov were beginning to think of home. The star of the month-long handover mission was Frenchman Jean-Loup Chrétien, who grabbed the headlines by participating in the first spacewalk by a non-Soviet or American citizen. However, the presence of six men in the cramped conditions of Mir–Kvant put a strain on power and life support systems, and there was general relief all round when Titov, Manarov and Chrétien departed on 21 December 1988.

Poliakov remained on board with newcomers Alexander Volkov and Sergei Krikalev, but the end of Mir's first period of manned operations was in sight after two years of non-stop occupation. Delays in completion of the station's new modules meant that the station was mothballed in April 1989 and it remained empty for the next 4½ months.

The initial indication that its solo run was about to end came with the launch of Progress M-1, the first automated craft to arrive at the front port of a Soviet space station. With Mir's fuel tanks topped up, Alexander Viktorenko and Alexander Serebrov set off for the station on 6 September. The main purpose of their mission was to receive and check out Mir's second add-on module.

Once again, unexpected difficulties arose after launch, with an automated approach aborted just 20 m (67 ft) from the station. However, Kvant 2 eventually docked at the forward transfer section on 6 December 1989. Two days later, a manipulator arm on the 20-tonne module grappled with the docking node and pivoted Kvant 2 into its permanent slot on Mir's upper lateral port.

Among the facilities provided by the newcomer were a metal shower cabin which could also be used as a sauna; six more gyrodynes to aid the orientation of the expanding station complex; a system for processing urine so that the recycled water could be electrolysed to produce oxygen by an Elektron generator similar to the one on Kvant; and a 1-metre (39-in) diameter EVA hatch that would become the main exit into outer space. Kvant 2 also carried various cameras and spectrometers for Earth observation, and an incubator for hatching and rearing Japanese quail.

The new year began with a series of spacewalks to tidy up the outer hull of Mir. The second of these was the final excursion through one of the hatches in the station's docking node. Three more EVAs followed, during which Viktorenko and Serebrov tried out an upgraded version of the Orlan pressure suit, then twice practised acrobatics on a special backpack that enabled cosmonauts to fly untethered around the station.

The next expedition got off to a bad start when the incoming crew's Soyuz TM-9 spacecraft arrived with three loose thermal insulation blankets. While mission control worked on methods of repairing the ferry, the station was manoeuvred in order to reduce condensation and temperature contrasts.

The next 20-tonne module, Kristall (Crystal), eventually set off for Mir on 31 May 1990, but the unfortunate precedents set by the previous modules were repeated when a thruster problem aborted the first docking attempt. Kristall eventually pulled in at Mir's front port on 10 June, and the next day it was transferred to a lateral site directly opposite Kvant 2.

The third permanent addition to Mir carried two detachable solar panels and two docking ports that could be used by the Soviet Shuttle Buran. In fact, it was the US Shuttle which eventually pulled in at Kristall five years later. As its name suggests, it was also packed with materials processing furnaces and other experiments, including two astrophysical telescopes, a greenhouse and an Earth resources camera. Six more gyrodynes improved attitude control.

It was now necessary to carry out the unusual task of reattaching the flapping thermal blankets. The first stage went according to plan, when the Soyuz TM-9 was relocated at Mir's forward port. However, in their hurry to begin the complicated work, Anatoli Solovyov and Alexander Balandin opened the EVA hatch on Kvant 2 before all of the air had been evacuated. The metal door slammed back on its hinges.

This unusual picture shows Mir with its first three modules. At the front end are Kvant 2 (top) and Kristall (lower middle). Also docked are Progress M-17 (right) and Soyuz TM-16 (bottom). At extreme left is Progress M-18, which has just undocked from the station. (NPO Energia photo)

Blissfully unaware of what they had done, the duo edged along the module to the Soyuz, but found it impossible to tie the blankets back in place. The only alternative was to fold down two of the blankets and hastily retreat to Kvant 2 before their air supply ran out. To their dismay, they found that the EVA hatch would not close. They had to float into the module's central compartment and use it as an emergency airlock before they could remove their suits and breathe a sigh of relief. The exhausting EVA had lasted 7 hours 16 minutes.

Following a week of rest and reassessment of the situation, Solovyov and Balandin depressurised Kvant 2's mid-section and, after several attempts to close the hatch, they eventually succeeded in securing it and recovering the use of the EVA airlock.

Two weeks later, the next two-man crew and a batch of quail arrived at Mir's front door. However, Gennadi Manakov and Gennadi Strekalov found that the damaged hatch was beyond their capability to repair. The main highlights of their four-month sojourn were the first use of a Progress return capsule to send film and samples back to Earth, and the brief visit by a chain-smoking Japanese journalist on board Soyuz TM-10.

However, it was the long-term residents, Victor Afanasyev and Musa

Manarov, who stole the headlines in the weeks to come. On 4 January 1991, they succeeded in repairing the damaged hinge on the Kvant 2 airlock, thus clearing the way for further spacewalks over the next three weeks. In preparations to relocate Kristall's solar panels, the pair installed a 14-m (46-ft) movable crane on the exterior of the core module and then readied support structures on the Kvant module.

However, even as the crew prepared to expand the station's capabilities, the Soviet media began to take advantage of its new-found freedom to complain about the amount of time being spent on repairs and maintenance, even expressing doubts over the economic return from the multi-million rouble Mir.

These critical comments almost became superfluous as Mir entered its sixth year of operation, when the space station avoided destruction by a hair's breadth. Two days after a failed automatic approach on 21 March, Progress M-7 made a second attempt to dock with Mir. When further problems arose during the final stages of its inward run, ground control ordered the 7-tonne cargo ship to veer away from the station at the last minute. The relieved cosmonauts, watching from their Soyuz spacecraft, stared in amazement as the huge, cargo-laden ship miraculously skimmed past the station's outstretched panels and disappeared into the black background.

During a relocation manoeuvre of the Soyuz from the forward to the aft port on Mir, the crew were able to pinpoint the cause of the problem to a faulty Kurs antenna which had lost one of its dishes after being inadvertently kicked during an EVA. Fortunately, the resupply of the station was successfully achieved when the errant Progress eventually pulled in at the evacuated port.

The eventful 175-day expedition concluded with the arrival of a space rarity – British cosmonaut Helen Sharman, only the third woman ever to set foot on a Soviet space station – accompanied by the replacement crew of Anatoli Artsebarski and Sergei Krikalev. The publicity potential of the Briton's week-long visit was marred somewhat by occasional power blackouts as Mir's solar panels drifted out of alignment with the Sun. Even the financial aspect proved a disappointment, since the Soviets had to pay for Sharman's entire stay after a commercial sponsorship deal fell through.

The new team's programme got under way in spectacular style with six spacewalks in little more than a month. With only one axial port operational, the first priority was to bring the damaged Kurs antenna back online. This was achieved on 24 June, clearing the way for a series of EVAs to construct a 14.5-m (48-ft) girder on the Kvant module. This would eventually house a propulsion system that would augment Mir's existing attitude control thrusters.

Politics came to the fore towards the end of August, when the cosmonauts had to listen helplessly to news accounts of a failed attempt to oust Soviet President Gorbachev. Although the government was restored to power,

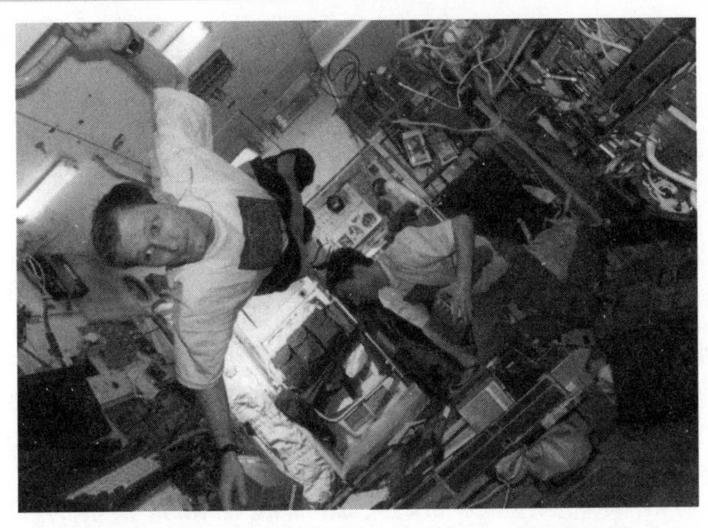

Sergei Krikalev and visitor Klaus-Dietrich Flade inside the crowded Mir base module. (DLR photo)

Gorbachev's position was fatally weakened and the Soviet Union rapidly fragmented into separate republics.

With financial and political considerations increasingly affecting the space programme, two missions were merged into one when veteran Alexander Volkov arrived on 4 October with two "guests" – one from Austria and the other from newly independent Kazakhstan, home of the Baikonur launch centre. In the absence of a flight engineer, Krikalev drew the short straw and was asked to extend his stay on Mir by another six months. Headlines in some Western media delighted in describing the plight of the "stranded cosmonaut".

As a result of the reshuffle, Artsebarski was given the task of shepherding the visitors back to Earth while Volkov joined Krikalev for a prolonged, frustrating session of repairs, communication problems and power shortages. Even the ground controllers staged industrial action in support of demands for pay rises. On the positive side, Krikalev was able to carry out his seventh EVA of the mission, although even that was complicated by a malfunction in Volkov's pressure suit. Krikalev eventually touched down on the Kazakh steppes on 25 March 1992, after 312 days in orbit.

After a 4½-month holding mission by cosmonauts Alexander Viktorenko and Alexander Kaleri, efforts to upgrade the station's attitude control were continued by Anatoli Solovyov and Sergei Avdeyev. A new thruster unit delivered by Progress M-14 was eventually installed during three spacewalks in early September 1992. In a symbolic gesture, the duo also replaced the now-tattered Soviet flag with its equivalent from the new Russian Republic.

The next six-month sojourn was by Gennadi Manakov and Alexander

A pale Sergei Krikalev receives a hero's welcome on his return to the steppes of Kazakhstan after 312 days in space. During this time, the Soviet Union had ceased to exist after a failed coup and the resignation of President Gorbachev. Alongside Krikalev are Alexander Volkov and German Klaus-Dietrich Flade (right). (DLR photo)

Poleshchuk, who arrived in Soyuz TM-16, the first spacecraft to pull in at one of the Shuttle docking ports on the nose of Kristall. This brought the overall mass of the seven spacecraft currently making up the Mir complex to more than 90 tonnes.

Of particular interest during their stay was a test of a revolutionary solar reflector/solar sail that was deployed from Progress M-15, and a subsequent trial of a manual system that allowed the cosmonauts on Mir to control the flight of the unmanned craft. The new capability was envisaged for use as a back up in case of future docking problems with the ferry craft. Further undocking and redocking manoeuvres were carried out before the Progress was commanded to self-destruct in the atmosphere.

Just as important were the two EVAs to fit solar array electric drives to Kvant. A replacement handle for the Strela crane was delivered by Progress M-18, the first time that two of the robotic craft had occupied Mir ports simultaneously.

Mir's 14th prime expedition arrived on 3 July 1993 in the form of Vasili Tsibliev and Alexander Serebrov. A month later, mission control prepared the crew for an emergency return to Earth during an appearance of the Perseid meteor shower, but despite a peppering of the station's windows and exterior, the precautionary measures proved to be unnecessary.

September saw three spacewalks to inspect the station's exterior and erect another experimental girder that could have multiple applications in the construction of future space stations. Two more EVAs in October completed the scrutiny of the station's entire hull for impact damage. Apart from 10 mini-craters on Mir's windows and numerous pockmarks on the thermal insulation and metal exterior, the most notable flaw was a 5-mm (0.2-in) hole through one of its solar panels. Whether it was excavated by a meteoroid or man-made debris was impossible to determine.

As further evidence of Russia's economic woes, the mission was extended by two months due to a shortage of rocket engines for the Soyuz booster. Yet further problems faced the outgoing crew when their Soyuz TM-17 hit Kristall with two glancing blows during the customary fly-around prior to deorbit. Fortunately for the new arrivals – Viktor Afanasyev, Yuri Usachev and Dr Valeri Poliakov – both spacecraft remained unaffected and life returned to the normal routine of repairs, experiments and exercise.

SHUTTLE–MIR

By this time, following an agreement signed in June 1992, plans to combine Mir and Shuttle operations were well under way, with the long-term aim of preparing both sides for joint efforts in building and operating an international space station. Indeed, one of the reasons for the crew's carelessness during the Soyuz TM-17 mishap was their eagerness to inspect a NASA-provided rendezvous and docking target for use in forthcoming Shuttle missions.

In the coming months and years, Mir became the focal point of the world's human spaceflight programmes as erstwhile enemies and ideological competitors struggled to learn how to cooperate and share a common goal. As a preliminary step in breaking down barriers, Sergei Krikalev flew on board the Shuttle Discovery in February 1994, with fellow cosmonaut Vladimir Titov following suit a year later.

However, before the cash-strapped Russians could take advantage of their new source of revenue, the rookie crew of Yuri Malenchenko and Talgat Musabayev had to overcome an unexpected challenge. Shortly after joining Poliakov in July 1994, Progress M-24 was scheduled to arrive with much-needed supplies and equipment for a forthcoming European Space Agency mission. After two aborted docking attempts using the Kurs automatic rendezvous system, the manual override control system on Mir proved its value on the third occasion when Malenchenko guided the Progress ship home with the aid of a TV monitor.

Euromir 1 began on 6 October with the arrival of ESA astronaut Ulf Merbold alongside Alexander Viktorenko and female flight engineer Elena Kondakova – the wife of Salyut hero Valeri Ryumin and only the third Russian woman to fly in space.

The world space endurance record holder, Valeri Poliakov, admires the view from a window on Mir during the fly around by Shuttle Discovery on 6 February 1995. Poliakov lived on board the Russian space station for 14 months – long enough to fly to Mars. (NASA photo STS63-711-080)

Merbold's 30-day stay marked the beginning of a series of long-term residencies by non-Russians. Next in line was NASA astronaut Dr Norman Thagard, the first American ever to fly in a Soyuz or enter a Russian space station.

As one era commenced, so another drew to a close. On 22 March 1995, the record-breaking crew of Viktorenko, Poliakov and Kondakova returned to a heroes' welcome. Kondakova had set a new space endurance mark for women of 169 days, but it was her medical colleague, Poliakov, who astounded the world by surviving in orbit for 438 days. In a journey equivalent to a return trip to Mars, the intrepid traveller had battled fatigue, homesickness and weightlessness as he watched the Sun rise and set more than 14,000 times.

Thagard's target was rather more modest: to break the existing US space endurance record set more than two decades earlier by the final Skylab crew, and to pave the way for a new era in international cooperation.

Chapter Three

Handshake in Space

Remarkable though it was, Thagard's mission to Mir was not the first manifestation of close cooperation between the space programmes of the United States and the former Soviet Union.

Certainly, from the late 1940s onwards, an Iron Curtain separated East from West, communist nations from capitalist nations. However, although relations between the two superpowers seesawed considerably over the post-war period – plumbing new depths during the 1962 Cuban missile crisis – there were periods of relative *rapprochement*, particularly after the end of the decade-long Vietnam War.

A temporary thaw in Soviet–American frostiness during the early 1970s saw a historic agreement to fly a joint manned space mission, which culminated in the 1975 docking of Soyuz and Apollo spacecraft and the famous first handshake between astronauts and cosmonauts.

NASA officials attempted to cement this encouraging example of burgeoning cooperation by putting forward proposals that would take advantage of the relative strengths and weaknesses of both programmes.

Even before the successful implementation of the Apollo–Soyuz Test Programme in 1975, NASA Deputy Administrator George Low was proposing exploratory studies involving future joint space stations, possibly preceded by a Space Shuttle–Salyut docking mission and Soviet use of the Shuttle for cooperative projects of mutual value. In return for using the Shuttle to ferry crews and cargo to and from the Salyuts, NASA was hoping to conduct long-term research on the Soviet stations until it could build its own version.

However, despite a renewal of the US-Soviet Space Cooperation Agreement in 1977, accompanied by an agreement between NASA and the USSR Academy

Artist's impression of the Apollo (left) and Soyuz spacecraft docked during the 1975 ASTP mission, which marked the first thaw in US–Soviet relations. (NASA S-74-24913)

of Sciences for a Shuttle–Salyut programme, the era of joint space operations was drawing to a close. The displeasure of the Carter Administration with the Soviet record on human rights and their intervention in Afghanistan meant that the momentum was lost. The Cooperation Agreement was allowed to lapse in 1982, and relations only began to improve again after Mikhail Gorbachev's rise to power three years later.

By 1987, dialogue between the two governments led to an agreement to exchange of biomedical data from human space flights. However, the major advances took place after the collapse of the Soviet Union in 1991. This was followed by the creation of a Russian Space Agency, under the leadership of long-term bureaucrat Yuri Koptev. In June 1992, a promising first meeting between Koptev and new NASA boss Daniel Goldin was followed up by a Joint Statement on Cooperation in Space signed by Presidents George Bush and Boris Yeltsin.

Human spaceflight was at the heart of the new *détente*. Among the items listed for further consideration were crew exchanges on forthcoming Shuttle and Mir missions, a rendezvous/docking between the Shuttle and Mir, and the use of Mir for long-term medical experiments. NASA also agreed to order

Russian Soyuz TM craft as interim crew return vehicles for Space Station Freedom.

The Shuttle–Mir programme expanded still further after the new incumbent, President Clinton, called for another redesign of Freedom and increased international participation in an effort to reduce the ballooning cost of the station.

On 5 September 1993, the human space programmes of the two nations were inextricably linked – something that would have been regarded as impossible during the height of the Cold War – by an agreement signed by Vice President Gore and Russian Prime Minister Viktor Chernomyrdin. A November addendum to the Space Programme Implementation Plan detailed the three-phase plan for US–Russian space cooperation.

This 1994 artist's impression shows the Shuttle docked to the Kristall module on Mir. Note that the Spektr module, which arrived in 1995, is shown (bottom). However, the Priroda module, which docked to Mir in 1996, and the Shuttle docking module have been omitted. (NASA S-94-30837)

Phase One comprised combined operations with the Shuttle and Mir. Up to 10 Shuttle dockings with Mir were anticipated, with Russia providing two years of astronaut flying time on Mir, opening the door to the first long-duration flight experience for US astronauts in more than 20 years. A wide-ranging scientific research programme would be created by outfitting the unflown Russian Spektr and Priroda modules with US experiments. In return, NASA agreed to pay $400 million to Russia – breaking its traditional policy of not exchanging funds in cooperative endeavours.

Phase Two foresaw the construction of a joint interim space science facility, based upon a second-generation Mir module mated with a US laboratory. Successful implementation of this initial structure could then lead to construction of a truly international space station in Phase Three.

PHASE ONE

The first phase of the International Space Station programme, known as Shuttle–Mir, had four main objectives: learning how to work with the Russians; reducing technical risks for Phases Two and Three; advancing studies of long-duration human spaceflight; and conducting scientific research.

Phase One began modestly in February 1994, when Mir veteran Sergei Krikalev became the first cosmonaut to fly on a Shuttle. One year later, Vladimir Titov repeated the experience, this time enjoying the opportunity to wave to his compatriots on board Mir as Shuttle Discovery spent three hours in close proximity to the orbital complex.

Having demonstrated the capability to approach Mir from below and to coordinate American and Russian tracking and communications, the way was clear for the first Shuttle–Mir docking during the summer of 1995.

First, however, it was the turn of four-time Shuttle astronaut Dr Norman Thagard, to make history by acting as a guinea pig for US–Soviet collaboration. Thagard had the distinction of being the first American ever to be carried into orbit by a Soyuz vehicle, to live in a Soyuz capsule (if the short visits by the ASTP crew in 1975 are discounted) or enter a Russian space station.

Having experienced at first hand the difficulties in becoming proficient in an unfamiliar language, the very different training techniques used at Star City near Moscow, the idiosyncrasies of Russian technology and attitudes to authority, Thagard found his task to be harder than anticipated. The slim physician from Florida succumbed to a feeling of cultural and physical isolation, exacerbated by two Russian-speaking comrades who tended to sideline their American colleague, and a diet which was very different from that to which he was accustomed.

Equally frustrating was his secondary, caretaking role as he was left alone inside the station on five occasions while cosmonauts Dezhurov and Strekalov floated around Mir's exterior. Their task was to prepare the way for the next 20-tonne module, known as Spektr (Spectrum), to dock at Mir's forward end.

On 12 May, the men collapsed one of the 14-m (46-ft) long solar panels on Kristall and moved it 35 m (115 ft) to the far end of the station before relocating it on the Kvant module. Thagard was able to contribute to the success of the operation by recycling the motor until the panel completely withdrew into its container. A second panel failed to retract completely, but it was judged to be no threat to an incoming Shuttle. There followed a complicated reshuffling of spacecraft as Progress M-23 vacated the forward axial port on Mir's node on 23 May, then Kristall was moved, with the aid of its in-built Ljappa arm, to the front end and then to another lateral port. This rearrangement left the forward docking site clear for the arrival of Spektr.

To everyone's delight, the newcomer became the first Mir module to dock without needing a second bite at the cherry. Thagard was particularly pleased because Spektr carried 880 kg (1,940 lb) of US equipment, in addition to its home-grown multispectral sensors that would be used to observe the Earth's surface and atmosphere. Among the American equipment now available to the physician and his astronaut successors was a bicycle exercise machine (officially known as an ergometer), a freezer, a centrifuge and several laptop computers.

Spektr was soon moved to its permanent slot opposite Kvant 2, once more freeing the forward port for the highly mobile Kristall. However, despite these successes, the enlarged station was not without its problems. In particular, Spektr's full electricity-generating potential was not fulfilled because one of the solar panels in the unusual V-shaped pair on the module's nose refused to deploy.

A proposed EVA to unfurl the solar array had to be cancelled when flight engineer Strekalov balked on the grounds that it was too dangerous and the crew had no suitable equipment. Mission managers refused to meet his demand for a legally binding "insurance" document, then, in the face of this almost unprecedented revolt, slapped a $10,000 fine on the stubborn cosmonaut.

With Kristall and its androgynous docking port now in position to accept a Shuttle, the Atlantis orbiter lifted off from Florida on 27 June. The first Shuttle–Mir docking mission was to bring home Thagard's crew and deliver their Russian replacements, Anatoli Solovyov and Nikolai Budarin.

Robert Gibson, who had resigned as NASA's chief astronaut in order to command this prestigious mission, was at the controls when the Russian-built docking system on Atlantis meshed with its counterpart on Kristall's nose. The

100-tonne delta-winged orbiter was now linked with the 100-tonne Mir, creating the largest space structure ever assembled.

Once the air pressure in both craft was equalised, Gibson floated into the airlock for the long-awaited photo opportunity – the symbolic handshake with Mir commander Dezhurov. Soon afterwards, 10 people squeezed into Mir's core module for the group picture. While the retiring crew endured a programme of biomedical studies in the Shuttle's Spacelab, the crew of Atlantis transferred cargo that included food, half a tonne of water generated by the American spacecraft's fuel cells, and bolt cutters to be used to free Spektr's jammed solar panel.

As the four-month stay on Mir drew to a close, the crew became increasingly irritated. Dezhurov, who had recently received news that his mother had died, refused to allow scientists to monitor his sessions in the lower body negative pressure unit, a device designed to help cosmonauts prepare for the rush of blood to their legs on their return to normal gravity. Both he and Thagard complained that they were being given too much work, with insufficient time allocated to complete it.

As he celebrated his 52nd birthday on board the Shuttle, Thagard also confided to reporters his distaste for Russian space food – particularly canned perch – and his cravings for those all-American favourites, hot dogs, hamburgers and ice cream. His self-imposed diet and rigorous exercise regime on Mir were blamed for the astronaut losing some 10 lb in weight. The lean astronaut also confessed to suffering from homesickness.

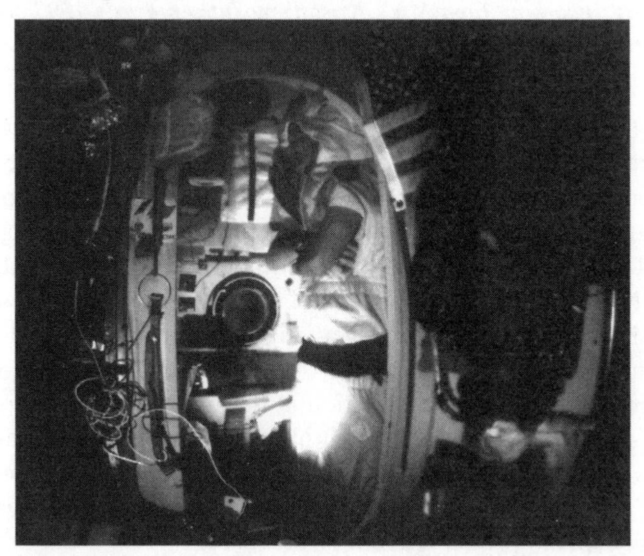

The ultimate room with a view. Astronaut Norman Thagard zipped into his sleeping bag in one of Mir's crew cabins. (NASA/RSA transparency 95-HC-399)

Shuttle Atlantis undocking from Mir's Kristall module on 4 July 1995. The partially retracted solar panel on Kristall can be clearly seen. (NASA photo 95-HC-679)

"The cultural isolation is extreme," he confided. "For an American on board a Russian space station, you're the only English speaker on board in general. There were times when I went 72 hours without speaking to an English-speaking person." The lack of communication with his colleagues and the outside world, particularly his family, took its toll. "All of those things start to weigh heavily after a while," he said. "If I had been looking at six months, I would have been real worried that I wasn't going to make it."

He then urged medical researchers to probe beyond the physical problems of weightlessness by peering in addition at the impact of prolonged spaceflight on human psychology. "I think we still have to find out what's happening with mineral loss from bone, for instance, and obviously radiation is an ongoing problem as long as you're in space," he said. "But my impression is that psychological aspects probably loom largest."

Before the tired astronaut could return home, he had to attend a farewell ceremony, where the Shuttle crew handed over a few parting gifts –

commemorative pins, wrist watches, fresh fruit and a bag of flour tortillas – to the incoming Russian duo. Solovyov was delighted to receive a personal present of a Houston Rockets basketball shirt.

The farewell was marked by a risky PR exercise, with Mir left on autopilot while Solovyov and Budarin entered their Soyuz to capture unique pictures of Atlantis undocking from Mir. During the 45 minutes it was left unoccupied, the station's main computer shut down, temporarily extinguishing Mir's external lights. The Russian crew hastily returned on board to regain control.

Lying back in special contoured seats installed on the mid-deck of Atlantis, Thagard and his two Russian colleagues finally returned to Florida on 7 July after clocking up an American record of 115 days in space. Apart from a difficult re-acquaintance with normal gravity, the crew could not escape from earthbound bureaucracy. Once the US Customs Service discovered that two Russians would be arriving on American territory, NASA had to hastily arrange for visas to be sent up to the two aliens from outer space.

THE EUROPEAN TOUCH

While mission managers on both side of the Atlantic tried to absorb the lessons from the first joint Mir mission, preparations were already underway for further international flights to the Russian station. Things were looking good after several spacewalks by the resident crew succeeded in freeing a non-rotating solar panel on Kvant 2 and the undeployed array on Spektr (apart from two segments at the tip which tilted at right angles to the remainder of the panel).

On 25 September 1995, German Thomas Reiter of the European Space Agency arrived on board Soyuz TM-22 with Yuri Gidzenko and Sergei Avdeyev. After the previous month-long excursion by Ulf Merbold, his ESA compatriot, Reiter's extended stay was intended to provide important experience of long-term spaceflight for one of the International Space Station's major contributors. The highlight of his planned 135-day mission was to be a rare spacewalk by a European astronaut.

Three days before the EVA took place, Reiter's crew was informed that its spell in orbit would have to be extended by six weeks due to the parlous state of the Russian economy – a theme that regularly reared its head in the years to come. Lack of funds to pay rocket assembly workers would delay completion of the Soyuz booster to be used by the next Mir crew.

On 20 October, Avdeyev and Reiter used the Strela crane to swing around Mir's exterior and mount cassettes on Spektr's exterior in order to sample the man-made and natural meteoritic debris surrounding the station. After the

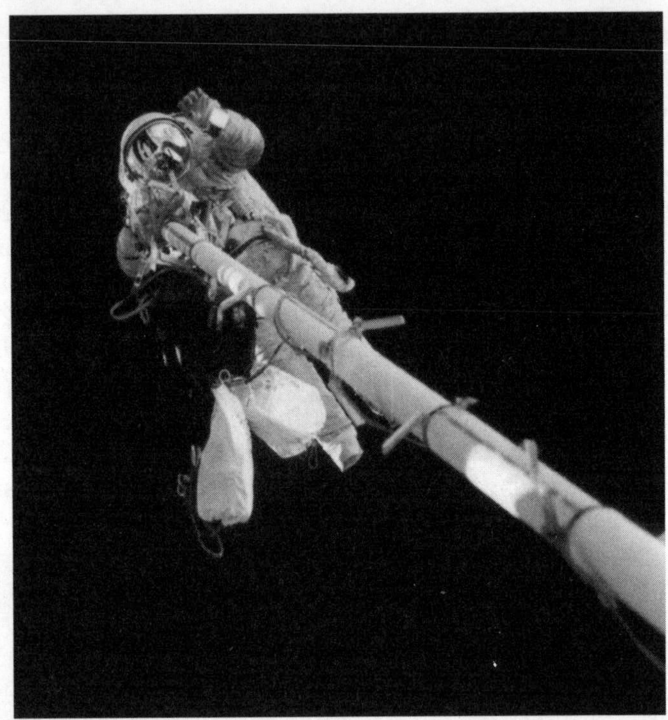

Thomas Reiter performs the first spacewalk by a European astronaut on 20 October 1995. (ESA photo)

exhilaration of walking around the world at 26,000 km/h (17,000 miles/h), normal routine resumed as Reiter continued his biomedical and materials processing experiments.

Three weeks later, the Shuttle Atlantis set off for its second mission to Mir. This time, instead of delivering a US astronaut, the orbiter brought a 4.2-tonne Russian-built docking module. Fitted at both ends with an androgynous docking port, the orange module was designed specifically to ease future link-ups with the Shuttle. On the occasion of the previous Shuttle–Mir docking, it had been necessary to relocate Kristall to Mir's forward longitudinal axis so that Atlantis could safely avoid protruding panels and other appendages, but this complicated manoeuvre was accepted as totally impractical on a regular basis.

On 14 November, mission specialist Chris Hadfield expertly used the Shuttle's remote arm to unstow the new module from its berth in the payload bay. In a preliminary rehearsal for similar activities on the future International Space Station, Hadfield took the opportunity to try out a Space Vision System that could be used as a docking aid when direct viewing was impossible. After soft dock was achieved by firing the Shuttle's thrusters, the procedure was completed when the docking mechanism retracted, drawing the module into a tight embrace with Kristall.

The following day, the module began its active life after Atlantis docked and the crew disembarked, passing through it to greet Mir's occupants. With representatives of four nations on board, the station had become truly established as a staging post on the road to more lasting international cooperation in space.

Apart from the docking module and the now usual supplies of food, water and research equipment, Atlantis also delivered a guitar and two solar panels, including a jointly developed array that used the latest American components. The value of the Shuttle as a space truck – something that had never been available to the Russians – was also shown by its ability to return loaded with the results and samples from experiments together with broken or worn-out hardware.

This wide angle IMAX image was taken seconds after Shuttle Atlantis undocked from Mir on 18 November 1995. The newly delivered Shuttle docking module (bottom) can be seen on the nose of the Kristall module. (NASA photo S95-22141)

Normal service resumed after the departure of Atlantis, with the continuation of maintenance by the cosmonauts and scientific research by Reiter. More supplies and Christmas presents arrived with Progress M-30. As a reward for his patience in enduring the extension to his life aloft, Reiter enjoyed a second spacewalk in the company of Gidzenko on 8 February 1996. Apart from retrieving some of the cassettes they had installed on Spektr, the pair cleared the multi-million rouble EVA backpack – which had only been used on two trial flights – from the Kvant 2 airlock and dumped it outside the station.

Twelve days later, Mir passed its 10th anniversary – double its planned lifetime – but, with the station more in demand than ever and proving to be a most effective source of foreign currency, there was no suggestion that it should be retired. Indeed, the Russian authorities were already considering ways of commercialising Mir while continuing to cooperate on the International Space Station programme. Only three weeks before, the US and Russian space agencies had announced that space veterans William Shepherd and Sergei Krikalev would be the first long-term occupants of the ISS.

LUCID

On 23 February, the next crew, which comprised Yuri Onufrienko and Yuri Usachev, arrived in Soyuz TM-23, clearing the way for Reiter and his companions to head for home after six months aloft. The newcomers were looking forward to a busy tour, including five scheduled spacewalks and commissioning of the sixth and final Mir module.

Their first excursion on 15 March completed installation of a new crane that would be used to mount the additional solar panels that were due for delivery on the next Shuttle flight. Also on board Atlantis when it duly docked with Mir on 24 March was the first Spacehab, a commercially developed pressurised module that could be used to transport cargo or scientific experiments. However, the most precious cargo was Dr Shannon Lucid, an experienced Shuttle astronaut and the oldest woman to fly in space.

The mission notched up another first even before Lucid was able to swap places and settle down inside the Spektr module on Mir. Astronauts Linda Godwin and Michael Clifford carried out the first spacewalk undertaken while a Shuttle was docked to the Russian station. Since Atlantis would not be free to chase after one of them if a tether became accidentally detached, the duo's backpacks carried newly developed SAFER manoeuvring units.

The much-delayed Priroda (Nature) module was eventually launched on 23

Shannon Lucid settles down to read a book in the Spektr module. She went on to break the space endurance record for female space travellers. (NASA photo 96-HC-659)

April. Three days later, despite some concern over a partial loss of battery power, the 20-tonne module successfully docked at Mir's front node. Once it was transferred to a side port opposite Kristall, the crew began the task of stripping out the now-redundant batteries and unloading its cargo. Only then would it be possible to begin the ambitious experimental programme with the advanced Earth observation equipment provided by Russian and European institutes and microgravity experiments provided by NASA, but first they had to unload yet more supplies from the incoming Progress M-31.

Once normal activities were resumed, Lucid was able to spend most of her time concentrating on experiments, although she was left in charge of the station when her Russian partners floated outside to continue the programme of construction work. During their first EVA on 21 May, the duo employed the new crane to transfer the US–Russian cooperative solar array from the docking module to Kvant 1. There followed a publicity stunt that would have been anathema in the days of Communist supremacy – the filming of an inflated Pepsi can for a TV advertisement. Four days later, the solar panel was extended with a hand crank.

The busy programme continued with three more excursions over the next three weeks, during which they deployed various experiments, dismantled and

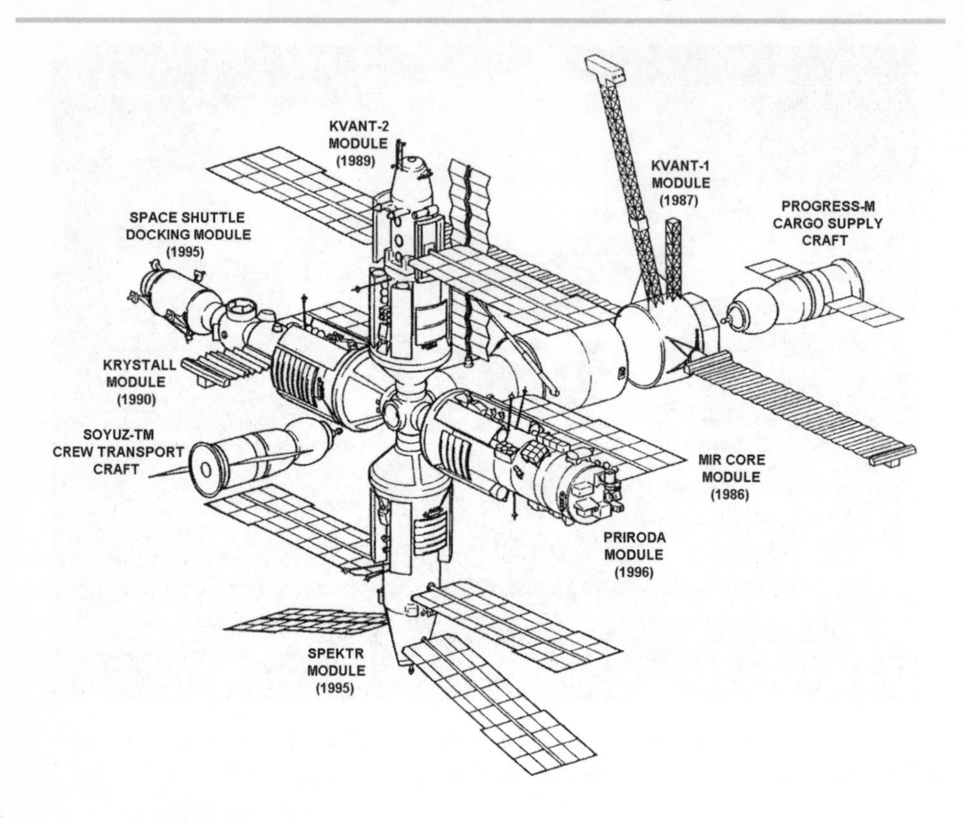

The completed Mir station as it appeared after the arrival of the Priroda module in 1996. (Energia Diagram.)

replaced the elderly Rapana truss with a more modern version, and freed a jammed dish antenna that would be used by a radar sensor on Priroda.

Russian financial difficulties surfaced once more when it was announced that construction of the rocket carrying the next Soyuz crew was behind schedule. Although this did not directly affect Lucid, she was advised that her trip would also have to be extended due to problems with hot gas leakage from the Shuttle's solid rocket boosters. Her reward would be the opportunity to break all existing endurance records for women and US astronauts in general.

In the absence of a Shuttle resupply mission, the Russians had to send up another Progress cargo craft. After a rapid unloading, the Progress vacated its port to make way for the next crew, cosmonauts Valeri Korzun, Alexander Kaleri and French "spationaute" Claudie André-Deshays. For a brief period, the male-dominated Russian station enjoyed the female touch, with André-Deshays based in Priroda and Lucid's living quarters established in Spektr.

The brief French connection ended on 2 September with the departure of Onufrienko, Usachev and André-Deshays, but the national space agency,

CNES, was soon booking further, more prolonged flights on Mir. Lucid later described her farewell with the two Yuris as "maybe the most difficult time I had while I was on Mir".

Indeed, after the difficulties experienced by her predecessor, Norman Thagard, one of the major successes of her record-breaking mission was the establishment of a friendly personal and professional relationship with her two long-term companions. "It dawned on all three of us at once how remarkable it was that here we were, three people who grew up in totally different parts of the world, mortally afraid of each other," she said. "Here we were sitting in an outpost in space together and getting along just great." The space veteran had little time to get to know her new crewmates. On 19 September, Shuttle Atlantis docked with the station. Two and a half hours later, the hatches were opened, allowing the usual round of speeches and souvenir exchanges to take place as nine people crammed into Mir's core module.

The visit of STS-79 was marked by the first astronaut hand-over in space. Lucid did her best in the few days available to brief her friend and replacement, John Blaha, on the whereabouts of everything he would need and how to survive the long confinement. "I told John he was going to have a great time and to take each day as it comes and not to get too hung up on a lot of details," she said.

With the transfer of the Russian Sokol suits and couches that were stored in

STS-79 astronaut Terence Wilcutt floats through the Shuttle docking module towards Kristall, preceded by a water bag. (NASA photo STS079-335-001)

readiness for a sudden emergency evacuation, the handover was complete. Lucid returned to a heroine's welcome – the most-travelled woman ever to fly in space and the new US endurance record-holder.

Blaha's period of occupation in many ways mirrored that of the remarkable Lucid as he continued the ongoing US experiments in Spektr and Priroda, and initiated a few more. Samples of the air, water and exposed surfaces in Mir showed that the enclosed environment in the aged station was still remarkably hygienic. However, the friendly camaraderie of Lucid's flight was unlikely to be repeated after a suspected heart problem for Gennadi Manakov grounded the prime Russian crew. Blaha was obliged to spend four months isolated aloft with two virtual strangers.

Power supply was augmented in early December when Korzun and Kaleri completed the link between the cooperative solar array and the station's main system. Around the same time, Blaha harvested the wheat originally planted by Lucid – the first time that grain had been produced on the station.

The now-inevitable delays caused by cash problems and deteriorating relations between state enterprises involved in building rockets appeared again when the Russian Space Agency announced that the next supply ship would be arriving three months late. Only three Progress craft would be launched that year, compared with the previous annual rate of five.

A more ominous sign of the deterioration in the once-proud Russian human spaceflight programme was the announcement that the Kurs automatic rendezvous apparatus, normally compulsory on any manned mission, was to be omitted from Soyuz ships. Cosmonauts would be required to complete dockings under manual control. Rather more concerning were the implications for future unmanned Progress missions to Mir.

The long-delayed Progress M-33 eventually arrived on 22 November, loaded with some 2 tonnes of supplies. Apart from the usual fuel, food, water, oxygen and equipment, the unmanned craft brought eagerly awaited messages and Christmas presents from home. Almost as welcome were the spare parts for the sewage system that had been out of action for the past month.

FIRE!

Two more spacecraft docked in the first months of the new year. Blaha's replacement, physician Jerry Linenger, was on board Atlantis as it docked on 15 January 1997. In a sign of things to come, the traditional press conference had to be postponed after a master alarm warned the mixed crews that Mir's batteries were running dangerously low. Once again, the value of a crew change-over was demonstrated as the experienced Blaha briefed his NASA colleague Linenger on the finer points of Mir operations and personal hygiene.

Linenger later wrote, "All of this information was very important because, to be honest, much of the housekeeping and science equipment on the then 11-year-old space station had been modified or jury-rigged over time, so much so that the ground training I had received in Russia was outdated at best, and in many cases no longer usable."

Blaha was also blunt about the difficulties of dealing with the Russian partners, particularly those running the mission on the ground. He complained to Linenger about the "atrocious" space-to-ground communications that severely restricted conversations with family members and NASA support teams in Moscow. As far as the Russian ground controllers were concerned, he had been "an inconvenience, a nuisance to them". Not only did the Russian ground control team regard him as a liability, but the Mir commander was also accused of treating him as a second-class citizen.

It was an inauspicious introduction to life on Mir for the newcomer, and one from which Linenger never really recovered. The first few weeks were full of hectic activity as the Mir crew – commander Korzun, flight engineer Kaleri and Linenger – said a temporary farewell to Progress M-33, then relocated their elderly Soyuz at the Kvant 1 module to clear the front port for a fresh craft.

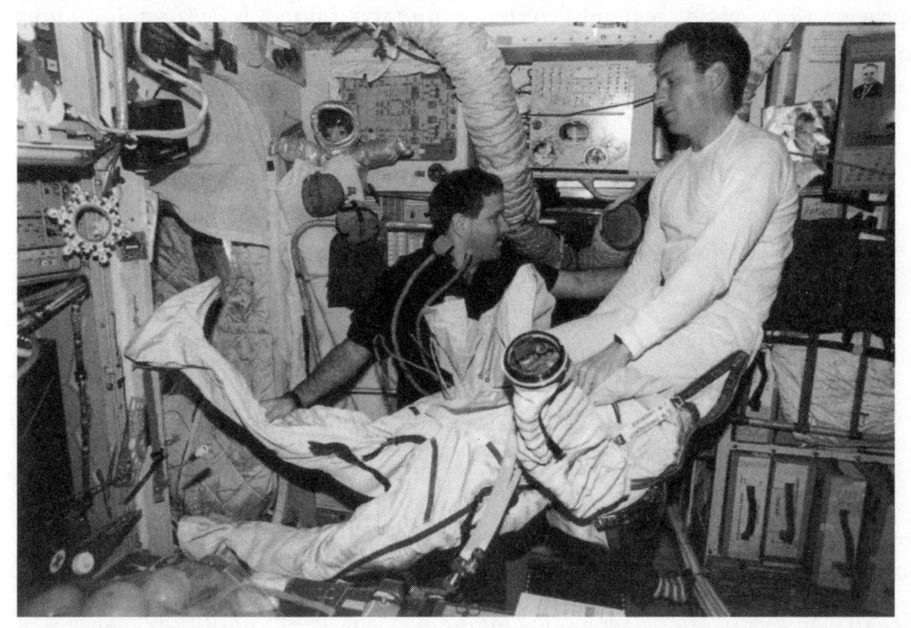

Assisted by STS-81 Shuttle astronaut John Grunsfeld, Mir newcomer Jerry Linenger floats into his Sokol pressure suit shortly after arrival on the Russian space station. (NASA photo STS081-372-021)

Soyuz TM-25 arrived at the station on 12 February, carrying Vasili Tsibliev, Alexander Lazutkin and German "guest" Reinhold Ewald. The docking was completed at the second attempt after the automatic Kurs system failed.

The blows just kept on coming. On 24 February, a black day for the Russian space programme, Yuri Koptev, director of the Russian Space Agency, told the world that the start of the International Space Station programme would be postponed for six months due to delays in building the Russian Service Module.

That evening, the six-man crew was settling down for a meal and a press conference in the core module when the peace was shattered by the clang of a master alarm. As Linenger emerged from the Spektr module to find out the reason for all the noise, he almost bumped into a frantic Tsibliev. "Fire!" came the abrupt warning. Looking along the length of the main module, Linenger's worst fears were confirmed. The Kvant 1 module seemed to be ablaze and black smoke was rapidly billowing into the body of the station, blocking all visibility of the terrible events taking place just a few metres away.

The life-threatening conflagration had occurred because Mir's life support systems were severely stretched by the presence of six humans on board. In order to supplement the oxygen supply, the cosmonauts burned chemicals in a metal canister to release additional supplies of the precious gas. On this occasion the canister had split, transforming itself into a blowtorch hot enough to melt metal.

If the fire could not be extinguished, the station's occupants would have

The six-man crew of Mir gather around the dining table. Shortly after, their enjoyment was interrupted by a near-catastrophic fire in the Kvant 1 module. Front row: Alexander Kaleri (left), Jerry Linenger, Valeri Korzun (right). Rear: Vasili Tsibliev (left), Reinhold Ewald, Alexander Lazutkin (right). (DLR photo)

little choice but to abandon ship and head for home. Unfortunately, a Soyuz can only carry a crew of three. With one of the lifeboats cut off by the fire, the remaining three cosmonauts would be stranded on Mir with no method of escape.

After they succeeded in grabbing oxygen masks that worked – at least one that was seized by Linenger failed to operate – the men were able to collect their thoughts and pursue a plan of action. While the newly arrived crew activated the one available Soyuz, Kaleri shut down the ventilation fans, leaving Korzun and Linenger working in unison to fight the fire.

This proved much harder than expected. Fuelled by the flow of oxygen from the canister, the flames continued to rage, despite the foam fountain squirting from a succession of extinguishers. Only after 14 minutes of desperate endeavour did the inferno burn itself out.

Doubts about their future still remained. With thick smoke filling the station, the men had little choice but to inhale through their oxygen respirators and hope that Mir's filtering system would clear the air. After an hour of tense inactivity, Korzun removed his empty inhaler and joyously declared the air to be breathable. Everyone stripped off their contaminated clothes, and, still naked, began to wash down the bulkheads before cleaning themselves from head to toe. The human crisis was over and the Mir station was saved to fly for several more years. Meanwhile, to no one's surprise, the Russian and American space agencies insisted that Mir was safe for future occupants.

Physician Linenger was called upon to treat some minor and second-degree burns, but none of the crew suffered any long-lasting effects. On 2 March, the number of occupants of the station was reduced once more to three when Korzun, Kaleri and Ewald ended their stay in orbit.

Two days later, another Russian money-saving effort flopped when an attempt by the crew to redock Progress M-33 with the aid of the TORU remote control system ended in failure. Assisted only by an image of Mir transmitted onto a video screen in the core module and two joysticks to command the engines on the Progress, Commander Tsibliev was to have taken control of the robot ship and brought it safely to port. In reality, the TV monitor remained blank, leaving a panic-stricken crew to search vainly for the approaching metal monster that could spell doom for their entire mission if it collided. Fortunately, the runaway missile slid past the station's protruding panels, leaving the crew grateful to be alive. While ground controllers dumped the rubbish-laden Progress in the Pacific, the crew continued in the vain hope that this reckless venture would not be repeated.

The next day, the second Elektron oxygen-generating unit shut down, leaving the crew with the dubious pleasure of relying solely on the fire-prone solid fuel canisters. One canister would have to be burned per person per day.

Fortunately, no more conflagrations occurred before the next Progress arrived on 8 April with a fresh supply of "candles" and spare parts for one of the broken-down Elektrons.

However, the crew's patience was tested once more when the attitude control computer failed. Alarms rang again to tell them that the station's power was so low that the gyros stabilising Mir ceased to operate. Plunged into darkness, the trio seized their penlights, switched off as many systems as possible and waited for the sunlight shining on the solar panels to recharge the flat batteries.

By late April, the crew and ground controllers were growing concerned over leaks of alcohol and ethylene glycol from pipes in Mir's cooling system. As the cosmonauts tried to trace the sources of the toxic leakages, parts of the cooling system had to be shut down, causing the cabin temperature to soar to 30 °C, and the carbon dioxide scrubbing system was also temporarily switched off. Not surprisingly, the men began to show signs of stress and had to resort to medication to enable them to sleep.

As a reward for all his hard work and endurance, Linenger was allowed to experience his first spacewalk, stepping outside Mir for five hours to retrieve and deploy some experiments. However, the homesick astronaut was happy to see Atlantis arrive on schedule on 17 May. On board was his replacement, Shuttle veteran Michael Foale, and a mass of new equipment, including a gyrodyne and a pristine Elektron unit.

Foale's tour of duty began with a prolonged headache, possibly due to oxygen deprivation and poor ventilation in his new home. Linenger's comment to him, "Mike, you're going to have some challenges ahead," must have also given him some pause for thought. However, little more than a month after the Shuttle's departure, the eager astronaut and his two weary colleagues had to endure a much more serious headache – the near-destruction of Mir.

COLLISION!

On 24 June, the Progress M-34 cargo ship, laden with rubbish, pulled away from the Kvant 1 port as dozens of its predecessors had done before. Only this time the 7-tonne ship was to make a brief return trip to Mir in a repeat of the near-disastrous link-up attempt by Progress M-33. Not surprisingly, cosmonauts Tsibliev and Lazutkin were displaying signs of nervous tension.

Once again, the station was out of touch with ground control, and commander Tsibliev's only aids were a pair of joysticks and a TV screen. All seemed to go smoothly at first, but the crew became increasingly concerned as the incoming craft refused to react to the cosmonaut's commands. As the

Astronaut Michael Foale exercising on the Mir treadmill. By the time this picture was taken on 30 September 1997, Foale had narrowly survived a spacecraft collision, and endured numerous power cuts and computer failures. (NASA photo STS086-E-5358)

Progress veered left and swept along the side of the station, the crew braced for impact and prayed. The next thing they felt was a shudder, followed by a warning klaxon. The Progress had slammed into a solar panel and radiator on the Spektr module, puncturing its outer skin and allowing precious air to seep into the vacuum of space. "I remember thinking that I was probably going to die," said Foale.

With their ears popping as the cabin pressure dropped, the crew's first thought was to activate their Soyuz lifeboat. Clearing the air tubes and power cables from the hatches of the docking module, they succeeded in sealing off Spektr with minutes to spare. The immediate emergency had been overcome, but with their main power supply now severed, the station had begun to spin out of control.

They had little choice but to switch off all non-essential systems, including the temperature control, lights and toilet, then try to find a way of recharging the station's flat batteries. Fortunately, they managed to find a way of using the thrusters on the Soyuz to turn the rotating structure towards the Sun. Gradually, normality began to return, and Michael Foale learned to adjust to the loss of his sleeping accommodation and personal belongings in the cold, airless Spektr.

Things began to improve when the next Progress arrived on 7 July, loaded with new power cables, a special hatch with cable adapters, a medical kit, tools, computer spares and a replacement hygiene kit for Foale. Initial plans for a dangerous internal spacewalk to assess the damage to Spektr were postponed when Tsibliev developed an irregular heart rhythm and had to be prescribed medication.

Still, the exhausted trio were not out of the woods. On 16 July, Lazutkin accidentally disconnected a cable linked to Mir's attitude control computer. Once again, his colleagues were awakened by the sound of alarms as the station drifted out of position and plunged into darkness. The discomforts and frustrations of another prolonged period of power shortage had to be endured.

As ever, the Russians tried to portray the shambles that purported to be an operational space station in a positive light.

A close-up view of the buckled and punctured solar panel on Spektr after the collision with an incoming Progress craft on 25 June 1997. (NASA photo STS086-387-014)

"We have had fires before, but not such effective ones. We have never had depressurisation of a compartment before. We have never had to repair the station's heat regulation system before. There is a very great deal of useful information which we are constantly applying to the International Space Station project ...," pronounced Mir flight director Vladimir Solovyov.

Meanwhile, it was clear that the two long-term cosmonauts had reached the end of their endurance, so the task of restoring some semblance of normality to Mir was entrusted to a replacement crew. With mixed emotions, Foale watched as the long-suffering Tsibliev and Lazutkin handed over to spacewalk specialist Anatoli Solovyov and engineer Pavel Vinogradov. However, even their homecoming turned out to be more bumpy than expected when the retro-rockets failed on their Soyuz capsule just before touchdown in Kazakhstan.

Despite further frustrating power cuts and a crashed computer, conditions slowly improved. Condensation was reduced by circulating warm air through the Priroda and Kristall modules. Foale was able to resume his science experiments, including the first ever planting of seeds that had been produced in space.

On 22 August, the new Spektr hatch was installed and 11 cables running from the module's undamaged solar arrays were reconnected, enabling a gradual restoration of power. No obvious signs of the collision were found inside Spektr, though there were some white crystals floating around – possibly from soap or shampoo – and there was a thin layer of frost on the inner walls.

The crew continued to work hard to improve the power situation before the Shuttle arrived in September. Then, with his eventful stay on Mir nearing an end, Foale's patient forbearance during times of extreme stress was rewarded when he joined Solovyov for a six-hour session outside the station. Once again, a visual inspection revealed no obvious sign of a puncture in Spektr's outer hull, but the duo did succeed in realigning one of the solar arrays to allow it to capture more sunlight.

Despite continuing glitches – the main computer crashed again on 8 and 14 September – NASA officials decided to continue with the Shuttle–Mir missions.

"The importance of this programme cannot be overestimated," said Shuttle–Mir manager Frank Culbertson (later to command ISS Expedition Three). "This is where theory meets reality, where the practical lessons we learn aboard Mir are already paying large dividends as we prepare to start construction of the Space Station in less than a year."

Michael Foale was also happy to contradict suggestions that the collaborative programme had fulfilled its main objectives and that little science had been returned from his traumatic four and a half months aloft.

The forward transfer node on Mir linked the four add-on modules – Spektr, Priroda, Kristall and Kvant 2. This view of Valeri Korzun almost buried inside the cluttered compartment gives some idea of the problems that faced the crew in sealing the hatch to Spektr after the collision. (NASA photo STS079-353-034)

"We're fairly resourceful in recovering as much as possible ...," he said. "The things that were lost in the impact were a lot of the life sciences equipment and most of the samples – blood, urine, things like that. ... Also, we lost the power to drive a lot of the power-hungry microgravity and materials sciences experiments, but with some judicious planning, a lot of hard work by the folks on the ground as well as the crew, alternate ways were found to do a lot of the kind of studies. ... My gut feeling is that when it's all said and done, when the dust settles, we're going to be looking at 70–80% of a science programme accomplished on this mission."

The seventh docking between a Shuttle and Mir took place on schedule on 27 September 1997. Among the welcome items delivered by Atlantis were a new computer – which was swapped while the Shuttle was on hand to control the attitude of the station – and more than 4 tonnes of supplies and water.

Highlights of the six-day link-up were a spacewalk involving Vladimir Titov – the first Russian to use a US spacesuit – and a final flyaround in which the Shuttle crew surveyed Spektr while the Mir occupants blew air into the damaged module. Particles seen seeping from the base of the dented solar array suggested that this was the site of the hull breach.

Foale's successor, David Wolf, was a physician who had only been promoted to prime position the previous July when astronaut Wendy Lawrence was suddenly deemed too small to wear a Russian Orlan EVA suit. His welcome to the vagaries of life on Mir was marked by a failure of the Progress M-35 cargo ship to undock from the station. Once the problem was solved – a clip had inadvertently been left on the docking module – the redundant craft cleared the rear port to make way for its replacement.

With the station now replete with fresh parts and supplies, Wolf was free to conduct his biomedical and protein crystallisation experiments in the Priroda module. On 20 October, he retired to the safety of the Soyuz ferry while his colleagues conducted an inspection of Spektr and connected cables that linked the automatic steering system to two of the module's three working solar

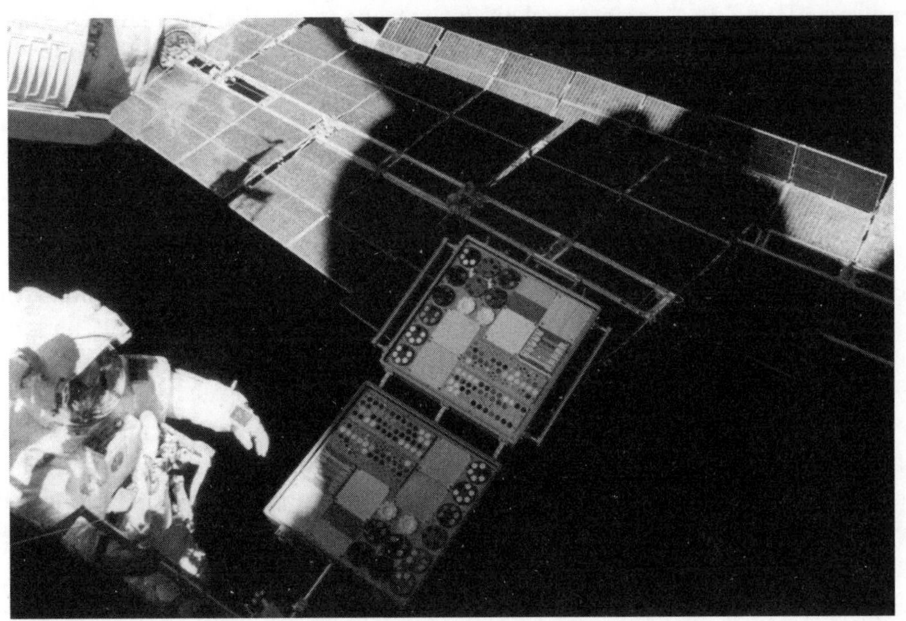

On 1 October 1997, cosmonaut Vladimir Titov made history by becoming the first Russian to use an American pressure suit during a spacewalk. Here he is shown alongside an experiment which exposed various materials to the space environment for six months. (NASA photo STS086-332-005)

arrays. Three days later, Yuri Koptev, head of the Russian Space Agency, appeared on TV to explain that Mir would be withdrawn once the new International Space Station was operational – probably in the second half of 1999.

Two more EVAs followed in quick succession as Wolf helped the Russians to fold and replace a solar panel on the Kvant module, then install a device that would enable an extra carbon dioxide removal system to be fitted inside the station. It was yet another display of their remarkable ability to upgrade aged hardware in orbit. They even found time to launch a small replica of Sputnik to mark the 40th anniversary of its historic flight.

Despite an occasional blip, such as a short power loss on 14 November and several days of "tropical temperatures", the expedition progressed relatively smoothly. Tass news agency even reported, "for the first time in four and a half months the crew are working full time on scientific experiments and not being distracted with damage-limitation jobs".

This assessment proved to be premature, with another computer crash and loss of attitude control on 24 November. Work overload meant that a spacewalk set for 5 December was postponed for a month and the next Shuttle docking was also delayed by five days. Then a trial of an innovative new German robot known as Inspektor – a potential aid for astronauts on the International Space Station – came to a disappointing end when the small, manually controlled spacecraft went out of control.

With power levels almost back to pre-collision levels, the Russians began confidently to look forward to further visits by paying customers. However, the first priority was to repair a leaking hatch in the Kvant 2 module which had left the airlock depressurised and unusable.

Despite the inconvenience of yet another computer malfunction, cosmonauts Solovyov and Vinogradov stepped outside the station to examine the cause of the air leak. Five days later, on 14 January 1998, Wolf had the opportunity to experience his first spacewalk in the company of Solovyov, who was taking part in a record-breaking 16th EVA.

It was then time for Wolf to finish packing as Shuttle Endeavour blasted off with his replacement, Andrew Thomas, the seventh and final NASA astronaut to live on Mir. The newcomer's four-month sojourn got off on the wrong foot with his Russian overseers when he complained that his "Sokol" pressure suit was too small.

No sooner had the change-over and resupply mission been completed than Soyuz TM-27 arrived with cosmonauts Talgat Musabayev, Nikolai Budarin and French "guest" Leopold Eyharts. During the three-week handover to the new resident crew, Eyharts and Thomas cooperated as they both tried to acclimatise and carry out their respective science programmes.

Anatoli Solovyov, Andy Thomas and Leopold Eyharts enjoy a meal in the Mir core module. Solovyov is the world's most experienced spacewalker, with 16 EVAs. (CNES/Eyharts)

Solovyov, Vinogradov and Eyharts bade their farewells on 19 February. The next day, the 12th anniversary of Mir's launch, was marked by a routine relocation of the fresh Soyuz to the rear of the station in order to clear the front entrance for a second docking by Progress M-37. This time, the redocking took place in automatic mode. A minor panic ensued on 26 February when a trace contaminant system overheated, generating a small quantity of smoke, but more important business soon loomed for the busy cosmonauts.

A spacewalk scheduled for 3 March had to be abandoned when Musabayev and Budarin broke three spanners in trying to release a secondary latch on the exit hatch of the Kvant 2 module. The frustrated cosmonauts had to redirect their attentions to the replacement of a broken air conditioner after ground control informed them that all EVAs and efforts to brace the damaged Spektr array would have to be postponed until the next Progress arrived on 17 March. A slight deviation in the craft's final flight path meant that the docking was completed under manual control by Musabayev, but Russian controllers reassured reporters that there was never any chance of a collision.

On opening the door to the cargo ship, the crew reported their pleasure at smelling fresh apples. Also on board were new airlock hatches, a set of wrenches, a replacement propulsion unit, a computerised photo album for Thomas, a CD player and three two-volume sets of Beatles music.

With the over-tightened latch now unscrewed, the way was clear for spacewalks to resume. In the first three weeks of April, the two cosmonauts stepped outside the station on no less than five occasions. During more than 30 hours of spacewalks, they successfully braced the damaged solar panel on

Spektr, discarded an old propulsion unit that had run out of fuel and fitted a replacement on the 'Sofora' boom.

Despite this upgrading of Mir, Russian officials had now begun to openly discuss details of how the aged complex might be decommissioned by December the following year.

The final weeks of Thomas's four-month sojourn were spent in concluding his 27 experiments, packing his scientific hardware and personal belongings, and conducting an inventory of US equipment on board the station. His Russian colleagues also found time to take part in a series of biomedical experiments, particularly after Progress M-39 arrived with 15 newts and some 80 snails.

However, the venerable station still had a trick or two up its sleeve, and the last days of the US presence on board were marked by another main computer crash which sent Mir spinning out of alignment with the Sun. A substitute computer developed a "glitch", so once again the crew was obliged to suffer a powered-down spacecraft for two days until stability was restored – just in time for the arrival of Shuttle Discovery to transport Andy Thomas.

The ninth and last Mir docking by the American vehicle took place on 4 June 1998. On board was veteran cosmonaut Valeri Ryumin (now deputy director of the Energia corporation which was responsible for Mir operations) who had come to observe the condition of the station for himself. Despite several failed attempts to identify the leaks in Spektr by filling the damaged module with coloured nitrogen-acetone gas and searching for tell-tale whisps, Ryumin declared himself very satisfied with Mir's overall state.

With Thomas, four other astronauts and more than a tonne of American equipment on board, Discovery touched down in Florida on 12 June. It was the end of the first phase in the International Space Station programme. Officials on both sides of the Atlantic were quick to list the advantages gained and lessons learned from some three years of close cooperation, while downplaying the major financial hurdles facing the future of the Russian space effort.

"I believe that having an unbroken presence like that has been extremely valuable to our awareness of what it takes to fly a station, to maintain an outpost on the frontier like that on a continuous basis," said Shuttle–Mir programme manager Frank Culbertson.

> If a problem occurs you've got to make it a twofold solution: one is, "Is it safe now and are we doing OK," and then the other is, "How do we fix it so it never recurs in the future or that the impact is minimised?" We've learned that we have to take that approach to any problem that occurs or any situation ... you always have to have somebody who's paying attention to what's going on from an operational standpoint and psychological and all the other aspects of having somebody deployed and continually operating. That's a different mindset for

The generation gap. Salyut veteran Valeri Ryumin, now director of the Shuttle–Mir programme, greets Kazakh cosmonaut Talgat Musabayev moments after the opening of the hatches between the Shuttle and Mir. (NASA photo STS091-379-008)

> NASA from what we've ever operated before. Even the Skylab missions had a distinct beginning and end and a break in between the missions. To have a continuous presence like this – I've got a lot of respect for the Russians for maintaining the presence they have on orbit and continue to support it the way they have and keep the mission rolling. We've got to learn how to do that and do it in a way that doesn't grind our people down and produces productive missions.

Culbertson went on to elaborate on the severe difficulties that faced some of the US–Mir crews.

> The events that occurred in 1997, the fire, the collision, the systems problems ... they were part of the character building of the programme that we didn't anticipate. They were part of the forcing together of the two sides to work together on problems and solve them ... and I believe we've gone much further in Phase One than we thought we would in terms of building a partnership and building trust than we might have otherwise.
>
> I think we were well on our way to a good programme, but I believe these things forced us to drop the façade ... To deal with the real problems underneath, and to look more closely at each other on how we do operate and what the infrastructure is like. ...

NASA's ISS programme manager, Randy Brinkley concurred.

> We tend to learn more from our difficulties and our failures and our problems than we do from our successes, and whereas people may look critically at some of the problems on Mir – the fire, the collision – those things have had tremendous benefit for us in terms of understanding not only the causes, but also the corrective action and ensuring that we're able to apply those to the International Space Station. Clearly that has tremendous value to us, and from a working relationship, having worked through those difficulties together has created a significant bond between the American and Russian team in Phase One, and that relationship and confidence in one another, understanding, and trust have certainly not been limited just to Phase One. It has flowed over into the International Space Station, and our working relationship with our Russian counterparts in the International Space Station has dramatically benefited from, not necessarily the problems, but having to overcome those difficulties together.

MIR: THE FINAL YEARS

With the departure of the final Shuttle, the Russians were left to decide how best to conclude Mir's glorious, and sometimes inglorious, chapter of space exploration and endurance. The signs were not good. On 18 June, the mission control centre announced that, because of the expense of keeping five people in space, the next hand-over of Mir crews would be curtailed by a week – despite the presence of former Defence Council secretary Yuri Baturin.

A week later, Russia's chief spacecraft designers met to draft a letter to Prime Minister Kiriyenko, bewailing the lack of government funding for the station and warning that, if no further crews were sent aloft, the station could crash to Earth out of control. The Russian Space Agency chief, Yuri Koptev, described the situation as "tense and extremely complicated". He added that, if no money could be found "a situation will arise where in August we shall have to take the crew from Mir and close the station".

The questions over the future of the Russian space programme's aged flagship spilled over into a series of sour comments involving the role of the United States. When "sources" at the Russian Space Agency suggested that NASA might provide a Shuttle to deorbit Mir, the US agency was obliged to issue a strong denial. This, in turn, only increased the determination of some Russians to delay Mir's demise.

"It would be a sin to drown the station in the ocean," said veteran cosmonaut Alexander Serebrov. "If this happens, not only Russia but the whole world will be left without space research within the nearest six to seven years before the International Space Station begins to operate. Sending a Shuttle there would cost several million dollars [sic], but Mir's end would mean that only the US flag will flutter in space."

The Mir space station as it appeared in its heyday. This view of the 130-tonne station was taken from Shuttle Endeavour in January 1998 – three years before its destruction on re-entry. (NASA STS-089-716-019)

Discussions with Finance Ministry representatives ended with an agreement to fund the programme from "non-budgetary sources", including the sale of part of the government's stake in the Energia company. Despite these efforts, Yuri Semenov, boss of the cash-stricken industrial giant, complained that the state owed Energia about 800 million roubles. He then announced, "I'm going ahead with the (next) launch after taking on loans of over 200 million roubles".

Although this temporarily tided over the programme, it was now foreseen that Mir's mission would end even earlier than expected, in the summer of 1999. With paying customers and funds in short supply, the outstanding visits by Slovak and French astronauts were to be merged into one. Similarly, plans to send up five Progress craft during Mir's final months were scaled down to include just three.

After a delay of 10 days, the next long-term crew finally lifted off to Mir. Gennadi Padalka, Sergei Avdeyev and "guest" Yuri Baturin reached the station on 13 August. It was anticipated that Avdeyev would stay on board long enough to host the foreign visitors before closing down Mir for good the following June.

With Mir in a better state than it had been for years, the newcomers settled down to a programme of scientific experiments and routine maintenance. Power supplies were improved still further by an excursion into the depressurised Spektr to improve control of the module's three operational solar arrays.

With the Russian economy suffering a major financial crisis due to the collapse of the rouble and rampant inflation, it was hardly surprising that the launch of the next Progress supply ship was delayed 10 days, due to a shortage of funds to purchase the Soyuz rocket. Yet, despite these obvious signs of a programme verging on collapse, the Russian parliament, the Duma, appealed to President Boris Yeltsin to prevent the termination of the Mir complex before the "complete deployment of the International Space Station".

"The termination of works with the Mir complex will result in the expulsion of the Russian Federation from the international piloted space programmes and other large-scale scientific projects," said the appeal. This was expected to lead to a cut of over 100,000 jobs for highly skilled scientists and engineers and might also create social tensions and result in closure of companies.

When Progress M-40 finally arrived on 27 October, the Russian media announced that its successor would be the first of Mir's gravediggers. Once again, emotional appeals were broadcast to the Russian people, explaining that the Americans "have us by the throat". Not only was NASA asking Russia not to divide its sparse funds between Mir and the new International Space Station, but it was also taking advantage of cheaper construction costs to build large parts of the ISS.

Curiously, at the same time, Yuri Semenov openly touted a proposal to transfer some of Mir's unique research equipment to the International Space Station – preferably by means of a US Shuttle. Not surprisingly, NASA did not take the bait, although the agency did agree to pay $60 million in return for leasing room on the yet-to-be-completed Russian Service Module.

November saw a noticeable shift in emphasis as Russian Space Agency chief Yuri Koptev declared for the first time that "it is possible and worthwhile to keep Mir as a unique long-term space unit", while continuing to insist that the International Space Station was the only long-term show in town.

A six-hour spacewalk on 10 November installed experiments to monitor the Leonid meteor shower and launch another Sputnik replica, but, at long last, the International Space Station was about to steal the limelight. The launch of Zarya, the American-financed, Russian-built first section of the ISS, was rapidly followed by a successful docking with NASA's Unity module.

However, with the Russian Service Module dropping further behind schedule and the scientific exploitation of the new station languishing in its wake, Russian plans to deorbit Mir took a sudden U-turn. On 23 December, Yuri Semenov announced that plans were being drawn up to continue Mir's life until the year 2000, based on proposals from foreign investors and government guarantees. The first practical step towards putting this plan into operation came a day later when the engines on Progress M-40 were used to raise the station's orbit.

Within a month, Prime Minister Primakov had signed a government declaration agreeing to prolong its operation until 2002, at a cost of $250 million per year. RSC Energia was given exclusive rights to use the proceeds from Mir and sell scientific and technological programmes to anyone willing to invest in the station. Paying visits by foreign cosmonauts were once again to be actively encouraged.

Up aloft, as Padalka neared the end of his time in orbit, the cosmonauts suffered a disappointment when an experiment to deploy a solar sail failed. It was then time to transfer the old Soyuz to the rear port in preparation for the arrival of the next crew.

Soyuz TM-29 blasted off from Kazakhstan on 20 February 1999, carrying a highly unusual crew – possibly the penultimate group of Mir occupants. Alongside Viktor Afanasyev were two foreign "guests". Slovak Ivan Bella, whose trip was arranged as part-payment of long-term Soviet-era debt, barely had time to settle in during his six days on the station. In contrast, Frenchman Jean-Pierre Haigneré, who had originally expected to stay on Mir for just 35 days, was now expected to remain on board for six months at no extra cost. His $20 million mission would continue experiments undertaken by previous European astronauts and would include at least one spacewalk.

After a frustrating week in which he struggled vainly to keep alive 14 new-born Japanese quail chicks, Bella returned to Earth with Gennadi Padalka and the remaining trio settled down for a long haul. Only three weeks into their mission, mission control announced that the number of scheduled spacewalks would be cut from five to two because of a shortage of money. Nevertheless, when Progress M-41 arrived on 4 April, the media announced that it was carrying more food than usual – including supplies for the next replacement crew. An optimistic space agency director Yuri Koptev announced that two more Mir crews were in training in case additional investment could be found.

On 16 April, Haigneré and Afanasyev took part in a six-hour EVA to retrieve Leonid meteor samples and other experiments, as well as launch yet another mini-satellite. Meanwhile, the search for foreign investment seemed to have reached a successful conclusion when an Energia spokesman announced that Welsh entrepreneur Peter Llewellyn was prepared to invest at least $100 million to finance Mir operations.

The penultimate crew to visit Mir. Slovak Ivan Bella (left), Russian commander Viktor Afanasyev, and ESA astronaut Jean-Pierre Haigneré (right) prior to launch on 20 February 1999. (ESA photo)

Alas, although he turned up for training at Star City, the apparent saviour of the Soviet-era station failed to live up to his promises, and Mir's future in orbit remained doubtful. Indeed, there was no guarantee that sufficient cash would be available to cover the flight of the deorbiting crew, despite the creation of a fund-raising charity by former cosmonaut Vitali Sevestyanov.

While the debate over the station's future raged down below, the three men on board Mir continued their programme of regular maintenance and scientific experiments. Remote-sensing equipment on the Priroda module was used to scan the Earth's oceans and land surface, in addition to studies of the effects of weightlessness and space radiation on amphibians. Then, as the crew prepared for the arrival of the next Progress, the world's media discovered that there was an oxygen leak from the Kvant 2 module. At the same time, the Kazakh authorities slapped a ban on all launches from Baikonur cosmodrome after a Proton rocket exploded over their territory.

When the Russians reluctantly agreed to pay the money they owed for the lease of the huge launch complex, Progress M-42 was free to lift off – two days late. On board was a Georgian-built foldable antenna and a system for controlling Mir when it was no longer occupied. Shortly after, Afanasyev and Avdeyev failed to fully erect the Reflector antenna during a six-hour EVA, but their efforts met with success at the second attempt on 29 July, in what was billed as the last Mir spacewalk.

August began with the inadvertent shut-down of the main computer by mission control. Taking advantage of the error, the crew set about installing a back-up computer in preparation for the abandonment of Mir. Then, on 28 August, came the moment everyone in the Russian space programme had been dreading – the crew shut down the station's life support systems and headed for home. "We are leaving with a heavy heart," declared flight engineer Sergei Avdeyev. "We are leaving a little piece of Russia."

Until the last, Mir had continued to make its mark as yet another space endurance record was broken. On completion of his 379-day flight, Avdeyev had spent a total of 747 days in orbit – more than anyone in the history of spaceflight.

Now, for the first time in almost 10 years, there was no one living on board the giant structure. On 7 September, the station's gyros were slowed to a stop, leaving Mir in an uncontrolled, drifting mode. Left to the vagaries of atmospheric drag, the station's altitude gradually fell over the next few months from 380 km to 340 km. Then, as the year 2000 dawned, Mir seemed to receive a reprieve. Financial backing from US businessman Walt Anderson paved the way for the creation of an international consortium, known as MirCorp, that would invest funds in its modernisation. Under an innovative leasing agreement, the station would be available for multiple commercial uses, ranging from tourism and advertising to industrial production and scientific research.

To the mighty annoyance of the Americans, two Progress craft and a Soyuz ferry were speedily sequestered from the production line for the ISS programme and reassigned for the renaissance of Mir.

"How can the Russians maintain a serious involvement in both Mir and the International Space Station?" asked US Congressman Dave Weldon, vice-chairman of the House Space and Aeronautics Subcommittee. "They are so cash-strapped, I don't see how they can do it." Even NASA boss Dan Goldin declared himself "not pleased with the performance and attitude" of RSC Energia. Company spokesman Nikolai Pronin countered, "The International Space Station is still non-existent and if someone wants to conduct long-term experiments in space, Mir would be the only place to do it."

With Mir's main computer once more operative, the orbit of the 130-tonne complex was raised and stabilised before a new generation of ferries was launched on 1 February. Progress M1-1, which was equipped with a more powerful propulsion unit and increased fuel reserves, also delivered a large supply of oxygen to raise the internal air pressure of the leaking station.

Just who would be the first to set foot on the newly commercialised station remained uncertain. An announcement that Russian film actor Vladimir Steklov would be shooting a movie about the efforts of a cosmonaut commander to save a doomed space station proved to be premature when financing for the project failed to materialise. So it was that the resurrection of Mir was entrusted to veteran Alexander Kaleri and rookie Sergei Zaletin.

To prepare for the opening up of the station, little time was wasted in using the modified Progress to raise the station's orbit still further, eventually lifting it above 350 km (220 miles) once more. As the new crew completed a manual docking with Mir on 6 April and began to reactivate the slumbering station, MirCorp and Energia announced their plans to send a 29th expedition aloft later in the year.

Although all systems were restored to full working order, the crew had less success in tracking down the station's small air leaks. After a painstaking process of disconnecting cables and air ducts, then closing and pressurising each of the station's modules and compartments in turn, the men did manage to track down and seal a faulty hermetic plate at the entrance to Spektr.

A second Progress M1 vehicle arrived on 28 April, loaded with propellant that would be used to raise the station's orbit and prolong its life. A spacewalk two weeks later revealed that a US-built solar panel had failed because of a cable short circuit. Back on Earth, the Russians could still not decide on Mir's future. While space agency officials merely agreed to maintain its orbit with the aid of further unmanned ships, MirCorp enthusiastically declared that two further manned missions would take place by spring 2001.

It was eventually revealed that American millionaire Dennis Tito had agreed to pay $20 million for a week-long trip to Mir in early 2001. A subsequent announcement stated that a US TV programme called *Survivor* was also prepared to run a competition with the ultimate prize of a trip to the aged complex.

But it was all too late. Kaleri and Zaletin once more mothballed Mir and returned home on 16 June after spending just 70 days on board. This time, there was to be no reprieve. With the much overdue launch of the Zvezda module on 12 July, it became clear to even the most enthusiastic Mir supporter that a new show had hit town and the old-timer could no longer compete.

Although MirCorp continued to insist that Mir had a commercial future, the consortium was struggling to find funds to keep the station aloft. In October, Progress M-43 arrived at the station with fresh food, water and oxygen that could be used by another expedition. Although its engines were fired to raise Mir's orbit, serious discussions had resumed about how best to dispose of the huge structure without endangering human life or property.

On 19 November, the government finally decided that, since the continued safe operation of Mir in an automated state could no longer be guaranteed, its mission must be terminated in late February 2001. As if to prove the wisdom of this decision, mission control lost contact with Mir on Christmas Day, leading to speculation that a rescue mission would have to be launched.

Although the emergency was soon over and control was re-established, the warning was there for all to see. Further alarms were raised on 18 January when the station lost power and drifted out of alignment, forcing the launch of Progress M1-5 to be postponed. Only by firing Mir's thrusters was the situation stabilised to enable the cargo ship to be delivered safely.

Outfitted with eight fuel tanks instead of the usual four, Progress M1-5 was given the task of lowering the station's path and sending it to its doom. Aided by the effects of atmospheric drag as enhanced solar activity caused the atmosphere to balloon outwards, Mir's orbit began its inexorable decline.

As Nature took its course, specialists in orbital dynamics around the world revised their calculations on a daily basis. The general consensus was that the structure would fragment into perhaps 1,500 pieces at it encountered atmospheric friction on its plunge into the Pacific. Meanwhile, Russian controllers insisted that the destruction of their proud flagship would provide valuable experience for deorbiting the International Space Station at the end of its operational life.

Tension mounted as the rate of descent increased and the date for the critical manoeuvres approached. The moment of truth came on the morning of 23 March, with the first of a carefully choreographed series of rocket firings by the engines on Progress. Mir's kamikaze plunge was accelerated by two more burns during the following five hours.

Less than half an hour after the third and final firing, which took place at an altitude of just 80 km (50 miles), spectators in Fiji observed trails of flaming wreckage streaking across the night sky. Fortunately, the last-minute insurance policy taken out by the Russians was not required, as the remnants of the once proud space station were scattered over a broad, uninhabited area of the southern Pacific Ocean.

It had been a remarkable success for the hard-pressed Russian managers, but no one was cheering in the silent control centre just outside Moscow.

Fireballs over the Pacific mark the end of Mir on 23 March 2001. (Digital photo)

Chapter Four

Metamorphosis

Today's ISS involves no fewer than 16 countries – the United States, Russia, Canada, Japan, Brazil and 11 members of the European Space Agency – making it the most ambitious global programme ever undertaken. However, the concept of an international space station created under the leadership of America is far from new.

As early as October 1969 – only three months after the triumphant Moon landing by Apollo 11 – NASA administrator Tom Paine was visiting Europe to present the elements of a future US space programme, including a multipurpose space station, with a view to encouraging offers of participation. By March 1970, Paine had also undertaken preliminary discussions with the Japanese government.

In a March 26 report to President Nixon, Paine commented on his recent meeting with representatives from 17 countries and the three European space organisations: "It seems clear that our proposed space station – Space Shuttle systems would obsolete many of their proposed developments before they became fully operational. For this reason, our proposals for international participation are receiving thoughtful attention."

Paine's initiative largely fell on stony ground, since Nixon was unwilling to fully fund the ambitious, $6–10 billion post-Apollo programme. The disillusioned Paine resigned in September 1970.

However, despite savage financial cutbacks, NASA continued to lobby for a 12-person space station as the obvious destination for its Space Shuttle. Without the orbital facility, the argument went, one of the main justifications for building the heavy-lift Shuttle had disappeared. An interim programme, Skylab, was already under way, but the Apollo-derived space station was a technological dead-end and would be abandoned long before the Shuttle was operational.

At first, agency officials continued to promote the concept of a space station that could accommodate up to a dozen crew and would be launched by a single Saturn V rocket left over from the Apollo programme.

When it became obvious that this would be too costly, two of America's leading space companies, Rockwell and McDonnell Douglas, put forward proposals for a cheaper, more flexible, modular space station. The plan they presented in early 1972 described a station that would be assembled in orbit like a giant construction kit. The first sections would provide living quarters, a dormitory, and a control module. Additional units containing scientific equipment and experiments would be delivered by the Shuttle.

Further downscaling followed as designers struggled to put together a programme that might overcome the Administration's tight financial constraints. Proposals to reduce costs by launching free-flying, unmanned Research and Applications Modules were countered by criticisms that they would still need to be serviced and controlled from a nearby space station. The third alternative was a so-called Sortie mission, which involved flying a single module in the Shuttle cargo bay. Although it was the cheapest of the research scenarios, it also had less to offer from a scientific point of view.

With the Nixon Administration's approval of the Shuttle in January 1972 offset by its refusal to fund a modular space station, NASA decided to opt for a stop-gap solution - Spacelab. Even better from the American viewpoint, the Europeans were eagerly waiting in the wings to build the pressurised laboratory.

THE INTERNATIONAL ASPECT

The European Space Research Organisation (ESRO) and the European Launcher Development Organisation (ELDO) were the forerunners of the European Space Agency. They also represented the most important of NASA's potential partners (and competitors) apart from the Soviet Union. In 1971, NASA approached ESRO and ELDO with three options for contributions to the post-Apollo programme.

In an unprecedented move for an American human spaceflight endeavour, NASA suggested that Europe might develop parts of the Shuttle orbiter, including the tail and the payload bay doors. In addition, the Europeans also agreed to study a reusable "Tug" upper stage that could transport satellites between the Shuttle's low orbit and much higher operational orbits. The third option was a Research and Applications Module that could be delivered to a space station, used as a free-flying, unmanned laboratory, or carried piggyback in the Shuttle's cargo bay.

While ESA spent some $20 million on preliminary Phase A studies for these projects, American political support for cooperation with the Europeans was superseded by more practical considerations. Concerns arose over the ability of European industry to develop leading edge technology for the Shuttle or Tug. NASA and officials from the US military also expressed concern that the space programme would become reliant on foreign technology.

The wish list was soon whittled down to just one – a reusable science laboratory (later to be known as "Spacelab") that would be carried in the Shuttle's payload bay. Europe would have to take it or leave it. In return for building the laboratory, Europe would gain access to Shuttle flights for its astronauts as well as valuable experience in building human-rated space hardware and carrying out pioneering research in microgravity.

Hampered by delays in Shuttle development and cost overruns – the final cost of the laboratory spiralled to $1 billion, twice the original forecast – the first Spacelab eventually flew in November 1983. On board Shuttle Columbia

View of the European Space Agency's Spacelab in the Shuttle payload bay. (ESA photo)

was Ulf Merbold, the European Space Agency's first astronaut. A further 32 Spacelab flights (some involving only unpressurised pallets) took place over the next 15 years before most of the hardware was retired in 1998. This extended experience was eventually used as the basis for the major European contributions to the ISS – a laboratory module and a logistics supply module.

Meanwhile, NASA officials had not abandoned their hopes for a future station. Behind the scenes, they were still looking forward to a time when the Shuttle would be operational and a more supportive President would occupy the White House.

A fascinating insight into agency thinking is provided by a memorandum sent by Arnold Frutkin, the NASA Assistant Administrator for International Affairs, to George Low, the NASA Deputy Administrator, on 7 June 1974. Frutkin wrote:

> I have assumed that it would be easier to get domestic clearance to explore a space station internationally than to get domestic approval for a space station per se before inviting international participation, Therefore, we propose an approach here in which all elements of the project would be attacked on an international basis: justification, definition, design, construction, operation and use.

The memo continued,

> The three plausible partners for an international space station effort would be the US, the Soviet Union, and the European Space Agency. All have space experience, will have had manned flight experience, and have the necessary resources. (Canada is extremely limited in resources and Japan has shown no disposition to contribute to a non-national space purpose.)
>
> We could approach either the USSR or Europe bilaterally, but I believe that each would be reluctant to enter into a strictly bilateral arrangement in the foreseeable future, Europe because of conservative space funding views and current space commitments, the Soviet Union because of political and security considerations. I do not think we should put the USSR forward as the senior partner since Europe would be quite offended (in view of the Spacelab agreement). Moreover, Europe might really be a better partner operationally, technically, financially and politically.
>
> Our best bet would be to approach both Europe and the USSR simultaneously, holding over each the possibility that we might be going ahead with the other. On this basis, we may be able to motivate *both* to work with us on a tripartite basis. This would give us the strongest basis for a large undertaking – economically and politically. Our approach is calculated to make the USSR and Europe feel they have very little to lose and perhaps something to gain by entering into the particular procedure we are proposing.

These comments have particular relevance in light of future developments that brought both the former Soviet Union and Europe (as well as Canada, Japan and others) on board the current International Space Station programme.

Although an anticipated approach to the Soviet Union as early as 1974 may seem surprising, it must be remembered that a temporary thaw in East–West relations was under way. The 1975 docking of Apollo–Soyuz and the subsequent handshake in space seemed to open the door to future collaboration. NASA even offered the Space Shuttle for carrying crews and cargo to and from the Salyut space stations. In return, it was hoped to conduct long-term research on the Soviet stations until America could build its own orbital facility. These efforts ended with the collapse of the US-Soviet *détente* in 1979.

That same year, with Shuttle development delayed but well advanced, NASA and contractors began conceptual studies of a space station that could be delivered into low Earth orbit and assembled by the reusable vehicle. In order to increase its appeal to the nation's lawmakers, potential users and the general public, this so-called Space Operations Center would serve as a laboratory, a satellite servicing centre and a construction site for large space structures.

THE BIRTH OF FREEDOM

Despite all of these early manoeuvres, the story of today's International Space Station really began in 1981, with the success of veteran Hollywood movie actor, Ronald Reagan, in the US presidential election. Unlike his predecessors, Reagan had big plans for space – both civil and military – and was not about to be obstructed by a small matter of finance.

Within a few weeks, the new incumbent had chosen James Beggs – former secretary of transport in the Nixon Administration during the 1970s and one-time senior executive with some of America's largest corporations – as the new NASA administrator. His deputy, also chosen by the White House, was Hans Mark, a senior Air Force staffer for President Carter and a keen supporter of links between NASA and the Department of Defense.

Despite their different backgrounds and perspectives on the direction that the agency should take, the two top men at NASA agreed that construction of a space station should be given priority. It was a good time to choose, not only because the much-heralded Space Shuttle had finally blasted off from Cape Canaveral on its maiden flight, but also because the new President seemed open to persuasion that America could lead the "Free World" in regaining the high ground of space from its Communist rival, "The Soviet Evil Empire".

"It is the logical next step," argued Beggs.

With the Space Shuttle now flying, it seemed sensible to provide a space station that the reusable space truck could regularly service – after all, this had been the original justification for the Shuttle when it was first proposed in the 1960s.

NASA argued that space station technology development and construction would offer a worthy successor to the Shuttle programme, guaranteeing money and jobs to an agency that was striving to find a vision for the late 20th century. Furthermore, America would be able to compete with the long-duration efforts of the Soviet space stations, learning how to utilise the space environment for development of new metal alloys, drugs and other materials. In the long term, a US station might even serve as an intermediate staging post for exploration of the Moon and planets.

The two men began to marshal support from any agency that might have a possible use for a permanent orbiting structure. Mark approached his contacts in the defence community, even suggesting to the Central Intelligence Agency (CIA) that the station could be a suitable spy platform.

However, their efforts initially came to naught, partly because Defense Secretary Caspar Weinberger was vehemently opposed to the station on the grounds that it would deprive the DoD of its regular access to Shuttle flights. It was also clear that an automated reconnaissance satellite would be a much cheaper and more stable asset than a multi-billion dollar manned facility in which a crew was continually bouncing around and carrying out exercises. If large sums were to be invested in space-based enterprises, the military and intelligence communities had much higher priority projects to pursue.

Little progress was made, despite the creation in October 1982 of a special Inter-Agency Working Group to discuss space station issues, under the leadership of British-born NASA veteran John Hodge.

Undeterred, Beggs established a Space Station Task Force at NASA headquarters in May 1982, opening the door to the first conceptual designs for a space station. Furthermore, returning to a policy discussed almost 10 years earlier, Beggs began to make informal advances to potential international partners once more, with the aim of encouraging them to take part in the station's development, construction and operations. This farsighted move, designed to pre-empt criticisms of excessive cost and increase political support while providing some protection against future cancellation, would stand the programme in good stead during later periods of crisis.

Since NASA still offered the only viable means of sending astronauts into orbit (at this time, visits to Soviet space stations were generally limited to residents of communist countries), the suggestion had a certain attraction. However, some caution was in order, since Europe, Canada and Japan all had

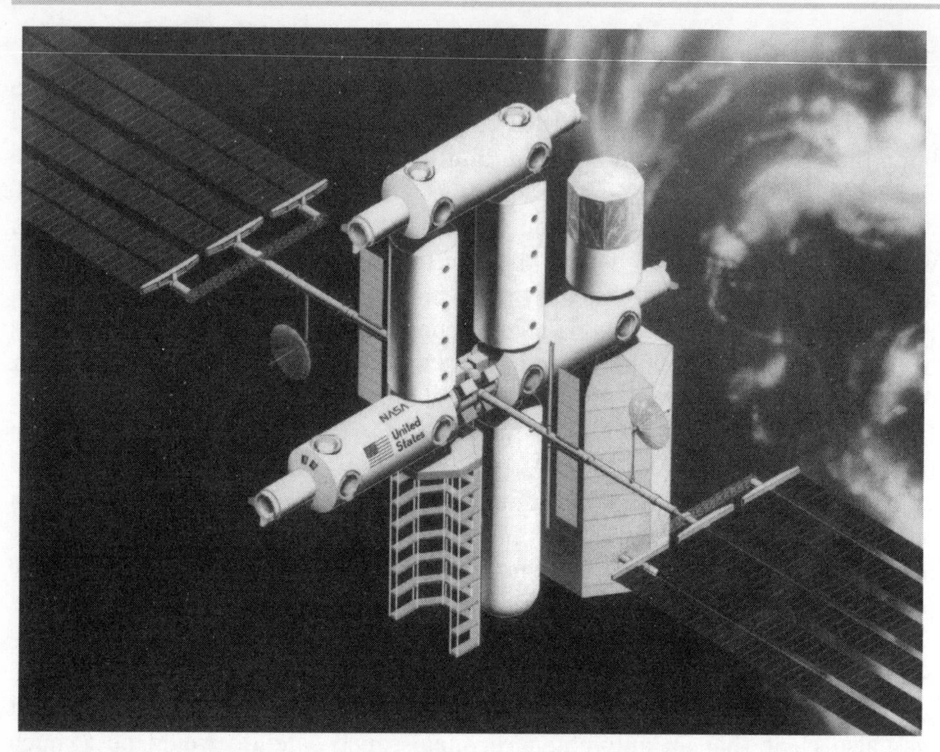

This 1981 space station design was produced by the Space Operations Center in Houston, Texas. It was envisaged that the station could be completed after seven or eight launches and carry a permanent crew of 8 to 12 astronauts. Two command centres are attached to the central energy truss, with two habitation modules above and a logistics module to their right. A docking tunnel lies across the top of the habitation modules. Lower structures included a propellant storage tank (centre) and an unpressurised hangar (right). (NASA photo S81-39234)

mixed experiences from previous cooperative efforts with their American counterparts. The US giant rarely hesitated to override its so-called partners if national priorities were at stake.

Their concerns were not assuaged when NASA convened an international space station workshop in the summer of 1983. In his keynote address, Beggs informed the delegates that a space station was required "to maintain our leadership".

The inference was clear when he went on to state, "If we can attract international cooperation, then other nations will be cooperating with us in the resources that they spend on space, rather than competing with us."

While the other space-faring nations mulled over their options, President Reagan accepted NASA's vision for the future. In his State of the Union Address on 25 January 1984, Reagan gave a ringing endorsement to the space station programme by declaring that one of the great national goals was to develop a new frontier based on the pioneer spirit. "I am directing NASA to develop a permanently manned space station and to do it within a decade," he enthused. "A space station will permit quantum leaps in our research in science, communications, in metals, and in life-saving medicines which could be manufactured only in space."

The President went on to describe an $8 billion NASA programme, without any military involvement, that would lead to the completion of a fully functional space station by the early 1990s. He made it clear, however, that the door was open for outside contributors to help with development of an even larger, more capable, station. "We want our friends to help us meet these challenges and share in their benefits," he said. "NASA will invite other countries to participate so we can strengthen peace, build prosperity, and expand freedom for all who share our goals."

NASA chief James Beggs followed up Reagan's declaration of intent with a tour of Europe, Japan and Canada to sound out the opinions of potential partners. On his return, he reported to Secretary of State George Schultz, "The reaction so far to the President's call for international cooperation has been both strongly positive and openly appreciative."

His upbeat assessment of the attitudes of the various governments is summarised in these extracts from the report:

> Prime Minister Nakasome and other Japanese officials, while still cautious in public, made it obvious that Japan will participate in a significant way. The Japanese believe they made a mistake in not joining us on the Shuttle, and are determined not to be left behind again.
>
> In Europe, Italian Prime Minister Craxi was openly ebullient about the prospect of cooperation and strong Italian participation is assured.
>
> The French will be tough bargainers, and obviously intend to pursue their own independent space programs, but I am confident that we can agree on mutually beneficial terms for cooperation. By the way, you will be interested that (President) Mitterrand observed to me that his recent proposal for a European military space station fell on deaf ears in Europe.
>
> Chancellor Kohl was in Washington during my stop in Bonn, but the relevant Ministers were quite clear that a major German contribution will be forthcoming. The British were more cautious, and while I believe they will participate, it will probably be on the same terms that have marked their recent space-related activities – relatively small-scale projects done on a multilateral basis.
>
> ESA will almost certainly play a key role in managing Europe's Space Station participation, just as it did in the highly successful Spacelab project.

With the benefit of hindsight, it is interesting to note how accurate this assessment was. All of the nations mentioned – with the exception of the UK – eventually entered enthusiastically into the Freedom, and later ISS, programmes. However, there was still a long way to go if the initial advances towards these potential partners were to culminate in a binding "matrimonial" agreement.

Events moved rapidly at first. In April 1984, NASA established its Space Station Programme Office, followed by a Request for Proposal to US industry in September.

Meanwhile, at the London Summit in June, President Reagan repeated his invitation to other countries to participate in the ISS programme. The response was very positive, with Japan, Europe, and Canada each signing a bilateral memorandum of understanding with the United States in the first half of 1985. In the following year, the hardware contributions of the three partners were set on a firm footing. ESA decided to offer the Columbus laboratory module as part of a larger programme which would include a small spaceplane, named Hermes, and a free-flying laboratory. Japan would also build a laboratory module while Canada agreed to provide an advanced robotic servicing system based upon its successful Shuttle robotic arm.

DESIGN AND REDESIGN

As early as the summer of 1982, NASA had sponsored studies from eight US aerospace contractors to design appropriate space station "architecture" that would fulfil all of the requirements for such an orbital base. For example, the space station must be utilised for a wide range of experiments in the fields of material and life sciences, taking advantage of such characteristics of the space environment as microgravity and high vacuum. Secondary considerations included using it as an intermediate base for exploration of the Moon and planets, or as a facility where malfunctioning satellites could be repaired.

Such was the complexity of the project and its various configurations that NASA officials decided to change the agency's usual practice of designating a single lead centre to take control of a particular project. Instead, the responsibility for developing different portions of the space station was divided into four work packages in April 1985, each of which was assigned to different NASA centres, based on their areas of expertise. (As costs increased and the design changed, this was consolidated into three work packages in 1991.)

In addition to awarding contracts to companies for design and development work, these field centres were responsible for overseeing the satisfactory completion of the contracts and for ensuring that any necessary programme

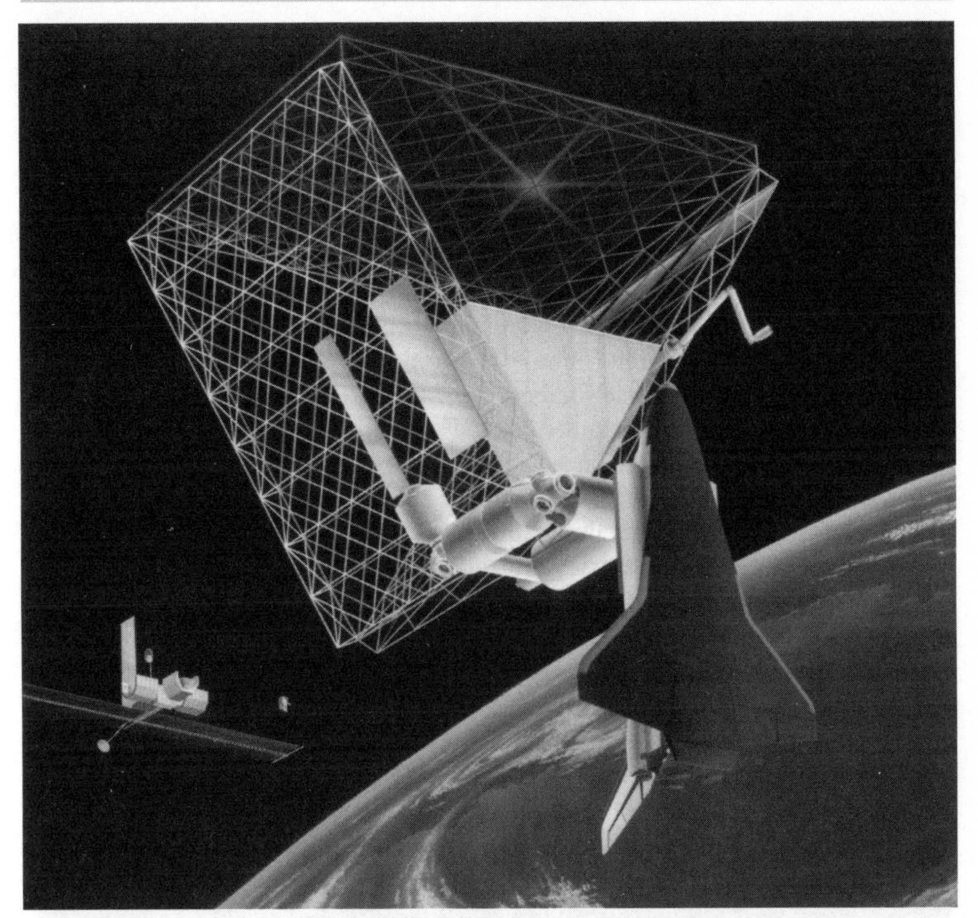

Concept for a triangular space station presented during a NASA industry briefing on 11 July 1984. The "roof" of the structure would be covered with solar cells for power generation. Rigid beams connect five accommodation and research modules. (NASA photo S84-30005)

changes were carried out. Each of the NASA field centres involved in the station was meant to be subordinate to a Space Station Project Office at Reston, Virginia, while overall management control and supervision was assigned to NASA HQ in Washington DC.

Work Package 1 was led by Marshall Spaceflight Center in Huntsville, Alabama. It covered pressurised modules, pressure shells for nodes, logistics modules, environmental control/life support, internal thermal control, internal audio and video.

Work Package 2, led by Johnson Space Center in Houston, Texas, included the truss, airlock, node outfitting, external thermal control, EVA support, data management, communications and tracking, propulsion, Shuttle interface and crew return vehicle.

Goddard Spaceflight Center in Maryland was responsible for Work Package 3 – the attached payload accommodation and telerobotic servicer – while Work Package 4, which included power modules and electrical power distribution, was run by Lewis Research Center in Ohio.

The first design studies – known as Phase A – for the station were presented at Johnson Space Center (JSC) in 1984, following preliminary discussion of various concepts involving contractors, JSC and Marshall Spaceflight Center. Discussions revolved around such factors as design feasibility, operations, payload accommodation, on-orbit assembly and technical feasibility. The eventual "reference configuration" which was given preference over four other concepts was dubbed the "power tower".

The name was derived from the station's elongated design, with eight solar arrays for generating 75 kW of electrical power at one end of a huge, central truss. At the opposite end were five pressurised modules and the Space Shuttle docking port. The modules to be occupied by the crew comprised two habitation units, two laboratories and a logistics resupply module. All of the modules, each of which had integrated airlocks and docking ports fore and aft, were attached in a rectangular "race track" arrangement.

The "power tower" design for the space station that was favoured in July 1984. Five modules are located at the end of a 122-m (400-ft) long central truss. (NASA photo S84-37255)

Although this design had a number of advantages in terms of power supply and flexibility, the need to improve the microgravity environment on board meant that it gave way during Phase B studies in 1985–86 to a "dual keel" configuration. In this version, the single beam was replaced by a rectangular structure with an additional "horizontal" truss aligned through its centre.

This meant that the station's mass would be distributed much more evenly, with four enlarged, pressurised, modules shifted to the middle of the station. Nodes with docking ports and airlocks on each of these would enable the station to be easily enlarged by adding further modules. Solar arrays were located away from the modules to reduce vibrations. Additional payloads and experiments could be attached to the two transverse keels, well away from the central modules.

Even before this design came to the forefront, NASA managers were faced with the trauma of the Challenger disaster. With the loss of seven crew and the grounding of the entire Shuttle fleet for 32 months, agency officials were faced by an entirely new list of operational requirements. The greater emphasis on safety had a number of serious implications for the space station. Fewer Shuttle flights per year, a lower payload lift capability and restrictions on astronaut extravehicular activity inevitably led to a downsizing of the station's structure. In particular, the "dual keel" configuration was scaled down to a single truss, with the option to add a second at a later date.

The "dual keel" space station design. (NASA photo S86-33300)

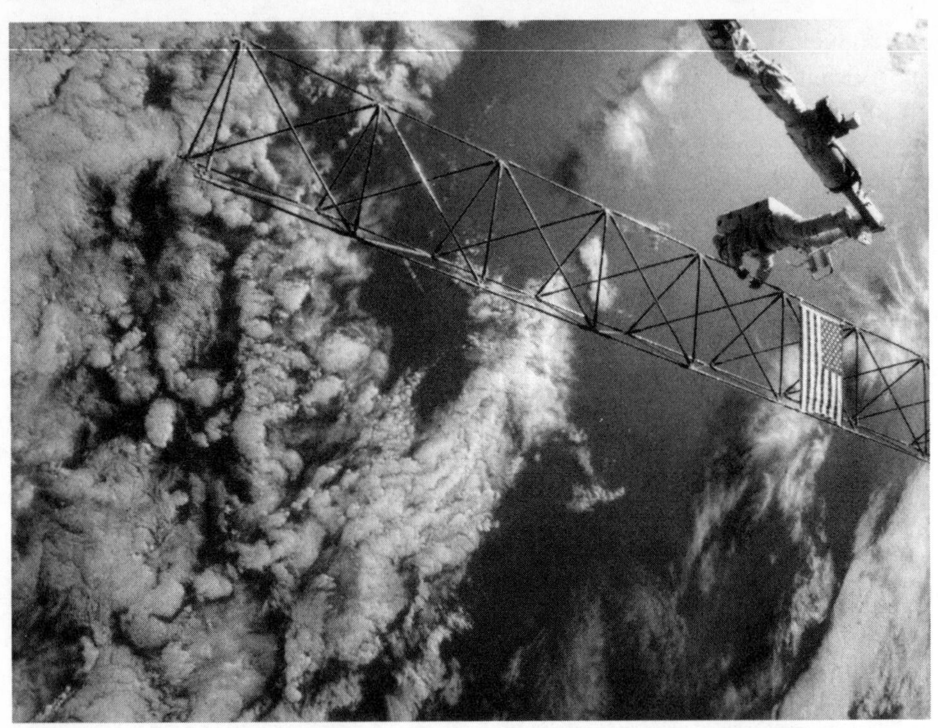

Space station construction techniques were tested on Shuttle mission STS-61-B in November 1985. Spacewalking astronauts attached beams, nodes and struts to evaluate different methods of assembling large structures in space. Here Sherwood Spring grasps the ACCESS tower. (NASA photo 61B-120052)

After a major cost review in early 1987, NASA decided to develop the station in two steps, Phase One and Phase Two, in an effort to reduce the cost of development and limit the number of spacewalks.

Against this background of a two-year hiatus in the Shuttle programme and modifications to the baseline design of the space station, negotiations continued between the USA and its overseas partners. These culminated in September 1988 with the signature of an Inter-Governmental Agreement among participating countries, just as the Shuttle returned to action.

At last, the way seemed clear for the station – now dubbed "Freedom" by President Reagan – to make progress towards commencement of orbital assembly in early 1995. However, although design and development of the station hardware got under way, a black cloud of exploding national debt and a tripling of costs hung over the programme. NASA's requests for more money coincided with increased criticism from Congress. Some of the programme's previous supporters, disappointed by the scaling down of the station's capabilities, also began to snipe at the limited science return to be expected.

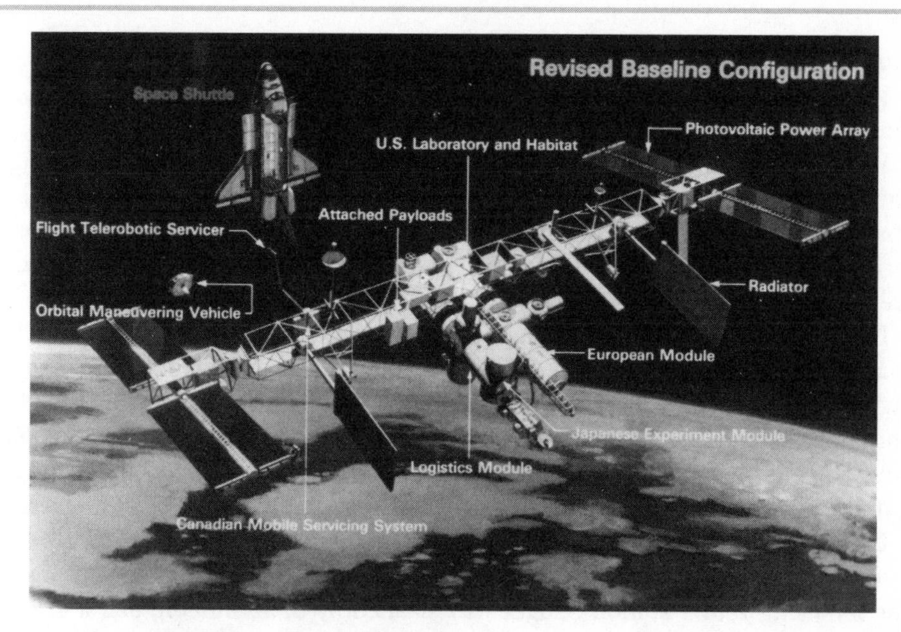

The 1987 revised baseline configuration for the Freedom space station. (NASA photo S87-33893)

These concerns were reflected in the 1990 report by the Advisory Committee on the Future of the US Space Programme, a blue-ribbon panel appointed by the recently elected President, George Bush. One of its recommendations called for "redesigning the Space Station Freedom to lessen complexity and reduce cost". It went on to suggest that "Freedom be revamped to emphasise life sciences and human space operations, and include microgravity research as appropriate".

CONGRESSIONAL CUTBACKS

Even before these damning conclusions, Congress had begun to question the space station budget. In the summer of 1989, the House appropriations committee, which set NASA's budget in the House of Representatives, recommended a $400 million cut in the funding request of $2.05 billion for Freedom. At the same time, Committee members were warning that they would recommend an additional $600 million reduction the following year.

As the programme's top three men decided it was time to quit, NASA administrator Richard Truly ordered a team of senior managers, project officials and astronauts to prepare for the worst – a funding cut of more than $1 billion over the next two years. Proposals produced by the workshop meetings, known as "Scrub 89". included a reduction in the station's crew size from eight to four, a halving of the power supply and numerous other changes to onboard facilities, attached experiments, oxygen supply and propulsion system.

When alarming rumours began to spread, the international partners clamoured for an immediate briefing. The downsizing was bad enough, but what concerned them even more was NASA's clear violation of international agreements, which stipulated that the Americans would consult their partners over potential changes.

Despite the inevitable ill feelings that reverberated around the globe, ESA Space Station Director Fredrick Engstrom tried to make a positive response by suggesting that the international modules be launched before the US laboratory. This cash-saving but rather humiliating suggestion fell on deaf ears. NASA's Space Station Director, Richard Kohrs, responded by issuing a memo to his 20,000 workforce to cut spending immediately.

Congress eventually allocated almost $1.8 billion for Freedom in 1990 – more than NASA officials had anticipated – but the problem would not go away. The following year's station budget was slashed by $550 million, and a drastic reduction of $8.3 billion over the lifetime of the Freedom programme was introduced.

Congress also told NASA to redesign the station so that it could be built and funded piecemeal. As a result, the truss that formed the station's spine was to be prefabricated on Earth rather than assembled in orbit, and its length was to be cut from 154 m (500 ft) to 107 m (350 ft). The US Laboratory and Habitation Modules were also to be outfitted on the ground and their lengths were reduced from 13.4 m (44 ft) to 8.2 m (27 ft) in order to bring their masses within the launch capability of the Shuttle.

The Freedom assembly schedule was also "rephased", which meant that the first launch would be delayed nearly a year, from March 1995 to early 1996. Man-tended capability – a term referring to short-term occupation of the station – would not begin until mid-1997, while permanent occupation would have to be postponed for three years until 2000.

When the new design was unveiled in March 1991, NASA chief Truly pleaded "Let's give it to the engineers to build". Representative Barbara Boxer scathingly responded that the new design was "a garage in space with nothing in it and nothing going on around it".

Matters worsened in May 1991 when the House subcommittee voted to scrap Freedom and spend the savings on social programmes. The following month, the future of the entire project hung in the balance as the entire House weighed the pros and cons of space station Freedom. At the end of the day, the implications of upsetting America's closest allies combined with throwing thousands of jobs (and possibly votes) into the melting pot saved the day for

The scaled down version of the Freedom space station that was unveiled in 1991. (NASA photo S91-20394)

George Bush, Vice-President Quayle and the NASA human spaceflight programme. Freedom survived by 240 votes to 173.

RUSSIA TO THE RESCUE

Despite spending some $8 billion on redesigns – the figure originally quoted for completion of a fully manned station – and producing mountains of paperwork over the past 8 years, the space station was not much nearer to becoming a physical reality when President Bush appointed a new administrator for the ailing space agency. The knight on a white charger who was expected to ride to NASA's rescue was Daniel Goldin, an outspoken product of a working-class Jewish family and a 25-year career in the aerospace industry.

Fully aware of the agency's precarious budgetary position and lack of direction, Goldin wasted little time in establishing his reforming credentials and his personal authority. After sacking or replacing half of the agency's field centre directors, he proceeded to slash the NASA workforce and thousands of contractor-related jobs. If NASA was to be obliged to struggle along with an annual budget of no more than $14 billion for the foreseeable future, so be it. The agency would have to find a way of achieving its goals by implementing a new "faster, better and cheaper" policy.

One of his first meetings was with his Russian counterpart, a huge bear of a man called Yuri Koptev. During an exploratory get-together in Goldin's Washington apartment, the two men came to realise the similarities in their predicaments – loss of prestige, financial cutbacks and an uncertain future for their respective space programmes. To Koptev's surprise, the Americans seemed to need his help as much as he needed theirs.

The response to this modest meeting was almost immediate. Within days, a more high-profile summit took place between US President George Bush and Russia's new leader, Boris Yeltsin, who announced their intention to initiate a joint human spaceflight programme that would culminate in the historic docking of a US Shuttle with the Mir space station and cosmonauts flying on the American orbiter. At the same time, the US agreed to purchase Russian Soyuz ferries as Freedom's lifeboats, an agreement which provided much-needed finance for the cash-starved Russians while allowing considerable savings in development costs for NASA.

The respite was short-lived. In the Presidential election later that year, Bush was defeated by William Jefferson Clinton, the first Democrat in the White House since Jimmy Carter, 12 years earlier. Space was far from the top of the new incumbent's list of priorities, but he was fully aware that tough times were ahead during his battles with the cost-cutting Republican majority in the Congress.

The figures told an unmistakable tale. Despite previous downscaling efforts, the price label on Freedom for the period to 1999 was still a whopping $30 billion, and the government's financial watchdog, the General Accounting Office, estimated that total expenditure over the station's 30-year lifetime would easily break the $100 billion mark.

Particular problems loomed in Work Package 2, which involved Johnson Space Center and prime contractor McDonnell Douglas. Criticism reached a new level when Democrat Senator Bob Krueger successfully called for the head of John Aaron, NASA's space station manager in Houston.

With this in mind, the new administration soon made it clear that it planned to slash NASA's budget by 15% over the next five years. Clinton also had the option to appoint his own NASA administrator, but, with Goldin still settling into the job and promising to revitalise the ailing agency, the President decided to maintain the status quo.

After two weeks of infighting within the White House – discussions from which Goldin was excluded – Clinton announced yet another redesign for the space station in an effort to reduce costs. Instead of complaining, the NASA chief openly supported the move. "I was frustrated that the programme was out of balance," he said. " It would be unconscionable if we proceeded with a fully funded space station – given the tremendous budget deficit problem this country has – at the expense of not starting any new activities in spacecraft."

Congress and the international partners were less enthusiastic. Goldin was also accused of undermining his top subordinates, including Space Station Director Richard Kohrs, by contradicting their testimony to the House space subcommittee.

The European Space Agency, which had only recently obtained agreement from member states to fund the Columbus laboratory, had to postpone its contract awards. A Japanese official declared that his country was becoming "fed up" with the frequent, sudden changes in the programme and technical specifications.

With all space station work on hold, Goldin brought back NASA veteran Joseph Shea to oversee the latest redesign of Freedom. Within a matter of weeks, Vice-President Gore announced the creation of the "Vest Committee". an expert advisory group under the leadership of Dr Charles Vest, President of Massachusetts Institute of Technology.

In early April 1993, the White House presented the NASA redesign team with its terms of reference. The Advisory Committee was told to consider three alternatives: a low-cost option of $5 billion, a mid-range option of $7 billion and a high option of $9 billion. These costs were to cover all space station expenditure 1994–98, including adequate reserves to ensure programme implementation. However, in order to encourage maximum frugality, any expenditure over $7 billion (i.e. the high option) would have to be clawed back by savings in the rest of the NASA budget.

Just as significant as the new budget framework was the announcement that "full consideration" would be given to the use of Russian assets during the station redesign process. Russian engineers would be asked to contribute to NASA's deliberations "so that the redesign team can make use of their expertise in assessing the capabilities of Mir and the possible use of Mir and other Russian systems".

As NASA officials began to investigate ways of bringing their erstwhile rivals on board, Russian Space Agency chief Yuri Koptev and Energia general designer Yuri Semenov began to make tempting offers that would have major financial implications for both sides. Apart from offering a core module for the redesigned station, they also suggested that the US station should be linked with their next generation Mir 2 space station. Although the state of near-collapse of the Russian economy meant that there was little chance of the country being able to afford a replacement for its ageing Mir complex, the space officials insisted that a decision on Mir 2 was imminent.

MIR 2

When the core module of the Mir space station was launched in 1986, Soviet designers gave it a five-year operational lifetime. It was a time of huge space infrastructure projects, culminating in the development of the monumental Energia booster and the Buran space shuttle.

In the late 1970s and early 1980s, Mir 2 was expected to be very similar to Mir 1. Indeed, the back-up core module for Mir 1 was intended to fulfil a similar role for its successor. The only modification was the addition of a large beam-like truss to which solar panels and other equipment could be attached.

This modest plan changed dramatically after President Reagan announced his government's intention to build Freedom. By the late 1980s, Mir's replacement was envisaged as an entirely new concept in space stations. The so-called Orbital Assembly and Operations Center would comprise a standard Mir module, which would act as living quarters for the first crew. To this would be added a 90-tonne core section that would be launched by the Energia rocket – the world's most powerful booster – which would also be used to deliver huge truss structures, solar arrays and parabolic solar concentrators.

Five-man crews were to be ferried to the Mir-2 in the Buran shuttle or a new Zarya spacecraft, which would be carried aloft by the recently developed Zenit rocket. The Zarya, although similar in shape to a Soyuz, would also be able to carry up to 4 tonnes of cargo instead of a crew, and it was to be protected during re-entry by heat-resistant tiles developed for Buran. Almost 6 tonnes of fuel and supplies could also be brought to the station by a 13-tonne unmanned craft known as Progress M2.

By 1991, the Russian space programme was entering a period of retreat, and officials openly considered an intermediate solution popularly known as Mir 1.5. In this scenario, the life of Mir would be extended by docking the back-up module with the original core section, then moving one or more add-on modules to the newcomer. The old core would then be deorbited.

However, the Soviets soon reverted to the plan that was favoured in the early 1980s, which entailed launching the Mir 1 back-up as the core of Mir 2, then constructing a cross beam for solar arrays, solar concentrators and rotatable science platforms. By adding universal docking modules, the Mir 1

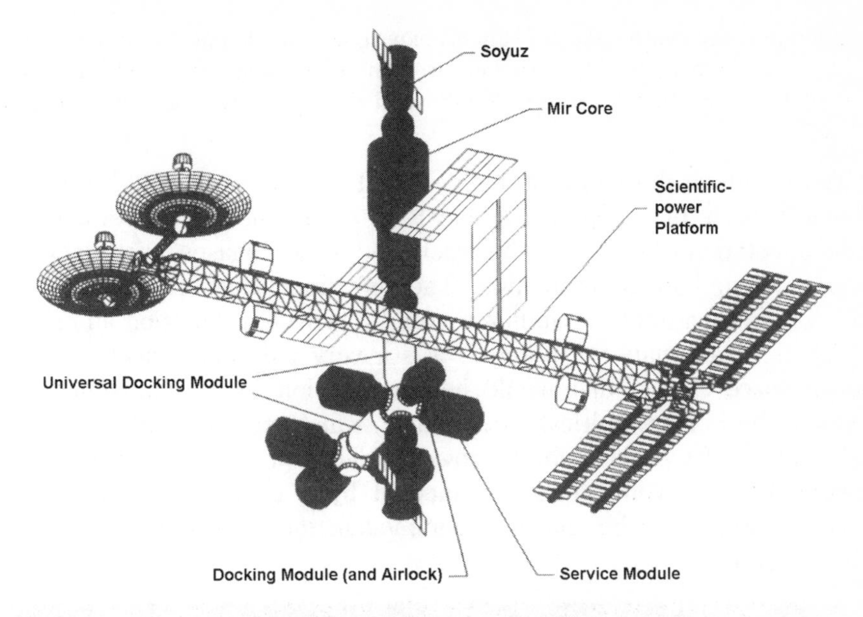

The Mir 2 space station as envisaged in 1992, just before the Soviet Union agreed to join the International Space Station programme. (NPO Energia diagram)

style station could be expanded to incorporate numerous small science modules.

By this time, Soviet expertise and apparent willingness to cooperate with other countries led to approaches from the European Space Agency. ESA was particularly interested in the opportunities presented by a 65-degree inclination orbit, as well as the option of an alternative partner if the US Congress decided to cancel Freedom. Even the Japanese expressed an interest in getting involved, should the political climate improve.

As it was, Russian Prime Minister Chernomyrdin and US Vice-President Al Gore signed an agreement in November 1993 to collaborate on a new International Space Station. Virtually all of the elements of Mir 2 were transferred to the Russian segment of the ISS, along with several new ones – most notably the FGB (Zarya) cargo block.

SPACE STATION OPTIONS

Bryan O'Connor, director of the station redesign team, explained why NASA was so interested in linking up with its Russian counterparts:

> We may not save too much by having the modules built there, since the price of the structure is not a big cost issue. But it might be a way to get to their

environmental control and life support expertise. For example, we might say: We like your carbon dioxide scrubber or your water system. So why not build modules there, install those systems, send the unit here where we finish the outfitting?

The results of the multinational deliberations were presented to the White House in early June. Options A and B were slimmed down versions of the Freedom station which was to be assembled in four phases, eventually leading to permanent human occupation. The only innovative suggestion was the possible use of a military satellite system to provide propulsion and guidance.

The third alternative, Option C, was a very different concept – a single-launch space station that would be carried aloft by an unmanned Shuttle system. The huge cylindrical core module, which measured 28 m (92 ft) long and 7 m (23 ft) across, would provide a large pressurised area which was separated into seven "decks" connected by a central corridor. Its seven docking ports would be able to accommodate the international modules and other add-ons.

Artist's impression of a fully evolved Option C space station. It includes a 28-m (92-ft) long core module and seven docking ports. Two Soyuz craft and a Shuttle are shown attached to the station. This concept was not selected during the 1993 design review. (NASA photo S93-35772)

Apart from Option C – disparagingly known as the "can" and generally slated by the international partners and by American industry – it was clear that the NASA team had deviated very little from the previous Freedom design. Most of the savings were expected to come from "streamlined management and a new operations approach". However, with 1994–98 cost estimates ranging from $11.9 billion (Option C) to $13.3 billion (Option B), none of the redesigns came anywhere near the $9 billion ceiling imposed by the Clinton Administration. Moreover, with total lifetime costs of between $41 billion and $87 billion, critics complained that the end product would still not justify the exorbitant expenditure.

Despite a recommendation from the Vest panel that serious consideration be given to Option C, the Clinton Administration decided to reverse its previous decision, opting for a hybrid of Options A and B that would utilise much of the Freedom hardware. However, although Clinton pointedly declared, "We are going to redesign NASA as we redesign the space station," he asked Congress to approve station expenditure of $2.1 billion for 1994.

The traditional opponents of the station (and big science) had other ideas. Less than a week after Clinton's endorsement of the programme, Democrat Tim Roemer posted an amendment to kill the space station. With many of the redesign details still to be defined, the opponents emphasised the uncertainties and weaknesses of the station, and almost prevailed. The amendment was defeated by the thinnest of margins: 216–215. "Today's vote is a clear signal that space station Freedom is on its way to being grounded," said a defiant Roemer.

A further sign of the direction in which Congress was moving was the decision to cancel another high-profile, but expensive, project known as the Superconducting Super Collider. Physicists were outraged at the loss of this groundbreaking facility, which they considered to offer much more scientific potential than the revamped Freedom, but all the calculations favoured the station. Thousands of jobs and votes were involved, and billions of dollars of expenditure would be thrown on the waste heap if Freedom was scrapped.

NASA Bets on Alpha

In the ensuing few months NASA was scrambling to revamp its much-criticised management structure and meet the President's requirement to save more than $4 billion over the next five years. In mid-August, Goldin announced that Johnson Space Center would become the host centre for the programme, taking over responsibility from the other field centres. At the same time, previous "Work Package" prime contractors Grumman, McDonnell Douglas, and the Rocketdyne Division of Rockwell International were subordinated to a single prime contractor – Boeing.

Meanwhile, discussions continued concerning the best way to involve the Russians in the project. On 2 September, Vice-President Gore and Russian Prime Minister Viktor Chernomyrdin, in the presence of Goldin and Koptev, signed a landmark agreement which ensured that Russia would play a major role in future developments. The new station – in deference to the Russians, "Freedom" was now renamed "Alpha" – was to be a unified structure that would include elements of Mir 2.

Details were sketchy, but the agreement envisaged an evolving programme that was divided into three phases. Phase One involved a number of rendezvous and docking missions between the Space Shuttle and Mir. The US would make a $100 million down payment to extend the life of Mir and further annual instalments of $100 million until 1997 to cover the costs of a joint Shuttle–Mir programme. The second phase envisaged the possible use of a Mir-type module flown in conjunction with a US laboratory and serviced by the Shuttle. More extensive Russian involvement in the international space station programme would depend on further discussions and consultation with the international partners.

In a joint statement, the leaders declared, "The parties are convinced that a unified space station can offer significant advantages to all concerned, including US partners."

Those partners who, once again, had not been consulted over the proposed deal, were less sanguine. Delays of up to four years for Canadian, Japanese and European elements would inevitably incur substantial additional costs. Furthermore, although this would allow more breathing space for the cash-strapped European Space Agency to complete its now-shrunken Columbus module, ESA officials were not too happy at the prospect of playing second fiddle to the Russian colossus after a decade of sitting at NASA's right hand.

It was also clear that the Russians had been obliged to eat a sizeable slice of humble pie in return for American political support for Yeltsin's unstable, cash-starved government and a promise to invest in their ailing space industry. The US had effectively employed its economic muscle to dictate terms to its new ally. However, both sides benefited from the political détente with their erstwhile rivals.

President Clinton spelled out his position in a letter written before a debate in the Senate to kill the station. "I wanted to convey to you my strong support for NASA's space station programme as an important science and technology investment for the United States, and as a symbol of peaceful international cooperation. We now have an historic opportunity to include Russia in this endeavour, thereby achieving an important step in putting the Cold War behind us."

"A prime motivation for this is in promoting stability in a nation that was once our enemy," said Jeffrey Manber, vice president for marketing at Energia USA. "We do have to spend money. But we will save untold billions by being able to avoid the expense of a Pentagon build-up."

As White House science adviser John Gibbons explained: "This initiative in space cooperation fits into the context of a much larger partnership with Russia, a relationship that will define the post-Cold War era. Our negotiations with the Russians produced a key understanding that Russia is committed to adhere to the guidelines of the Missile Technology Control Regime."

In effect, strict limitations were to be placed on Russian missile-related exports, with specific reference to a Russian agreement not to assist India in development of a home-grown cryogenic rocket engine. Furthermore, although the US government was now prepared to open the door to Russian competition in the international launch market, there were limitations to this grandiose gesture of goodwill. Russian companies would only be able to compete for a maximum of eight launches of telecommunications satellites until the end of 2000, and prices quoted would have to be comparable to those in the West.

In an attempt to pre-empt critics who might question the introduction of such a powerful aerospace rival to a US-led programme, NASA insisted that 75% of the hardware planned for the previous Freedom station would be used by the modified Alpha.

"With respect to potential Russian participation, however, no one should confuse the course we are charting as relinquishing control of the space station or exporting jobs out of the US," declared Gibbons. "In developing this cooperative programme, we are focusing on areas that will not negatively impact the US aerospace sector. We intend to proceed in a way that protects our vital domestic interests while maximising the benefit we can derive from fuller interaction with the Russians. In some areas, such as solar dynamic power and possibly closed life support systems, we believe the net gain of new technologies from the Russians could stimulate jobs in the US."

In the following months, the Russian and American space agencies firmed up the vague initial proposals. Under a plan presented to the White House in November, the Alpha/Russian station closely resembled the International Space Station of today, with a Salyut "space tug" and a Mir module attached to the laboratories of the other nations. The station would orbit the Earth at an inclination of 51.6 degrees (the same as Mir) rather than 28.8 degrees for Freedom. In order to compensate for the Shuttle's consequent loss of payload capability to a more highly inclined orbit, NASA would build a lighter aluminium–lithium external fuel tank for the orbiter. Command and control would be shared between the Russian ground control centre near Moscow and NASA's Johnson Space Center.

This 1996 artist's impression shows the International Space Station at completion of Phase Two, with key elements now provided by Russia. The schedule envisaged the first launch in November 1997, with Phase Two completed by March 1999. In reality, the first launch took place one year late, and Phase Two (without the Russian Science Power Platform and Universal Docking Module) was completed in September 2001. (NASA photo HQL-425)

Construction would commence in May 1997 with the launch from Baikonur of the Salyut FGB module. The first long-term expeditionary crew would arrive on a Soyuz vehicle in January 1998. Altogether, station assembly would entail 12 Russian missions and 14 Shuttle flights, with three further missions allocated for Japanese components and two for ESA hardware. A crew of six, including two Russians, would eventually be accommodated.

NASA's next challenge was to persuade Congress that the new joint enterprise would save up to $4 billion by bringing completion of the station forward by one to two years. In an effort to streamline management and save money, NASA boss Dan Goldin amalgamated the space station and Shuttle programmes. Boeing was confirmed as the prime contractor responsible for the design, development and integration of the space station.

However, the programme's political opponents remained sceptical, despite assurances from Yuri Koptev that Russia was committed to the station. He told Goldin in private, "You've had nine votes (in Congress) since you've been administrator. How do we know that *you're* not going to back out?"

Even station supporters such as Congressman George Brown, chairman of the House Science, Space and Technology Committee, began to question the wisdom of bringing the Russians on board. "I actually have more confidence in the ability of the Russians to fulfil their part of the deal than I do in the US," was Brown's sarcastic comment.

Tight budgets in Europe and Canada did little to help. The European Space Agency had decided it could not afford its Hermes spaceplane or free-flying platform, while the Columbus laboratory would have to shrink to a size compatible with the available funds. Meanwhile, the new Canadian government announced that it was trimming $US400 million over the next decade from its $1 billion contribution to the station. Dan Goldin promised to reschedule the programme in order to keep the Canadian contribution.

As the supposed savings dwindled and the start date of the revamped programme inexorably slipped to November 1997, Brown and Congressman James Sensenbrenner, the ranking Republican on the House space subcommittee, insisted on a back-up scheme for completing the space station if the Russians reneged on their promises.

Their lack of confidence was not helped by obvious problems with the former Soviet Union's space infrastructure. The previous year had seen launches delayed by lack of funding for rockets; a cosmonaut who had to stay in space an extra five months to accommodate a paying guest; a launch failure due to contaminated fuel; and ransacking of Baikonur cosmodrome by rioting Kazakh conscripted troops.

The European Space Agency's contributions to the Freedom space station programme were to have been an Attached Pressurised Module, a man-tended free-flying platform, an unmanned polar platform and the Hermes spaceplane (seen here docked to the space station). Of these, only a half-size Columbus laboratory and the polar platform for Earth observation survived. (ESA photo)

The political environment was hardly more promising, with open disagreements between the newly independent governments of the Ukraine and Kazakhstan and their former Russian overlords. On a number of occasions, the Ukrainians showed their displeasure by cutting off communications between mission control and the Mir space station – once during a spacewalk. With even less financial or political influence, the Kazakhs decided to threaten the annexation of the Baikonur launch complex. Russia responded by offering a $110 million annual rent, but the shambolic state of the cosmodrome was emphasised a few weeks later when a fire spread through a rocket assembly shed. The fire brigade stood by and watched helplessly – there was no water to extinguish the flames.

US–Russian relations were also on a knife edge, with accusations of a mole in the CIA, doubts about the future of President Yeltsin and anti-reform moves in the parliament. Worst of all, the potential instability of Russia's political system was emphasised when Yeltsin's tanks laid siege to the rebellious Russian parliament during an official visit to Moscow in early November. Goldin's NASA contingent was shaken in the conference room as they listened to the sound of tank shells blasting the supposed seat of Russian democracy.

Nevertheless, late February 1994 saw the first pair of American astronauts – Norman Thagard and Bonnie Dunbar – fly out to Star City near Moscow to begin training for the first of the Shuttle–Mir flights. Over the next year, they would learn about Soyuz and Mir systems, undergo splashdown and winter survival training and prepare for their programme of biomedical and materials sciences experiments.

Other signs of concrete progress were the commencement of the negotiations between the international partners that would formally establish Russia's role in the space station and the completion of the space station design review on 24 March. However, it also became clear that there were no easy ways to integrate the multitude of diverse designs and capabilities, especially when the Japanese made tentative suggestions about utilising their new H-II launch vehicle and ESA offered its unflown Ariane 5 with an Automated Transfer Vehicle as a supply ship.

The annual Congressional budget battle began in April with NASA's position already undermined by an admission that the inclusion of the Russians would not provide the previously indicated cost- and schedule-savings. This confession was compounded by an embarrassing Russian proposal that it should be paid $650 million for its contributions during 1994–97, rather than the initial figure of $400 million.

Shuttle safety concerns did little to help. NASA officials had conceded that some launch safety restrictions would have to be compromised if the orbiter was to meet the tight 5-minute launch window for space station missions.

Meanwhile, astronauts and the NASA–Congress Aerospace Safety Advisory Panel expressed fears that the three-vehicle Shuttle fleet might not be able to cope with the pressure of launching 23 assembly and utilisation missions in quick succession, particularly at a time of the first year-on-year cutbacks in the agency's budget since 1972.

Once again, station opponents in Congress, led by Representatives Tim Roemer and Dick Zimmer, introduced a bill that would terminate the space station. However, both NASA and the White House were at pains to emphasise that back-up plans were in place if the Russians failed to deliver their promised hardware.

In particular, NASA announced its intention to buy, rather than lease, the Salyut FGB module from the Russian Khrunichev enterprise. A Lockheed spacecraft propulsion known as Bus 1, which was originally built for a classified US programme, could also be used as a temporary substitute if the FGB failed to materialise. This Interim Control Module would also be available as a stopgap if the Russian-financed Service Module, the station's third component, was destroyed during launch or failed to dock with the fledgling station. Meanwhile, a cosmonaut crew was to be put on standby in case they were needed to carry out a manual docking of the Service Module to the other compartments. In a worst case scenario, a US Propulsion Module was proposed as a larger, more capable back-up if the Russians proved unable to meet their commitments to deliver a Service Module.

These assurances were sufficient to persuade key politicians such as George Brown and James Sensenbrenner to proffer their support. The amendment to terminate the space station was doomed.

Dan Goldin praised the "courageous decision" of the House of Representatives to continue to build the station. "It was a vote for America and for the American people, and a vote for our future," he declared.

Although criticisms of the International Space Station did not cease – indeed, they continue to the present day – the programme was never under serious threat of cancellation again. With the design more or less fixed and the funding on a fairly stable footing, the way was clear for NASA and its partners to start construction of the hardware.

Indeed, on 14 October 1997, while the first modules were still being built, a sixteenth contributor to the ISS programme was brought on board. Under an agreement between Brazil and NASA, the South American giant agreed to supply payload and logistics hardware in return for research opportunities and crew time on the station. The Brazilian contribution would include an unpressurised cargo container, a cargo-handling interface assembly and "Express" pallets to support external experiments.

However, even the late newcomer almost succumbed to the same economic jinx that had plagued the other space station participants. Little more than one year after signing up to the programme, the Brazilian currency nose-dived in value, threatening the ability of Brazil's National Institute for Space Research (INPE) to cover the costs of its contributions. It was a clear sign that there was still a long, rocky road ahead for all concerned.

Chapter Five

Coming and Going

An orbital space station may be compared to a desert island, cut off from the rest of civilisation. With no water supply or source of food, and no means of escape or rescue, would-be Robinson Crusoes would soon succumb to hunger and thirst. Similarly, no humans can exist on a space station without a method of delivering victuals and other supplies.

However, here the analogy breaks down, since the space station in question is actually a manufactured island. It owes its very existence to a laborious, dangerous and expensive building programme in which the pieces that comprise the isolated habitation are gradually delivered and assembled.

This orbital delivery service comes in many forms. In the former Soviet Union, Europe and Japan, the expendable rocket has traditionally taken precedence. For example, more than 1,600 Soyuz rockets have been launched from Baikonur Cosmodrome during the past 30 years, many carrying Soyuz capsules containing two- or three-member crews or Progress ships loaded with supplies. Unless there is a failure of the automated rendezvous and docking systems, these missions are completed without human intervention.

The space station modules themselves, each weighing up to 20 tonnes on the ground, are sent aloft by heavy lift rockets such as the Proton. While the launcher plunges into the atmosphere on a fiery path to destruction after releasing its payload into low-Earth orbit, the new module continues on its way with the aid of automated systems, eventually linking up with the rest of the expanding structure.

In contrast, NASA has consistently relied upon human participation in the assembly and resupply of space stations. Although the gigantic Skylab was inserted into orbit by an expendable Saturn V rocket and the three Skylab crews were launched by its smaller Saturn IB cousins, the final journey to and from the sole US space station was completed under manual control by the astronauts themselves.

In the case of the International Space Station, all American modules, equipment and supplies have been brought by the crewed Space Shuttle, with each stage of the Shuttle's docking, redocking and assembly activities controlled by astronauts.

So which method of operation is superior? In terms of reliability, the astronaut-based technique probably just has the edge, although failures during automated dockings by large Russian modules or Soyuz and Progress craft have always been overcome by modifying the software or using manual back-up systems.

On the other hand, the use of robotic systems avoids the huge expense of using a human-rated space vehicle – currently about $400 million per Shuttle launch. The need for human life support and almost total redundancy on spacecraft systems inevitably escalates the costs of each ISS assembly or supply mission.

Expendable rockets may be speedily processed at the launch pad and, even if the launch is a failure, there is generally no loss of human life. Mass production leads to economies of scale, and in the relatively low-wage Russian economy, a typical ISS Proton launch may cost little more than one-tenth that of a Shuttle.

Furthermore, if one of the three Shuttles capable of servicing the ISS was lost, the grounding of the fleet would lead to a major hiatus lasting several years that would impact the entire Space Station programme. The cost to the American taxpayer of replacing the vehicle, even if the White House and Congress granted the funds, would probably exceed $2 billion. If an unmanned Russian rocket explodes, an inquiry is held and the vehicle returns to duty in a matter of weeks. Even the loss of a new module or supply ship, while damaging to the assembly schedule or morale of the station crew, would soon be overcome. Only the destruction of a manned Soyuz rocket – something that has not happened since 1975 – would shut down the Russian programme for some considerable time.

The main characteristics of the various launch systems and the manned or unmanned spacecraft they deliver to the ISS are described below. Some have been in operation for decades, others are still in the development stage. However, until now, only two sites, one in the former Soviet Union and one in Florida, have witnessed the excitement and drama associated with launching humans into outer space.

Although the monopoly of manned spacecraft launches by Baikonur and the Kennedy Space Center will continue for the foreseeable future, ISS-related launch activity will soon spread to South America and Japan as the other international partners make their contributions to ISS assembly and supply.

Aerial View of Kennedy Space Center, Cape Canaveral with the Vehicle Assembly Building in the foreground and two Shuttle launch pads in the distance. (Nasa photo KSC-388C)

LAUNCH CENTRES

Cape Canaveral

Cape Canaveral is the most famous launch centres in the world. As its name suggests, it is located on a bend in Florida's east coast, a low-lying area of ridges and islands made from sand and shingle.

The site was selected as a missile range soon after World War II. This location enabled rockets to be launched over the Atlantic Ocean towards the east and northeast. Its southerly position was particularly favoured for eastward launches, since rockets and their payloads receive an extra boost from the Earth's west–east rotation. Other advantages were the frost-free weather and the availability of the Bahamas and other islands for tracking stations.

The first launch, a V2 rocket with a WAC Corporal second stage, took place on 24 July 1950. Almost eight years later, the Cape was the location for the launch of America's first satellite, Explorer 1. Since then, it has been the centre for all human spaceflight activity as well as launches by thousands of unmanned, expendable rockets.

Most of the site, known as the Cape Canaveral Air Force Station, is under military control, but the northern part of the complex on Merrit Island, which is named after President John F. Kennedy, is operated by NASA and used for Shuttle launches. Launch complex 39 was originally built for the Saturn V Moon rocket, but the two pads (39A and 39B) were modified after 1975 and are now used for Shuttle missions.

The Space Shuttle is attached to its booster rockets and external fuel tank in the 52-storey Vehicle Assembly Building, and then moved to the pad some 5 km (3 miles) away on top of a huge crawler transporter. Other specialised buildings are used to check out the Shuttle after it returns from orbit and to prepare the payloads that will be carried in its cargo bay.

When weather conditions permit, the Shuttle lands on the 4.5-km (2.8-mile) long runway at Kennedy Space Center at the end of a mission. This reduces costs and turnaround time.

Kennedy Space Center is now one of the foremost tourist attractions in Florida. Apart from its interest for space enthusiasts, the lagoons and swamps are also a nature reserve for a number of protected animals, including alligators and snakes.

Space Shuttle Discovery arrives at Launch Pad 39B after a six-hour journey from the vehicle assembly building at Kennedy Space Center, Cape Canaveral. (NASA photo S-88-42100)

Baikonur

Baikonur is the largest and best-known Russian launch centre, even though it is now deep within the independent republic of Kazakhstan. It lies in a sparsely inhabited, semi-desert region to the east of the Aral Sea, where the seasonal temperatures range from −40 °C in the winter to +45 °C in the summer. Although close to the southern limits of the former Soviet Union, its location (45.6° N, 63.4° E) is far less favourable with regard to gaining extra energy from the Earth's west–east rotation than sites that are closer to the equator, such as Cape Canaveral or Kourou.

The huge cosmodrome, which covers an area of about 70 km by 90 km (44 by 56 miles), is now a Russian-controlled enclave within Kazakhstan. Russia pays an annual rent to the Kazakh government of 115 million roubles per annum for the use of the complex. However, much to the annoyance of the Russians, the launch schedule has been interrupted after launch failures on several occasions through the intervention of the Kazakh government.

The name Baikonur is misleading. In an attempt to confuse its Cold War enemies, the former Soviet Union named its ICBM launch complex after a small mining town some 320 km (200 miles) northeast of the actual location, close to the town and railway junction of Tyuratam. This deception was soon uncovered, but the name has stuck. Also a secret was the nearby town of Leninsk, some 40 km (25 miles) away, where the numerous soldiers, engineering, design and support staff lived.

Construction of this huge centre began in 1955 and the first launch facilities were commissioned in 1957 for the R-7 launcher, the rocket that was used to launch Sputnik 1. Since Yuri Gagarin's 1961 flight in Vostok 1, all Soviet and Russian human-related missions have been launched from Baikonur.

Today, the cosmodrome has a runway and is also served by a dense network of railway lines and crumbling roads. Apart from the nearby town, the surrounding area is almost uninhabited. Two separate cultures exist side by side here as camel-owning nomads wander around beneath the rockets' flight paths.

There are currently 9 launch complexes with 14 rocket launch pads at Baikonur. Two launch complexes, each with two launch pads, are available for the heavy-lift Proton rocket - one for international launches and one for Russian military missions. Launch complex 333 (also referred to as "point 23") was used for the launch of the space station's Zarya module.

Some distance away are the two pads used for manned and unmanned launches of Soyuz rockets. These are used to send manned Soyuz and unmanned Progress craft to the ISS.

The huge 20-tonne modules built for the ISS have to be transported across country by rail to the cosmodrome, whereas the rocket stages are usually flown

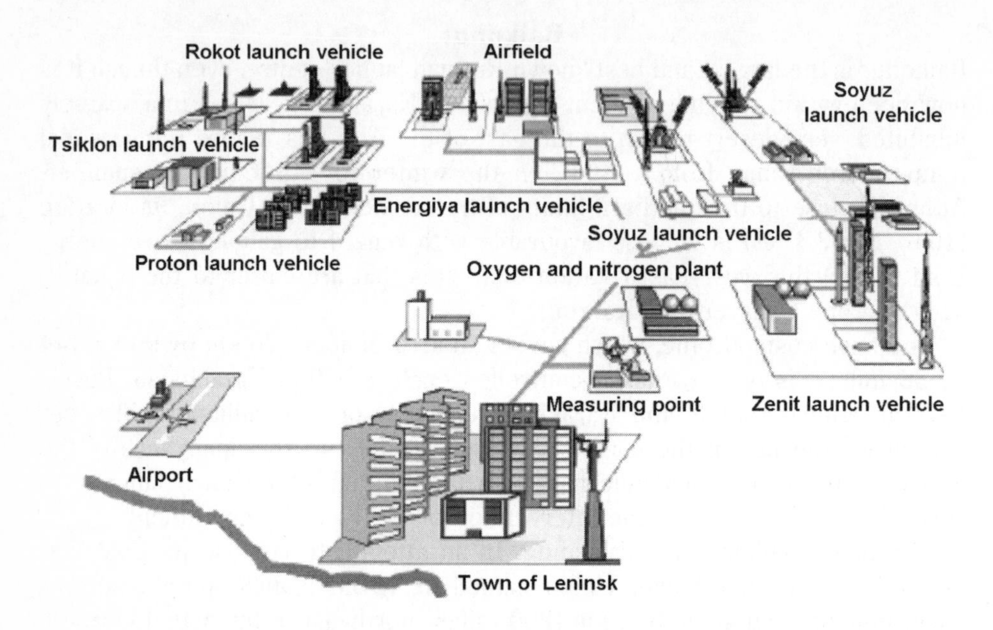

The location of the main launch facilities at Baikonur. (RSA diagram)

to Kazakhstan. Both the Soyuz and Proton rockets and their payloads are assembled and transported to the launch pads by rail car in a horizontal position, then raised upright prior to lift-off.

Final operations for the Soyuz vehicle begin six hours before launch. Filling with propellant starts three hours later. The cosmonauts enter the Soyuz spacecraft 2½ hours before the scheduled lift-off time. Access to the launch pad is much less restricted than in Florida, and the launch site is not evacuated until T − 15 minutes. Four pivoting trusses suspend the booster over the flame duct. When the engines ignite and the rocket begins to rise from the pad, the trusses are swung open by counterweights.

Although new hotels have been built nearby and some launch facilities have been upgraded – largely through foreign investment – the population of Leninsk (now known as Baikonur Town) has declined over the past decade from a peak of 120,000 to the current figure of 30,000. The complex has also been plagued in recent years by fires, theft and occasional riots by disaffected conscripts.

Until recently, Baikonur has been controlled by the Russian Military Space Forces, but the centre is now being handed over to the Russian Aerospace Agency. The site was off limits to Westerners for many years, and even during the Apollo–Soyuz programme, US astronauts were flown in at night, allowed to visit only the launch pad, then flown out again after dark.

Kourou

Although it is known as "Europe's spaceport", Kourou is actually located in French Guiana, South America. The French government decided to establish a rocket range there in April 1964 after detailed studies of 14 potential sites world wide.

One of Kourou's most important advantages is its near-equatorial location (latitude: 5.20° N), which means that rockets launched from there can gain maximum advantage from the Earth's west–east rotation. The coastal site also allows rockets to be launched towards the north or the east with relative safety for local inhabitants. Furthermore, the site is sparsely populated and free of hurricanes and earthquakes.

Today, the space centre is operated by the French Space Agency, CNES, on behalf of the French government and the European Space Agency. However, the commercial operation of the launch pads is the responsibility of a private company, Arianespace.

In the 1960s and 1970s Kourou was devoted to sounding rockets and small French or European launch vehicles, but since 24 December 1979, the centre has been devoted solely to the European Space Agency's series of Ariane rockets. Today, some 1,500 people are employed at the site, of whom 60% live in French Guiana.

As the highly reliable and successful Ariane 4 vehicle is phased out, the heavy-lift Ariane 5 is taking over as the primary launcher for commercial payloads, particularly communications satellites. A new launch complex, covering 21 km^2 (8 square miles), has been built to cater for this huge rocket. It includes a solid propellant plant, integration buildings for the solid fuel boosters and launcher, a final assembly building and launch pad 3 (known as ELA-3).

This is the site from which Europe's Automated Transfer Vehicle (ATV) will be launched to the ISS. A huge new payload preparation complex for the ATV, co-funded by Arianespace and ESA, was completed in 2001.

Tanegashima

Tanegashima Space Centre is Japan's main launch site. The launch facilities are located near the coast, in the southeastern corner of a long, narrow island known as Tanegashima. This means that launches towards the east or south take place over the ocean.

The site was selected to be as close to the equator as possible within the territory of Japan (latitude 31° N). It was also located where there would be minimal interference for the fairly small local population or for offshore fishermen.

Construction of the site began in 1966. It now consists of the Takesaki

Range for small launch vehicles; the Osaki Range for medium to large launch vehicles; the Masuda Tracking and Communication Station, situated 18 km (11 miles) to the north of Osaki Range; Uchugaoka Radar Station, situated 6 km (4 miles) to the west of the range; and the Kadokura Optical Tracking Station, 13 km (8 miles) to the southwest of the range. It is also equipped with other facilities, including those for ground firing tests of liquid engines and solid motors.

The H-IIA launch vehicle, which will be used to launch the H-II Transfer Vehicle to the space station, is launched from newly built or modernised facilities on the Osaki Range. These include a new vehicle assembly building, a mobile launch platform and a new blockhouse. With these upgrades, two H-IIA rockets can be processed and prepared for launch simultaneously.

Tanegashima's high-tech facilities, superimposed on a landscape of white, sandy beaches, subtropical climate and tropical plants such as hibiscus, banyan trees and cycad vegetation, make the island a popular tourist attraction.

LAUNCH VEHICLES

Space Shuttle

The Space Shuttle (for details, see Tables 5.1 and 5.2) has been America's sole means of sending astronauts into space since its debut on 12 April 1981. Five vehicles capable of flying in space have been built. One, Challenger, was destroyed in a catastrophic launch failure in January 1986 which resulted in the deaths of all seven crew members and the grounding of the Shuttle fleet for more than two years.

At present (2002) there are four operational Shuttles, three of which are equipped for ISS operations – Atlantis, Endeavour and Discovery. Columbia,

Table 5.1 Shuttle launch profile

Time (min: s)	Event	Altitude
T – 00:03	Main engine ignition	–
T – 00:00	Main engines at 90% thrust	–
T + 00:03	SRB ignition. Lift-off	–
T + 00:07	Begin pitch over	166 m (545 ft)
T + 01:02	Maximum dynamic pressure (Max Q)	13.4 km (8.3 miles)
T + 02:03	SRB separation	47.3 km (29.4 miles)
T + 08:24	Main engine cut-off (MECO)	117.5 km (73 miles)
T + 08:44	External tank separation	118.3 km (73.5 miles)

Table 5.2 Space Shuttle vital statistics

Length	
System	55.8 m (184 ft)
Orbiter	37 m (122 ft)
External tank	46.7 m (154 ft)
Solid rocket boosters	45.2 m (149 ft)
Orbiter wingspan	23.6 m (78 ft)
Payload bay dimensions	18.2 m (60 ft) long by 4.5 m (15 ft) wide
Lift-off weight	Approx. 2,040 tonnes (4.5 million lb)
Orbiter dry weight	Approx. 79.8 tonnes (176,000 lb)
Thrust	
SRBs (2)	1,497 tonnes (3.3 million lb) each in a vacuum
Shuttle main engines (3)	179 tonnes (393,800 lb) each at sea level at 104%
ISS maximum payload	16.4 tonnes (36,200 lb)

the oldest member of the fleet, is considered to be too heavy to reach the 51.6-degree inclination orbit occupied by the space station and make a useful contribution to assembly and supply missions.

The Space Transportation System (STS) – the formal name for the Shuttle – comprises three main components: a pair of solid fuel rocket boosters (SRBs), a huge fuel tank and the orbiter itself. Of these, only the external tank is not reusable.

The dart-shaped SRBs are like giant firecrackers packed with 450 tonnes of solid fuel – mainly ammonium perchlorate, mixed with aluminium, iron oxide and a binding agent. Once they are ignited, they cannot be shut down until their fuel is exhausted. They provide 80% of the enormous thrust that is needed to lift the 2,000-tonne Shuttle system off the ground. Once the SRBs are jettisoned, they parachute into the Atlantic Ocean, approximately 200 km (125 miles) from the launch site, where they are picked up and towed back to the mainland. They are then transported by rail to Thiokol in Utah for refurbishment.

The external tank, which is taller than a 15-storey building and contains over 2 million litres (530,000 gallons) of liquid oxygen and hydrogen in two separate sections, feeds the Shuttle's main engines during launch. Once the propellant has been consumed, the orange cylinder is jettisoned and burns up during its descent through the atmosphere.

Since 2 June 1998, the STS has used a super lightweight tank that is 3,400 kg (7,500 lb) lighter than the standard fuel tank. It is made of aluminium lithium, a lighter, stronger material than the metal alloy used in the previous version and also has an improved structural design. This modification enables the Shuttle to deliver heavier payloads to the ISS than would otherwise be possible.

Space Shuttle Discovery lifts off from Pad 39B at Cape Canaveral. (NASA photo STS095-(S)009)

After a burn of the orbital manoeuvring system (OMS) engines, the orbiter sets off in pursuit of the ISS. Further OMS and thruster burns over the next 40 hours raise and adjust the orbiter's path. Approaching from a lower, faster orbit, the Shuttle gradually catches up, closing the gap by 370 km (230 miles) during each 90-minute circuit of the Earth. A final intercept burn takes place about 2 hours 20 minutes before docking. The entire launch, rendezvous and docking procedure usually takes 42–43 hours.

During joint operations, the Shuttle is used to adjust the station's attitude and raise its orbit. Until the ISS received its own airlocks, all spacewalks associated with assembly and maintenance were carried out from the Shuttle. In addition, the orbiter's robotic arm has played a key role in the installation of the Destiny module and the Canadarm 2.

On most Shuttle missions, the orbiter remains attached to the space station for six to eight days (the longest that a Shuttle can remain in orbit is about 16 days). The ISS relies heavily on the Shuttle for delivery of new US hardware and replacement crews. Its capacious payload bay may house one or two Spacehab modules or a European-built Multi-Purpose Logistics Module (MPLM), as well as other units such as a Spacelab Logistics Pallet.

Three MPLMs have been built in Italy to carry several tonnes of cargo and experiments to the station. These high-tech, pressurised "moving vans" are lifted from the payload bay and docked to Unity for unloading by the crew. Towards the end of the Shuttle's visit, the module is returned to its Shuttle berth for the return to Earth.

Once the Shuttle has undocked from its port on the station, it is pushed to a safe distance by springs in the docking module before the thrusters can be fired. At a distance of 136 m (450 ft) from the station, the pilot manoeuvres the orbiter through a three-quarter circle for a final fly-around and photographic opportunity. A separation burn then moves the orbiter below and ahead of the station. Almost two days later, a deorbit burn by the OMS engines slows the Shuttle, causing it to re-enter on a gradual, circular descent path that brings it over the Pacific and the west coast of America.

Despite temperatures during re-entry that reach 1,500 °C (2,750 °F), the vehicle's thermal protection system dissipates the heat very rapidly. The orbiter returns to Earth as a glider, usually touching down on a runway at Kennedy Space Center in Florida or Edwards Air Force Base in California. However, the threat of high winds or storms at these sites may force the crew to stay aloft for an extra day or two. The landing typically takes place at a speed of about 345 km/h (215 miles/h).

With no replacement on the horizon, the Shuttle fleet is continually being upgraded to ensure that it will be able to operate well into the second decade of the 21st century. Apart from the super lightweight fuel tank, most of the orbiters have recently been fitted with an improved main engine, modified airlocks and docking ports located in the payload bay, a new "glass cockpit" with flat computer display screens, and a satellite navigation system.

Proton rocket

The three-stage Proton rocket (Table 5.3) played a key role in the early stages of the International Space Station's development. Not only was it used to place the first ISS module, Zarya, in orbit on 20 November 1998, but it subsequently delivered the first fully Russian contribution – the Zvezda service module.

The Proton has successfully flown more than 200 times since its introduction in 1965 as a launcher for Soviet heavy military payloads, lunar and planetary craft and space stations. It was designed by the Salyut Design

Table 5.3 Proton launch profile for Zvezda module

Time (min: s)	Event	Altitude	Speed
T − 0	Lift-off	–	–
T + 2:06	First-stage jettison	43 km (27 miles)	5,920 km/h (3,700 miles/h)
T + 3:03	Zvezda fairing jettison	77 km (48 miles)	7,520 km/h (4,700 miles/h)
T + 5:30	Second-stage jettison	138 km (86 miles)	15,840 km/h (9,900 miles/h)
T + 9:47	Third-stage jettison	184 km (115 miles)	27,040 km/h (16,900 miles/h)

Orbit at third-stage jettison: approximately 184 by 352 km (115 by 220 miles)
Orbit at ISS rendezvous: 384 km (240 miles)

Bureau and is currently manufactured by the Khrunichev State Research and Production Space Centre in Moscow.

The Proton can lift more than 20 tonnes into low-Earth orbit and is among the most reliable heavy-lift launch vehicles in operation, with a reliability rating of about 98%. With the Zvezda module, launch fairing and adapter in place on top of the booster, the Proton measures about 55 m (180 ft) tall, 7.3 m (24 ft) in diameter at its widest point and weighs about 700 tonnes (1,540,000 lb) when fully fuelled for launch. The engines use nitrogen tetroxide as an oxidiser, and unsymmetrical dimethyl hydrazine as the fuel.

The first stage includes six engines that are fed propellants from a single, central oxidiser tank and six outboard fuel tanks. At launch, the first-stage engines combine to provide about 860 tonnes (1.9 million lb) of thrust. The first stage, which is about 21 m (68 ft) long and 7.3 m (24 ft) across, eventually burns out and is jettisoned.

The Proton's second stage, which measures 17 m (56 ft) long and 4 m (13.5 ft) in diameter, then takes over. It was while the second stage was in operation that the fairing that protected the ISS modules during lift-off was jettisoned. This stage is powered by four engines that generate a total thrust of 215 tonnes (475,000 lb).

Zvezda was inserted into orbit by the Proton's third and final stage, which measured approximately 4 m in length and diameter (13.5 ft by 13 ft). This stage is powered by a single engine that creates 57 tonnes (125,000 lb) of thrust.

Once the third stage was jettisoned, almost 10 minutes after lift-off, Zvezda had reached an elliptical orbit with a high point (apogee) of 352 km (220 miles) and a low point (perigee) of 184 km (115 miles). Firings of Zvezda's engines during the following days manoeuvred it into an almost circular orbit at an altitude of about 384 km (240 miles), ready for the final rendezvous and

20 November 1998. Lift-off from Baikonur Cosmodrome of the Proton launch vehicle carrying the Zarya module, the first component of the International Space Station. (NASA photo S99-00883)

capture by the orbiting International Space Station, under the control of the Zarya module.

Soyuz rocket

The Soyuz launcher (Table 5.4) has evolved from the R-7, the famous rocket that was developed by Sergei Korolev in the 1950s. Originally designed as the Soviet Union's first intercontinental ballistic missile, the R-7 became the basis for the vehicles that launched the first Soviet satellites and cosmonauts into orbit.

Since its introduction in 1966, the Soyuz has become the most frequently used launch vehicle in the world. It is used to orbit many types of robotic satellites as well as all Soviet/Russian manned missions.

Table 5.4 Soyuz ascent timeline

Time (min: s)	Event	Altitude
T − 0	Lift-off	–
T + 1:58	First-stage separation	45 km (28 miles)
T + 2:40	Escape tower and launch fairing jettison	85 km (53 miles)
T + 4:58	Second-stage separation	170 km (105 miles)
T + 9:00	Third-stage cut-off	205 km (127 miles)

A Soyuz rocket lifts off from Baikonur on 31 October 2000, carrying the Expedition One crew to the International Space Station. (NASA photo ISS01-S-006)

The Soyuz rocket is described by the Russians as a three-stage vehicle. It is 49 m (162 ft) high and weighs almost 310 tonnes (680,000 lb) when carrying a crewed Soyuz spacecraft with an escape tower. Its payload lift capability to low-Earth orbit is almost 8 tonnes (18,000 lb).

The first stage comprises four strap-on boosters assembled laterally around the central core, which makes up the second stage. The four strap-ons are 19.9 m (66 ft) high and have a cylindrical–conic shape that tapers upwards. The upper, cone-shaped section houses the oxygen tank, while the lower, cylindrical part contains the kerosene fuel. Each of the liquid fuel motors generates 102 tonnes (225,000 lb) of thrust.

Ignition of the four lateral boosters and the core stage takes place at the same time. When the boosters run out of fuel and are jettisoned, the second stage continues to burn. This stage is 28 m (92 ft) high and its single engine generates 96 tonnes (212,000 lb) of thrust.

The third stage, which is 8.1 m (27 ft) high and 2.7 m (9 ft) in diameter, is used to insert the Soyuz spacecraft into orbit. Its single engine generates 30 tonnes of thrust. Once the required velocity is reached, 9 minutes into the flight, engine cut-off occurs and the crewed spacecraft separates.

A number of upgrades are in preparation. The first of these was introduced on 21 May 2001 when a modified Soyuz FG rocket, equipped with an improved fuel injection system, blasted off from Baikonur with Progress M1-6.

Ariane 5 rocket

The Ariane 5 (Table 5.5) is the largest and most powerful in the series of Ariane rockets that have been developed by Europe. After a launch failure on its maiden flight in 1996, the rocket was modified and began commercial operation two years later. The version to be used for ISS operations – in particular, the delivery of the European Automated Transfer Vehicle to low-Earth orbit – is the Ariane 5V (Versatile), which is still under development.

Table 5.5 Ariane 5V statistics

Height	55.4 m (182 ft)
Total launch mass	Approx. 760 tonnes (1,675,800 lb)
Propellant mass	Approx. 645 tonnes (1,422,000 lb)
Maximum diameter	5.45 m (18 ft)
Solid fuel boosters (2):	
Height	31.16 m (102 ft)
Diameter	3.05 m (10 ft)
Core stage thrust	117 tonnes (257,985 lb)
Solid fuel booster thrust	Max. 1,227 tonnes each (2,705,535 lb)

Launch of the third Ariane 5 from the launch site at Kourou in French Guiana, South America, 21 October 1998. (EADS photo)

Under current plans, the Ariane 5V will comprise two strap-on boosters, a cryogenic (liquid hydrogen and liquid oxygen) core stage, a reignitable upper stage, and a long fairing to protect the ATV during launch. At lift-off, 90% of the thrust is provided by the twin solid fuel boosters, with the single engine of the core stage providing the remainder (Table 5.6).

Once the strap-on boosters are jettisoned, the core stage continues to operate for another six and a half minutes. The second stage then takes over and carries the ATV to an elliptical, low-Earth orbit. After a long ballistic phase, the upper stage ignites once more to insert the ATV into a circular 300-km (188-mile) orbit. Its work done, the stage then brakes and re-enters for a splashdown in the Atlantic.

By using the upper stage in this way, the number of manoeuvres to be carried out by the spacecraft's own engines is reduced. This allows more time

Table 5.6 Ariane 5V/ATV launch timeline

Time (min: s)	Event
T – 0	Core stage engine ignition
T + 0:7	Solid booster ignition
T + 2:25	Solid booster separation
T + 9:00	Core stage shut-down and separation
T + 18:10	Second-stage shut-down #1
T + 64:45	Second-stage reignition
T + 65:25	Second-stage shut down #2
T + 69:25	ATV separation

for the ATV's systems to be checked out by the control centre in Toulouse, France, and to prepare for the rendezvous with the ISS.

H-IIA rocket

The most powerful rocket ever developed by the Japanese Space Agency (NASDA), the H-IIA (Table 5.7) is an improved version of the H-II heavy-lift vehicle that first flew in February 1994. Following two successive launch failures in 1998 and 1999, the H-II programme was cancelled and all efforts were transferred to development of the H-IIA.

After eight years and $1 billion of funding, the new version successfully completed its maiden flight on 29 August 2001. With launch prices slashed to less than half those of the H-II, Japanese officials are now hoping that the rocket will attract commercial customers and open the way for future access to the ISS.

The standard H-IIA, a two-stage vehicle fitted with a pair of strap-on solid

Table 5.7 H-IIA vital statistics (212/HTV version)

Overall length	53 m (174 ft)
Diameter of core stages	4 m (13 ft)
Solid fuel boosters (2):	
Height	15.2 m (50 ft)
Diameter	2.5 m (8 ft)
Lift-off weight	414 tonnes (912,700 lb)
Propellant weight	350 tonnes (772,000 lb)
First-stage and LRB engine thrust	112 tonnes (250,000 lb) each
Second-stage thrust	14 tonnes (30,900 lb)
SRB thrust	230 tonnes (515,000 lb) each
Payload to low-Earth orbit	Approx. 17 tonnes (37,000 lb)

rocket boosters, is able to lift a payload of 4 tonnes to geostationary transfer orbit. However, a more powerful "augmented" version will be required to lift the 15-tonne H-II Transfer Vehicle into low-Earth orbit.

In order to carry the Japanese cargo vessel, the H-IIA 212 vehicle will be equipped with two main engines, a side-mounted liquid rocket booster that incorporates two additional main engines, and an enlarged payload fairing. Like the engines of the two main stages, the 37-m (120-ft) long booster will use liquid hydrogen and liquid oxygen propellants. If all goes well, the enhanced version should be launched by 2003.

The H-IIA is launched eastwards across the Pacific Ocean from Tanegashima, one of the most southerly islands of Japan. During a typical launch sequence, the first-stage engines ignite before lift-off. Six seconds later, the SRBs ignite and the rocket leaves the pad. At 100 seconds into the flight, the SRBs burn out and are jettisoned. First-stage engine cut-off takes place 6½ minutes after lift-off. The first stage then separates, followed by second-stage ignition at 6 minutes 45 seconds. The burn time for the second stage, which has a restart capability, will depend on the type of payload and the orbit required.

SERVICE VEHICLES

Soyuz TM Spacecraft

The first manned Soyuz was launched on 23 April 1967 with cosmonaut Vladimir Komarov. Unfortunately, a planned docking with a second crewed Soyuz had to be abandoned when the spacecraft suffered a series of major technical malfunctions. Komarov died on the return to Earth when the spacecraft's parachute became tangled, and almost two years passed before a modified Soyuz was launched. This spacecraft was used for the first Soviet docking experiments, solo missions and the delivery of three-man crews to the Salyut 1 space station.

After a sudden decompression caused the deaths of the Salyut 1–Soyuz 11 crew in 1972, cosmonauts had to wear protective pressure suits, and the limited room in the Soyuz cabin meant that crew size was limited to two. Only with the introduction of the Soyuz T variant, first launched on 27 November 1980, could crews of three with pressure suits once again became possible. This version was used to carry cosmonauts to and from the Salyut 3–7 space stations.

The current version, known as Soyuz TM (Table 5.8), first flew on 21 May 1986. It was used by Mir space station crews and is now used for Russian-led crewed missions to the International Space Station. At least one Soyuz is

Table 5.8 Soyuz TM vital statistics

Launch mass (without shroud and launch escape system)	7.1 tonnes (15,655 lb)
Descent module	2.9 tonnes (6,395 lb)
Orbital module	1.3 tonnes (2,867 lb)
Instrumentation/propulsion module	2.6 tonnes (5,733 lb)
Delivered payload (with three crew)	30 kg (66 lb)
Returned payload	50 kg (110 lb)
Length	7 m (23 ft)
Maximum diameter	2.72 m (9 ft)
Solar array span	10.7 m (35 ft)
Volume of orbital module	6.5 m^3 (230 ft^3)
Volume of descent module	4 m^3 (141 ft^3)
Descent *g*-loads (1*g* = normal Earth gravity)	3–4*g*
Final landing speed	2 m/s (7 ft/s)
Landing accuracy	30 km (18.75 miles)

always docked with the ISS for use as an emergency crew return vehicle. Since its nominal orbital lifetime is 200 days, the spacecraft has to be replaced at six-monthly intervals. This takes place during 10-day taxi missions, which are available on a commercial basis for non-Russian passengers. Among those participating in such taxi missions have been the first space tourist, American businessman Dennis Tito, and French spationaute Claudie Haigneré.

The Soyuz TM weighs 7.7 tonnes (16,000 lb) and is designed to carry three crew members with a payload of 30–50 kg (66–110 lb). It comprises three modules. The 2.5-m (8-ft) long instrument module at the rear has an unpressurised section that incorporates the engines, manoeuvring thrusters and fuel tanks. A pressurised section includes the temperature control system, batteries, telemetry equipment, attitude control instruments and computer. Two solar 'wings', each more than 3 m (10 ft) long, provide power to the Soyuz for most of the flight. Batteries in the descent module are used for re-entry.

A 2.1-m (3½-ft) long descent module, in which the crew is seated during lift-off, orbital manoeuvres and re-entry, houses the main controls, radio and life support systems. An ablative heat shield at the base, parachutes and soft-landing engines are fitted for the return to Earth.

Above the descent module is an orbital module, which the cosmonauts use for eating, sleeping and recreation during quiet periods of the flight. It also houses the main atmospheric and purification system. The module is accessible via a narrow hatch, so it can be used as an airlock. On its nose are another hatch, opening inwards, and a docking probe.

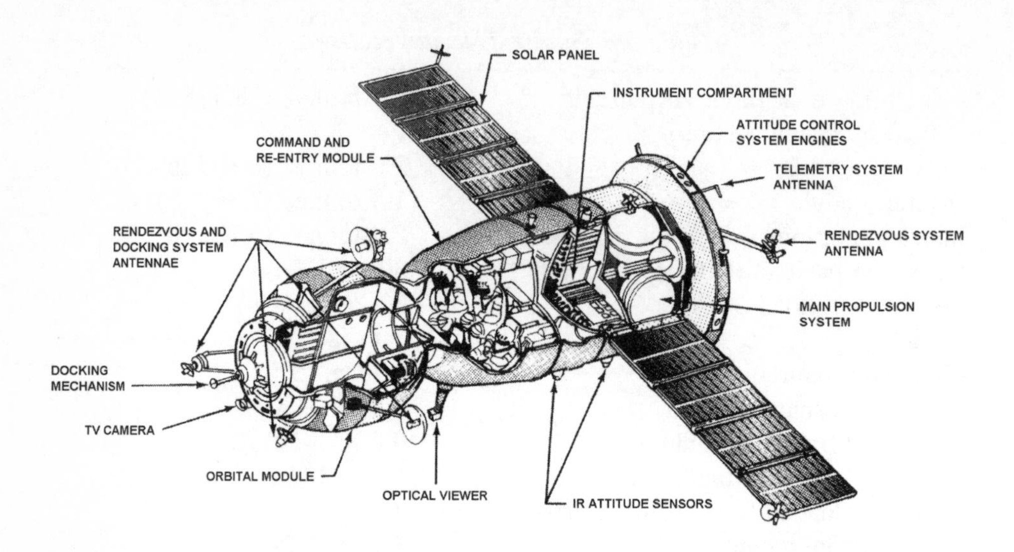

The Soyuz TM. (RKK Energia diagram)

Rendezvous with the ISS takes just over two days or 34 orbits. The Soyuz TM can dock to any of the Russian segments of the ISS (Zarya, Zvezda or Pirs). Docking is normally automatic, using the Kurs system, but the crew can use manual control if required.

After Soyuz separation from the space station, pyrotechnic charges split the spacecraft into its three sections. The crew returns to the flat steppes of Kazakhstan in the descent module, while the other modules burn up in the atmosphere.

If the number of residents on board the ISS eventually increases to six, as was originally planned, then two Soyuz vehicles will have to be attached to the station at any one time in case of emergencies. Since the current spacecraft are rather cramped, the Russians have redesigned the interior by raising the instrument panel to offer more leg room for foreign astronauts. However, if the reusable US Crew Return Vehicle is eventually built, the need for Soyuz rescue vehicles will diminish.

Progress M

Progress M is an unmanned cargo craft provided by Russia. It is based on the highly successful Progress craft that was developed by NPO Energia in the 1970s to resupply the Salyut 6 and Salyut 7 space stations. A modernised version (Progress M), with improved flight control systems, was introduced in 1989 for use with the Mir space station and has also been used to bring supplies, air and fuel to the ISS.

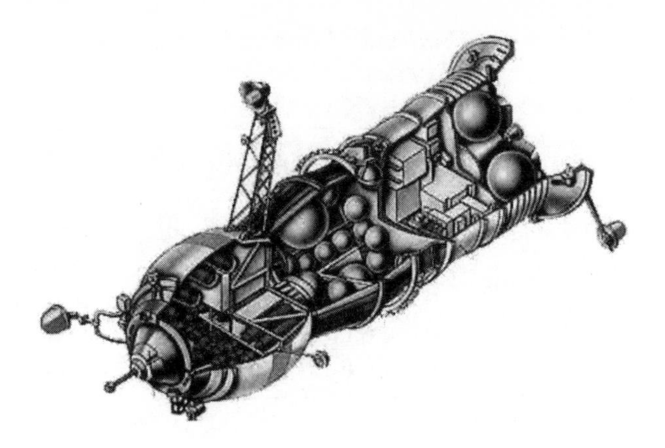

Cutaway of the Progress M spacecraft. (RKK Energia diagram)

The Progress M spacecraft closely resembles the manned Soyuz externally, but the pressurised orbital module is used for equipment and supplies, while the descent module has become an unpressurised section for fuel. A typical Progress M ferry will deliver approximately 2 tonnes (4,400 lb) of food, clothing, hardware (including spare parts) personal mail and gifts to the ISS. Once it has been emptied, the cargo section is filled with up to 2,000 kg (4,400 lb) of rubbish and jettisoned to burn up in Earth's atmosphere.

Loading and unloading by the ISS crew is carried out through the forward hatch on the pressurised cargo module. The propulsion and service equipment is at the rear of the spacecraft, along with two solar panels for power supply.

The Progress M has four propellant tanks (two each for fuel and oxidiser). Propellant is transferred to the ISS through connectors in the docking ring, but is also available to the thrusters on the Progress for attitude control and reboost of the station.

A larger Progress M1 craft (Table 5.9) has recently been developed by RKK Energia to allow the delivery of more fuel to the ISS. By rearranging the middle section of the spacecraft, extra propellant tanks were squeezed in, while the water tanks were moved to the front "cargo" section of the craft.

Twelve tanks with nitrogen and oxygen mix for the station atmosphere were placed on the exterior of the craft around the "neck" between the cargo and propellant modules. It is also equipped with a new digital flight control system and an upgraded rendezvous and docking system known as Kurs-MM.

The first Progress M1 spacecraft was launched to the Mir space station on 1 February 2000. The third in the series, Progress M1-3, was the first to dock with the ISS. It was launched on 6 August 2000. Others have followed every three to four months. If the automatic system fails, the Progress may be manually docked, using the TORU system in Zvezda.

Table 5.9 Progress M and M1 comparison

	Progress M	Progress M1
Total payload limit	2,350 kg (5,180 lb)	2,230–3,200 kg (4,917–7,056 lb)
Maximum pressurised (dry) cargo	1,800 kg (3,969 lb)	1,800 kg (3,969 lb)
Maximum water	420 kg (926 lb)	Up to 300 kg (662 lb)
Maximum air or oxygen	50 kg (110 lb)	40 kg (88 lb)
Maximum propellant for refuelling	850 kg (1,874 lb)	1,950 kg (4,300 lb)
Cargo volume	6.6 m^3 (233 ft^3)	6.6 m^3

A new, heavier version with an elongated cargo module has been proposed by RKK Energia for ISS resupply. Known as Progress M2, it was originally envisaged in the 1980s for use with the Mir 2 station. At the end of the 1990s, as the relationship between Russia and Ukraine improved, RKK Energia tried to revive Progress M2-based modules. While the enlarged Progress is unlikely to materialise in the current funding climate, the commercial Enterprise module and possible future Russian–Ukrainian modules for the ISS might use much of the hardware developed for this project.

Automated Transfer Vehicle

In October 1995, the European Space Agency's Ministerial Council gave the go-ahead for the development of the Automated Transfer Vehicle (ATV) as part of ESA's contribution to the ISS programme. The ATV (Table 5.10) is an unmanned supply vehicle that will be launched by an Ariane 5 from Kourou in South America. After insertion into an elliptical low-Earth orbit, the spacecraft's propulsion system will raise and circularise the orbit, prior to rendezvous and docking with the Russian segment of the ISS. Nine ATVs will be built to cover European ISS operations for 10 years (2004–2013) with a launch approximately every 12 months.

The cylindrical ATV is 9.8 m (32 ft) long and has a maximum diameter of 4.5 m (15 ft). It comprises two sections: the spacecraft itself and an integrated cargo carrier. Launch weight, fully loaded, will be about 21 tonnes (46,000 lb), of which approximately one-third will be cargo. This may consist of dry cargo, which is carried in a pressurised environment, water, air, oxygen or nitrogen, and up to 860 kg (1,892 lb) of propellant (fuel and oxidiser). There will also be up to 4 tonnes (8,800 lb) of fuel that can be expended on space station reboost and attitude control. Power during flight is provided by four solar arrays.

The ATV approaches the ISS. (ESA artist's impression)

Table 5.10 ATV vital statistics

Length	9.8 m (32 ft)
Maximum diameter	4.5 m (15 ft)
Solar arrays span	22.3 m (73 ft)
Mass at launch	20.75 tonnes (45,753 lb)
Spacecraft dry mass	5.3 tonnes (11,687 lb)
Cargo carrier dry mass	5.15 tonnes (11,355 lb)
Cargo upload capacity	7.7 tonnes (16,979 lb)
Dry cargo	1.5–5.5 tonnes (3,308–12,128 lb)
Water	0.8 tonnes max. (1,764 lb)
Gas (N, O, air)	0.1 tonne max. (221 lb)
ISS reboost and attitude control propellant	4.5 tonnes max. (9,923 lb)
Waste download capacity	6.5 tonnes (14,333 lb)

The ATV will be used to deliver dry and liquid cargoes such as experiments, food, compressed air and water; refuel the station by automatic transfer to the Zarya module; raise the station's orbit and provide attitude control during such manoeuvres; and remove waste for incineration during re-entry.

The vehicle carries a Russian-designed docking probe for linking up with the rear port on Zvezda. It is expected to remain attached to the ISS for periods of up to six months. During this phase, its hatch remains open to ease access for the crew. After departure, loaded with a maximum of 6.5 tonnes (14,333 lb) of waste, the vehicle will burn up in the atmosphere on re-entry.

H-II Transfer Vehicle

Like Europe's ATV, this Japanese-built spacecraft is designed to transport various cargoes to the ISS and for waste disposal through incineration during re-entry (Table 5.11). It will be launched into an elliptical orbit by a modified version of the Japanese H-IIA launch vehicle, then use its own propulsion and navigation systems to approach and hover beside the space station. Over a period of three days, the H-II Transfer Vehicle (HTV) will close to within 500 m (1,650 ft) of the ISS with the aid of data from satellites of the Global Positioning System. A laser-ranging system will then bring it to a distance of about 10 m (33 ft), where it will be captured and docked to the ISS by an astronaut-operated manipulator on the Japanese Experiment Module.

The HTV can carry 6–7 tonnes of cargo and equipment to the ISS with the aid of pressurised or non-pressurised logistics carriers. After the HTV's arrival at the ISS, an exposed pallet is taken out and cargoes are replaced by robotic arm. This exposed pallet has rollers on its sides to minimise friction and keep its position accurately when it is brought back to the HTV. The maiden launch of the HTV is currently scheduled for 2004.

HTV docking with ISS. (NASDA digital photo)

Table 5.11 HTV vital statistics

Length:	
Mixed Logistics Carrier Type	9.2 m (30.4 ft)
Pressurised Logistics Carrier Type	7.4 m (24.4 ft)
Diameter	4.4 m (14.5 ft)
Launch weight	About 15 tonnes (3,308 lb)
Payloads:	
Pressurised only)	About 7 tonnes (15,435 lb)
Pressurised/unpressurised mix	About 6 tonnes (13,230 lb)

H-II Orbiting Plane – Experimental

Japanese interest in shuttle-type space vehicles dates back to the late 1980s. As in the United States, it was argued that a reusable shuttle would dramatically reduce the cost of access to low-Earth orbit. After initial studies of manned spacecraft, the country's ambitions were scaled down to an unmanned vehicle that would be launched on top of the new H-II launcher - hence the name HOPE (H-II Orbiting Plane - Experimental).

Since 1992, Japan has followed a development programme involving small demonstrators that have tested various aspects of spaceplane technology. These included:

- Highly Manoeuvrable Experimental Space vehicle (HIMES), a 500-kg suborbital model that was launched from a high-latitude balloon.
- Orbital Re-entry Experiment (OREX), first flown on an H-II rocket in 1994 to test heat-resistant materials.
- Automatic Landing Flight Experiment (ALFLEX), a one-third size "lifting body" vehicle that carried out glide tests over the Australian desert in July–August 1996.
- Hypersonic Flight Experiment (HYFLEX), a small spaceplane that was launched from a J-1 rocket for supersonic flight testing.

Despite this modest progress, the programme's costs overran and the entire HOPE project was threatened with cancellation. In the end, a compromise solution was reached, with the inception of an interim programme known as HOPE-X (Table 5.12).

This small, unmanned experimental vehicle, was intended to demonstrate the technologies required for operational missions such as supply/recovery of cargo and equipment for the International Space Station. A High Speed Flight Demonstrator experiment was to be conducted in 2004, but, aside from a series of subscale-vehicle drop tests, Japan has put its HOPE-X development plans on hold.

If it is finally built, the spaceplane will be launched in a vertical position on

Artist's impression of HOPE-X spaceplane. (NASDA)

Table 5.12 HOPE-X vital statistics

Length	16 m (53 ft)
Diameter	10 m (33 ft)
Height	5 m (16.5 ft)
Launch weight	14.5 tonnes (31,973 lb)
Re-entry weight	10.5 tonnes (23,153 lb)
Launcher	Single-stage H-IIA rocket

the H-IIA rocket. After separation, the orbital manoeuvring system engines will inject the spaceplane into a 120-km (75-mile) by 200-km (125-mile) orbit. After one orbit of the Earth, the vehicle will brake and re-enter the atmosphere, protected by carbon and ceramic tiles. The mission will end with a landing on a 1,800-m (5,940-ft) long runway, possibly on Christmas Island. If successful, this would make Japan only the second country (after Russia) to fly an unmanned spaceplane.

Crew Return Vehicle

This winged re-entry vehicle is being designed by NASA and the European Space Agency to replace the Soyuz as the ISS "lifeboat".

One of the major limitations on expanding the number of ISS occupants from three to six or seven is the lack of a suitable rescue craft for a rapid return to Earth in an emergency. At present, this capability is provided by the Russian Soyuz TM vehicle, but at least two of these three-person craft would have to be docked to the station at all times for the use of an expanded crew.

In order to provide a cheaper, more flexible alternative, NASA and ESA initiated the development of a Crew Return Vehicle (CRV) that would be docked to the station and available to return a crew of seven to Earth at short notice.

In order to demonstrate the technology and reduce risks, the United States, Germany, Belgium, Italy, the Netherlands, France, Spain, Sweden and Switzerland have been developing a smaller prototype vehicle known as the X-38. This project combines proven technologies - a shape borrowed from the 1970s Air Force X-24 project - with modern aerospace technology.

A number of test vehicles have been built and flown from NASA's Dryden Flight Research Center in California since 1999. In each case, the small spaceplanes have been released from a B-52 aircraft and allowed to descend through the atmosphere before a drogue parachute is released, followed by the deployment of the world's largest parafoil. (The surface area of the parafoil is more than one and a half times that of the wings of a 747 jumbo jet.) Software developed by the European Space Agency then guides the uncrewed X-38 to a safe landing on the clay surface of Rogers Dry Lake at Edwards Air Force Base. The X-38 touches down at a speed of less than 64 km/h (40 miles/h).

An X-38 space test vehicle is currently under construction at the Johnson Space Center and an orbital test from a Shuttle is still pencilled in for 2005. In its latest variant, the X-38 has a base with a semicircular cross-section that could allow it to be compatible with launch on a European Ariane 5 rocket as well as on board the Space Shuttle.

If built, the CRV will be delivered to the ISS by the Space Shuttle and remain attached for up to three years. It will accommodate up to seven crew in a shirt-sleeve environment, and be capable of autonomous operation. The maximum solo mission duration after undocking is about nine hours.

The CRV must be able to separate from the station when it is at any attitude and tumbling at up to 2 degrees per second. During a typical CRV re-entry, the braking thrusters would be fired within three hours of departure from the station, then the deorbit propulsion stage would be jettisoned prior to re-entry. The vehicle enters the atmosphere at an altitude of about 120 km (75 miles), travelling at 27,000 km/h (16,875 miles/h). At first its attitude is controlled by thrusters, but rudders and flaps take over as air pressure increases. A drogue parachute is released at an altitude of 8 km (5 miles), followed by deployment of the parafoil. Automatic guidance navigation and control software takes over for the final descent and landing at a forward speed of just 10 m/s (33 ft/s).

The X-38 lands on Rogers Dry Lake in California during a test flight in November 2000. (NASA photo JSC2000-02453)

The first CRV was scheduled to fly in 2007. However, the entire programme is in doubt after major financial shortfalls announced by NASA in February 2001. Before the programme was shelved, it was expected to build four flight units and one spare vehicle.

ATTITUDE CONTROL

Apart from the Space Shuttle and Soyuz TM, both of which carry crews who can manoeuvre them to their particular docking ports on the ISS, all of the vehicles that will visit the station in the near future will rendezvous and dock under automatic control.

A stable attitude or orientation in space for the station is vital during all of these operations, but particularly in the case of automated manoeuvres. The attitude of the ISS is controlled in several ways.

The space station has small thrusters that shoot propellant into space to slowly rotate the craft for fine attitude adjustments. However, using this system alone would cost millions of dollars, so giant flywheels known as gyroscopes are also available to stabilise the station's attitude. The gyroscopes spin like heavy toy tops to maintain the station's proper tilt relative to the Earth without use of expensive, expendable fuel.

The first four American gyroscopes were delivered to the ISS, installed on the Z1 truss and tested by the STS-92 Shuttle mission. The status of the gyros can be monitored on the ground and alterations made by uploading commands or via modified laptop computers in the Destiny module.

During the early phase of construction, first Zarya and then Zvezda were responsible for ISS attitude control and reboost. Since then, spacecraft have always been docked to the station, and these are used for attitude control and orbit-raising manoeuvres. Two or three reboost operations will take place during a typical week-long Shuttle visit, each involving several hundred firings of the orbiter's reaction control system jets in order to raise the orbit by 3–8 km (2–5 miles).

DOCKING SYSTEMS

Two types of docking system are provided on the Russian segment. The traditional probe and cone mechanism, which has been operational since the early 1970s, receives the Russian manned Soyuz TM spacecraft and the unmanned Progress ferry, and will also be used by the European Space Agency's Automated Transfer Vehicle (ATV).

A similar system was used for the delivery of the Pirs docking module on 16 September 2001. The propulsion module and its payload approached the station from below and behind, then the station's thrusters moved it to the proper orientation for docking. The long probe, which protrudes from the cone-shaped "star" of the docking assembly, was driven into the drogue housing, and the two spacecraft were pulled together. The module was locked securely in place when 12 active latching hooks were driven to their closed position.

The arrival of a third docking port on Pirs (Docking Compartment 1) simplified considerably the Russian traffic congestion at the ISS. Without Pirs, Russian manned cargo ships could only dock either to the aft port of the Zvezda service module, or to the Earth-facing (nadir) port of the Zarya control module.

With only two regular docking ports available, the Progress ship had to be undocked from the station every time a fresh Soyuz spacecraft was sent up to replace its older predecessor. After the Soyuz exchange was completed, the Progress cargo ship had to be either discarded or redocked to the station with

The Pirs docking probe being packaged for reuse by cosmonaut Tyurin. (NASA digital photo)

the use of the TORU remote-control system operated manually from inside the ISS.

Now, with the Pirs compartment attached to Zvezda, the incoming Soyuz spacecraft are able to dock to the compartment's free Earth-facing port, so the Progress can remain in place even when two Soyuz craft are berthed on the Russian segment.

Zvezda's forward, zenith and nadir docking ports carry a different, hybrid, docking mechanism that is not compatible with the standard hardware installed on the Soyuz spacecraft. This Russian-designed Androgynous Peripheral Attach System (APAS) is used to attach larger and heavier Russian elements, such as Zarya and the Science Power Platform. It is also fitted to the Space Shuttle so that it can link up with the Pressurised Mating Adapters on Unity or Destiny. The ring-shaped APAS mates with an exact copy on the other spacecraft, and each can act as the passive or active unit.

A description of the STS-96 docking with the space station shows how the system works. About three hours before the scheduled docking time, the Shuttle Discovery moved to a point about 14 km (9 miles) behind the station. Then came the so-called Terminal Phase Initiation burn of Discovery's thrusters to close the gap over the next orbit. As the orbiter closed on its quarry, its rendezvous radar system provided the crew with information on the

View of the ISS as seen through the crew optical alignment system. (NASA photo STS105-707-055)

range and closing rate. Within a few hundred metres, this information was supplemented by a laser-ranging device mounted in the Shuttle payload bay, with additional checks from a handheld laser aimed through the Shuttle windows.

As the Shuttle moved closer, ground controllers commanded the station to turn perpendicular to the Earth's surface, with the Unity module pointing towards space and the Zarya towards the Earth. The solar arrays were then "feathered" so that they faced edge on to the Shuttle. Damage from the Shuttle's exhaust was also minimised by using Discovery's steering jets in "Low Z" mode, so that they fired at an angle rather than directly towards the station.

Final manoeuvres were manually controlled by commander Kent Rominger from the Shuttle's aft flight deck, where the approach could be monitored through the rear and overhead windows. From a point 150 m (500 ft) directly below the station, the Shuttle began to fly in a half circle that brought it 107 m

(350 ft) ahead of the station and then 76 m (250 ft) directly above it. While they waited to move into range of the Russian ground stations, mission specialist Tamara Jernigan powered up the docking mechanism, which already had its docking ring extended, and prepared it for use.

During the last stage of the inward leg, Rominger precisely aligned the docking ports on the Shuttle and the station with the aid of a "centreline" camera located in the centre of Discovery's docking mechanism. With the thrusters switched back once more to normal mode, contact was made at a speed of about 3 cm (1.2 in) per second. Absorbing springs dampened any relative motion as latches automatically held the two craft together. Jernigan then commanded the Shuttle's docking ring to retract and close the latches.

Since the arrival of Destiny, a modified approach to the station is undertaken by Shuttle crews. The ISS now remains oriented parallel to the Earth, with Destiny to the fore and Zvezda at the rear. When it reaches about 150 m (490 ft) below the ISS, the commander moves the Shuttle only a quarter-circle around the station, bringing it to a point about 100 m (330 ft) in front of Destiny and Pressurised Mating Adapter 2. The rest of the rendezvous and docking manoeuvres remain the same as those used earlier in the ISS programme.

Rendezvous and docking with the Russian segment are carried out using an upgraded version of the Kurs (course) automatic system that was developed for the Mir station. Special laser reflectors developed by ESA have been added to the aft end of Zvezda for eventual use with its ATV spacecraft.

Sometimes, the automatic docking system on an unmanned Progress fails to lock on to a comparable system on Zarya. In this case, a back-up manual docking system, known as the Telerobotically Operated Rendezvous System (TORU) may be brought into use. Using information from two antennas on Zarya and pictures from a camera on the Progress, a cosmonaut can take over control of the incoming spacecraft from a control panel in the station's Zvezda living quarters.

This system was used on a number of occasions on Mir, including one near miss and one disastrous collision between a Progress and the space station. It was first used on the ISS on 18 November 2000, when pilot Yuri Gidzenko used the hand controller to slowly bring a Progress supply ship to the Earth-facing docking port on Zarya.

ALTERNATE ACCESS

Not content with the space station's primary resupply vehicles – the US Space Shuttle, Russian Progress, European Automated Transfer Vehicle and the

Japanese H-II Transfer Vehicle – NASA has been seeking alternative means of access. As part of its Space Launch Initiative, the agency awarded 90-day contracts in August 2000, to enable four small businesses to study how emerging launch systems might provide alternative methods of ISS transportation. It was hoped that the contracts, worth a total of $902,000, would uncover a potential back-up capability to the government-funded vehicles.

"Alternate Access To Space Station is a potential market opportunity for emerging or established US launch companies," said Dan Dumbacher, manager of the 2nd Generation Reusable Launch Vehicle Program Office at NASA's Marshall Space Flight Center in Huntsville, Alabama. "These companies will develop concepts for alternate access to the Space Station, determine what a launch service needs to do to meet the requirements, and offer suggestions on specific development risk-reduction activities, such as technology development or business planning, that we need to perform. This potential alternate means of transportation could help us meet our commitments to the station."

The contingency resupply service would be capable of launching within a week if necessary and could enhance the Space Station's operational flexibility if the usual launchers were unavailable. Established launch services companies have also been studying the same idea under contracts managed by NASA's Kennedy Space Center in Florida.

However, the difficult economic climate, severe competition from large, established launch providers, and the reluctance of investors to finance space hardware projects seem likely to condemn this initiative to failure – at least for the foreseeable future.

Similar drawbacks are likely to affect a Russian proposal to develop larger freighters for the ISS. Arguing that production capacity at Energia, which makes the Progress, is already at the maximum, Khrunichev director general Anatoli Kiselov suggested constructing freighters with two to four times the cargo capacity of the current Russian workhorse.

According to Kiselov, the Khrunichev Centre has produced a design for a freighter that could carry up to 10 tonnes of cargo. "Thanks to the modular structure, the freighter can be fitted in various ways for dry and liquid loads, in any weight or size. The first such freighter could be launched in 2001," he said.

Less problematic is Khrunichev's intention to refit the Zarya-based FGB-2 so that it could carry 5 tonnes of freight to orbit. According to Kiselov, the FGB-2-based freighter could be converted fairly quickly. After delivery, it could be used as a storage module for the Russian section of the station.

FLIGHT CONTROL

Primary oversight of this busy traffic is now undertaken by a new mission control centre, located adjacent to the Space Shuttle flight control room at the Johnson Space Center in Houston. Support is provided by the Russian control room in Korolev, near Moscow, which also send up commands to the Russian segment. Meanwhile, a small NASA flight control team, designated the Houston Support Group, is permanently stationed in Korolev to facilitate communications and information exchange between the two control centres. However, responsibilities have changed since the programme began in 1998.

After the launch of the first module, Zarya, the primary command and control functions were undertaken from Korolev, using the Russian communications system. Once the Unity node was attached to Zarya, flight control was conducted from locations in both the United States and in Russia. While the Russian flight control team directed real-time ISS operations and performed command and control functions over the US systems, NASA flight controllers were able to oversee and approve all plans.

31 October 2000. Flight controllers in Houston's Mission Control Center follow the countdown to the Kazakhstan lift-off of the Soyuz carrying the Expedition One crew. (NASA digital photo) (http://www.spaceflight.nasa.gov/gallery/images/station/crew-1/html/jsc2000e27301.html)

The ISS mission control centre in Texas became permanently staffed at the time of Unity's launch in December 1998, about two weeks after Zarya's launch. However, it was not until the delivery of the Destiny laboratory and the primary communications system for the American segment in February 2001, that Station Flight Control in Houston assumed the direction of real-time flight operations, taking over the primary command and control functions. About a dozen flight controllers, led by an ISS flight director, are now on hand to monitor station activities.

In contrast to the short-lived Shuttle missions, the ISS now requires permanent, 24-hour monitoring. In order to meet the demand, NASA named 10 new flight directors in January 2001, the largest such class ever selected. This brought the number of US flight directors to 28. "Such a large class was needed to support around-the-clock operation of the International Space Station," said Jeffrey W. Bantle, chief of the Flight Director Office. "The first flight director, Chris Kraft, was selected during the Mercury era. Since that time, only 48 men and women have served as flight directors throughout the history of US human space flight."

A few months earlier, NASA opened a new training facility for flight control teams in the JSC mission control centre. Aware that the existing Space Station Flight Control Room and the Space Shuttle Flight Control Room would be permanently occupied with flight activities, the agency decided to outfit another room solely for training. The new facility includes 17 consoles and large front projection screens, identical to those in the main room of Mission Control. Like the rooms used for flight operations, the training flight control room can be linked to astronaut training facilities around the Johnson Space Center and other spaceflight control centres around the world.

These will eventually include control centres for the European Automated Transfer Vehicle (ATV) and the Japanese H-II Transfer Vehicle. The ATV control centre, which will execute all operations involving the unmanned spacecraft, will be located at the French Space Agency's site in Toulouse and is scheduled to become operational in 2004. The main tracking and control centre for all Japanese ISS operations will be at Tsukuba Space Centre, 60 km (38 miles) northeast of Tokyo.

Chapter Six

Construction Site in Space

More than 40 assembly missions spread over seven or eight years will be required to complete the International Space Station. With a mass more than four times that of Mir and an internal volume equal to that of a Boeing 747 Jumbo Jet, the ISS will be the largest artificial structure ever to orbit the Earth.

The mammoth task of constructing this monolith began one year late, in November 1998, with the launch of the first element – the Functional Cargo

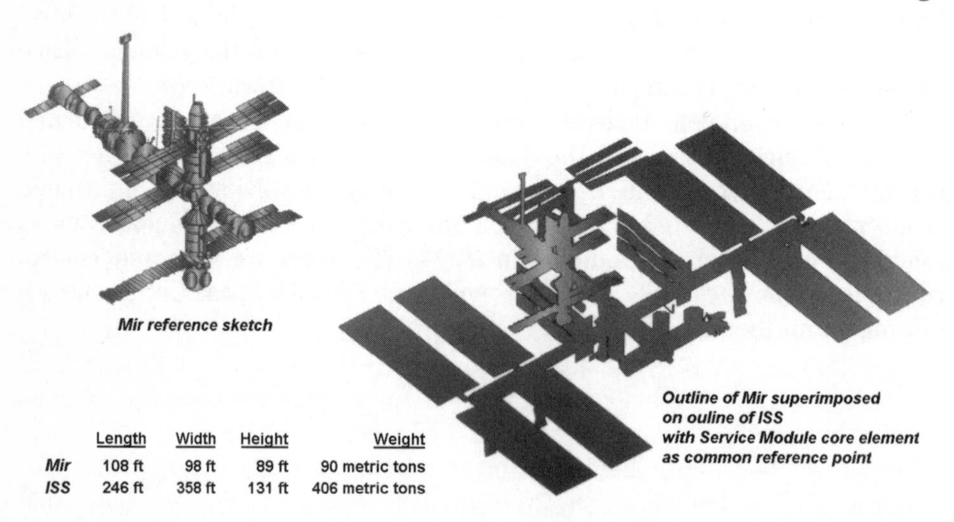

	Length	Width	Height	Weight
Mir	108 ft	98 ft	89 ft	90 metric tons
ISS	246 ft	358 ft	131 ft	406 metric tons

Size comparison of Mir and the completed International Space Station. (NASA photo 96E-00025)

Block (FGB in Russian) or Control Module, later to be known as Zarya (Sunrise). Since then it has continually encountered further delays and financial difficulties that threaten to truncate and modify its original design. This chapter tells the story of the progress made until the hand-over to the Expedition Four crew in December 2001.

THE RUSSIAN CORE

The first section of the ISS was built by Russia's Khrunichev Space Research and Production Centre, but it was actually financed by NASA under a $190 million contract to the American prime station contractor, Boeing. The bulk of the money was given to Khrunichev to build the FGB, with an extra $35 million added to allow the Russians to modify the module for docking by manned Soyuz and unmanned Progress craft.

Russia contributed the Proton launch, two Zarya simulators and the cost of ground control operations. A second FGB structure was built in case of a launch failure or a disastrous systems malfunction on the primary module.

Despite the supposed financial security offered by the NASA–Boeing contract, the FGB was not delivered on time. One of the first setbacks occurred on 9 December 1995 when the FGB was damaged during a hermetic test due to "operator error". Although the damaged section had to be replaced and further testing carried out, Khrunichev unrealistically insisted that the project schedule would not be affected.

In January 1996, the names of the first long-term occupants of the ISS were announced by NASA and the Russian Space Agency (RSA). Since the lion's share of ISS funding was being provided by America, overall command was given to William Shepherd, then serving as deputy manager for NASA's space station programme. The fact that most of the hardware they would occupy would be Russian-built was reflected in the selection of two Mir veterans – Anatoli Solovyov as Soyuz commander and Sergei Krikalev as flight engineer – to fly alongside Shepherd.

Another setback occurred on 19 February 1996 when the upper stage of a Proton heavy-lift vehicle, the rocket required to launch the first Russian-built segments of the ISS, malfunctioned, leaving a government-owned Raduga satellite stranded in the wrong orbit.

On 22 October 1996, Shepherd started training in Star City, near Moscow. The first stage of his course involved intensive Russian language classes and familiarisation with the systems of the Soyuz ferry craft. This was to be followed by joint work with Solovyov and Krikalev, learning about Zarya, the first segment of the multimodular ISS.

However, the Russian Duma, or parliament, could not resist the opportunity to score political points. Why, they demanded to know, was an American commanding a Russian space mission? There were even suggestions that Shepherd's appointment was illegal according to the Russian constitution! Unfortunately, cosmonaut Solovyov became involved in the dispute, indicating his displeasure that he was not to command Expedition One. Well aware that America was holding the purse strings on the entire ISS programme, the RSA had little choice but to replace the veteran cosmonaut with the more compliant Yuri Gidzenko.

LIFT-OFF!

The cornerstone of the monumental international building site, 20-tonne Zarya was based on the design of the Kvant 2 and Kristall modules used in the Mir space station. It provided an initial pressurised living quarters for visiting astronauts, as well as supplying some 3 kW of electrical power, communications, propulsion and attitude control. The attitude control system included 24 large thrusters and 12 smaller ones, while two main engines were available for major orbital manoeuvres. The module was launched with 4.5 tonnes of heptyl fuel, sufficient to operate for 430 days without refuelling. Its 16 fuel tanks would later be used to store propellant delivered by Progress cargo ships arriving at one of the module's forward docking ports.

"Sunrise" eventually soared into the sky above Baikonur Cosmodrome in Kazakhstan on top of a Proton rocket on 20 November 1998. By 24 November, it was in a 404 by 386 km (252 by 241 mile) orbit inclined at 51.6 degrees to the equator, and the module was circling the Earth once every 92 minutes. However, a number of teething problems surfaced during its first week aloft, including cabin humidity 10% above the norm, and a battery that would not fully charge.

Mission control in Moscow also reported that two antennas to be used by the Telerobotically Operated Rendezvous System (TORU) – a back-up manual docking system on the Zarya module – had not unfurled as planned. The antennas would be used to transmit range and closure rate information to approaching spacecraft heading for manually controlled dockings with the Russian section.

Back on the ground, the great debate continued about the future of Mir and its position with regard to Russian space priorities. Yuri Koptev, director-general of the Russian Space Agency, saw that the ISS offered the best prospects for the nation's future manned spaceflight programme. "If we find resources for both Mir and the International Space Station, that will be fine,"

The Zarya/FGB as seen from Shuttle Endeavour on 6 December 1998. (NASA photo STS088-719-059)

he said. "(However) in any respectable society, international obligations always have higher status than national ones. If we do not join this project, we will remain alone."

He went on to explain the very favourable terms that Russia had been given to participate in the ISS project. "By investing $6.8 billion, out of a total of about $100 billion, we have access to 35% of the station's capabilities and three out of seven seats on the crew. No need to feel nostalgic for the past. We should look into the future where the world community has decided to bet on international cooperation."

ENTERING THE OUTPOST

Only two weeks after Zarya became the keystone of the gigantic structure that would become the ISS, the Space Shuttle Endeavour blasted off from Cape Canaveral, carrying the second building block – a node or linking section named Unity. Also on board were five Americans and cosmonaut Sergei Krikalev, who was destined to be a member of the station's first long-term crew two years later.

NASA proudly declared that the first station hub, which had been under construction at the Marshall Space Flight Center since 1994, contained more than 9 km (6 miles) of electrical wiring, 216 lines for carrying fluids and gases and 50,000 mechanical units.

Two days after lift off, commander Robert Cabana completed a perfect rendezvous with the Control Module, clearing the way for mission specialist Nancy Currie to capture it with the Shuttle's robotic arm. Although the 20-tonne module was the heaviest object ever handled by the orbiter's remote manipulator system, Currie had little difficulty in successfully docking it with Unity in Endeavour's cargo bay. The newly joined modules towered 23 m (76 ft) above Endeavour's flight deck, like some outrageous creation of modern architecture.

Three spacewalks (EVAs) by Jerry Ross and James Newman were needed to prepare the station for independent flight. During the first EVA on 7 December, which lasted 7 hours 21 minutes, the astronauts completed some 40 connections involving electrical and data cables between the modules and

In this IMAX camera view, taken on 7 December 1998, the Shuttle's 15-m (50-ft) robotic arm grasps Zarya and prepares to attach it to the Unity module in the orbiter's payload bay. (NASA–ESA photo)

installed external handrails to assist future EVAs. They also inspected the two Russian TORU antennas that had not fully deployed following Zarya's launch two weeks earlier.

As Endeavour and the ISS passed over Russian ground stations, commands from the Moscow flight control centre activated a pair of Russian–American voltage converters, enabling power to start surging from Zarya to Unity for the first time. Unity's systems were then activated, including a pair of data relay boxes serving as the brain and nervous system for the US-built component. Near the end of the spacewalk, Ross removed thermal covers from the relay boxes after Unity's heaters began to control the module's temperature. By the time he re-entered Endeavour, Ross had clocked up 30 hours 8 minutes of spacewalks, a new record for American astronauts. ... And the ISS programme was only just beginning.

Preparations began for the first internal inspection of the station with the pressurisation of the Pressurised Mating Adapter 1 (PMA-1) between Unity and Zarya via remote commands from Moscow. While the tunnel was checked for leaks, flight controllers in Houston turned on air filters and fans and monitored temperatures inside Unity as heaters warmed up the module to a liveable temperature.

Yet another first for the ISS programme came on 8 December when commander Bob Cabana and pilot Rick Sturckow fired bursts from Endeavour's steering jets for about 22 minutes to gradually raise the highest point of the complex's orbit by about 9 km ($5\frac{1}{2}$ miles) to 397 km (248 miles).

Then, on 10 December, in the second EVA, Ross and Newman fitted two early communication antennas on Unity's outer skin and an external video cable linking Zarya to the starboard antenna. From then on, US ground controllers would be able to receive data and telemetry from Unity, instead of relying solely on information sent back when the station passed over Russian ground stations. At the end of the spacewalk, Newman was hoisted high above the Shuttle on the end of the robotic arm. With the aid of an extendable, 3-m (10-ft) long grappling hook, he was able to nudge free one of the baulky back-up rendezvous antennas on Zarya.

Then came the historic moment that NASA had been awaiting for many years – American astronauts floating into the International Space Station. The apparently straightforward transfer involved bringing the air pressure inside Endeavour to 1,013 mbar (14.7 lb/in^2), the same pressure as at sea level on Earth. The crew had then to equalise the air pressure on both sides of each hatch prior to opening them.

A total of six hatches had to be negotiated: the hatch on Endeavour's docking system; the hatch to Unity's forward mating adapter (PMA-2); the hatch to Unity; the hatch from Unity to the other mating adapter (PMA-1); the hatch to Zarya's spherical docking node; and finally, a hatch between the Zarya node and Zarya's main compartment.

In the spirit of international cooperation, Cabana and his Russian colleague, Krikalev, both opened the hatch to Unity at 1:54 p.m. Houston time on 11 December and floated inside. After turning on the lights and unstowing gear in the pristine module, the crew repeated the ritual for Zarya.

During their 28½ hours on board, the astronauts installed air ducts, completed the assembly of an early S-Band communications system and conducted a successful test of its videoconferencing capability by sending greetings to controllers in Houston. They also replaced a faulty battery current converter unit and stashed items of hardware behind panels in Zarya, ready for use by future visitors to the station.

When they shut the hatches behind them, they left tools, supplies and clothing to welcome the next visiting Shuttle crew, and the first long-term expedition, which was expected to arrive in January 2000. After turning off the lights and ventilation system, Endeavour's life support systems were used to increase the atmospheric pressure in the station and Shuttle to 1,078 mbar (15lb/in^2), a little above sea level pressure on Earth. Then, as each hatch in the station was closed, the crew lowered the pressure slightly to keep positive air pressure on the inside of each hatch, so assisting in sealing the doors. Dessicant bags were installed in Unity's portable, battery-operated fans to remove humidity from the module during its future unmanned operations.

Three members of the STS-88 crew inside the Pressurised Mating Adapter on Unity. Foreground, Jerry Ross; left, Robert Cabana; right, Sergei Krikalev. (NASA photo STS088-322-035)

The final spacewalk took place on 12 December. Ross and Newman stowed a tool bag on the exterior of Unity, disconnected umbilicals used during the docking with Zarya, installed a handrail on the Russian-built module and made a photographic survey of the 35-tonne station for review by engineers back on Earth. After Ross repeated his colleague's successful grapple of the other jammed rendezvous antenna, the pair wrapped up their third excursion by testing the Simplified Aid for Extravehicular Activity Rescue (SAFER) emergency backpacks. They had spent a total of 21 hours and 22 minutes outside the Shuttle.

The next day saw another ISS landmark as the new Flight Control Room in Houston issued its first wake-up call to orbiting astronauts – the song "Goodnight Sweetheart, Goodnight". A few hours later, Endeavour edged away from the station, its assembly mission successfully accomplished. For more than an hour, the Shuttle circled just 140 m (450 ft) from the abandoned complex, giving the crew ample opportunity to capture some spectacular camera shots before heading for home.

As the recently mated modules continued their solitary path around the Earth, mission controllers sent up commands to alter their orientation, pointing Zarya towards deep space and Unity towards the planet. The couple were then sent into a slow "barbecue roll", completing one revolution every 30 minutes in order to maintain an even temperature. During their long spell without occupants, Zarya's motion control system would be reactivated about once a week to ensure that it was working properly and its guidance system would be updated regularly with the latest orbital parameters.

THE LONG AWAKENING

Six months passed before the space station hosted another crew. During this hiatus, the ground control centres in Houston and Moscow took the opportunity to familiarise their teams with the complexities of their new baby, and to carry out tests on various systems.

By April 1999, the only hiccups to arise had involved the use of the American Early Communications System on Unity. At first, the right-side Omni (low-gain) antenna simply registered a persistently weak signal strength, but then communication via one of the high-gain antennas was completely lost during an evaluation test. Controllers suspected a faulty transmitter box. As a result, Russian ground stations once again became the primary ISS communication and command centres.

US confidence in the ability of their Russian colleagues to cope soon suffered a significant blow when the Russian flight control team sent an erroneous command to the ISS. The command was issued from a ground station when the direct link from the Mission Control Centre in Korolev, outside Moscow, was not working.

In order to send the command, the three-digit number identifying the command was incorrectly passed by telephone to the ground station, before being uplinked to Zarya's computer. This error caused the onboard computer to power up one of the module's solar array retraction motors. Fortunately, when the computer recognised the discrepancy, it automatically disregarded the command and shut off the power to the motor before the solar array could move. NASA was not amused, and issued a statement solemnly declaring that, "Meetings are under way to understand the sequence of events, to improve processes and prevent any reoccurrence."

Another minor hitch occurred at the end of April when flight controllers noted a false indication from one of the eight smoke detectors in Zarya. The faulty detector was powered off, but the rest of the smoke detection system continued to operate as normal.

Meanwhile, on 26 April, the next component of the station – the Russian Service Module – was at long last rolled out of the Khrunichev manufacturing plant and given its certificate of flight readiness. However, a lot of testing remained to be done at the Baikonur launch site in Kazakhstan before a launch date could be fixed. It seemed increasingly likely that Zarya and Unity would have to soldier on alone for some considerable time to come.

On the bright side, preparations for the arrival of the Shuttle Discovery continued to go well. With the station's main axis lying parallel rather than perpendicular to the Earth's surface, the Control Module had demonstrated that it could supply the 1,500 watts of power required to raise Unity's internal temperature to 18 °C – warm enough for the crew to enter.

THE FIRST SUPPLY DROP

Discovery and its multinational crew eventually lifted off from Cape Canaveral on the morning of 27 May 1999. At the time of launch, Zarya and Unity were cruising eastwards, 385 km (240 miles) above the outer banks of the Carolinas. The main tasks of assembly flight 2A.1 were to deliver some 2 tonnes of supplies and to continue outfitting the fledgling facility. Despite the additional room provided by a double Spacehab module, the Shuttle's crew quarters were even more cramped than normal. Stowed inside the cabin and the modules were clothes and food that would be stored on the station, together with various pieces of equipment. Outside in the payload bay were other pieces of hardware, the most significant of which were two cranes – one American and one Russian – that would be used during ISS assembly.

With Zarya pointing towards the Earth and Unity towards space, Discovery gradually closed the gap until it was just 180 m (600 ft) below the station. Over the next 80 minutes, Commander Kent Rominger flew the Shuttle in a half circle until it hovered directly above Unity. Once they passed within range of Russian ground stations, he gently guided the 100-tonne vehicle towards PMA-2, completing the docking exactly on schedule at 11.39 p.m. Houston time on 28 May.

The fourth spacewalk in the long saga of ISS construction – and the second longest in Shuttle history – followed in the early morning of 30 May. Tammy Jernigan and Dan Barry began by transferring components of the two cranes from a cargo pallet onto the two PMAs linked to Unity. Assisted by Ellen Ochoa on the Shuttle's robotic arm, they then installed two portable foot restraints that would fit both US and Russian EVA boots, and attached three bags filled with tools and handrails for use in future assembly operations.

A different form of hard labour followed the next day when the crew opened the hatch to Unity and began moving the cumbersome baggage into the station. Meanwhile, Canadian astronaut Julie Payette and Russian Valeri Tokarev worked on replacing faulty battery recharge controllers in the Zarya module. Other maintenance work over the remaining few days included fitting noise mufflers over the ventilation fans in Zarya and restoring the Early Communications System to full health. Crew fitness proved to be more of a problem as the visitors suffered from headaches, sore eyes and nausea.

By the end of the crew's visit, Unity and Zarya were crammed with packages containing clothing, sleeping bags, medical support equipment, hardware spares, and computer support and maintenance equipment. 318 litres (84 gallons) of water were also stashed to await the first resident crew.

With all outstanding tasks successfully completed, the crew closed the ISS hatches behind them on the morning of 3 June, after almost 80 hours inside the orbiting outpost. All that remained was to raise the station's orbit about 10 km (6 miles) with the aid of the Shuttle's Reaction System Control thrusters, then carry out a two and a half circuit fly-around photo opportunity.

Over the next few days, the crew deployed a small, mirror-encrusted satellite called Starshine and prepared for re-entry. The second mission to visit the ISS concluded successfully on 6 June with a night landing at Cape Canaveral. Up above, the abandoned space station was once more in its slow "barbecue roll", with Unity aligned towards the blue planet below.

THE DELAYED RUSSIAN STAR

As Discovery rolled to a halt on the Florida runway, hopes were still high that the Russians would be able to launch the core section of the ISS later that summer. These hopes were soon to be dashed and 11 months passed before another human set foot inside the mothballed station.

The blame for the major hiatus was placed fair and square on the shoulders of the Russians. On numerous occasions, Russian officials went on record to explain that financial shortages were holding up the Zvezda Service Module. For example, on 19 November 1997 Boris Ostroumov, Deputy Director of the Russian Space Agency, complained that funds for module construction, which were supposed to have been paid in January, had not actually arrived until May of that year. Even then, they took the form of promissory notes and mutual cancellation of debts rather than cash. "It is important to note that not all businesses can handle promissory notes, and the payment of the notes takes a year," he explained. "Mutual cancellations do not pay wages." As a result, suppliers of instruments and cables had not been able to deliver them on time, with the inevitable knock-on effect for the assembly teams of Khrunichev and Energia.

Inevitably, there were some angry exchanges between Russian engineers, NASA and various US politicians, who accused Russia of wasting time and resources on the 14-year-old Mir space station.

However, the saga of Russian delays and excuses continued unabated. The most serious hold up came from an unexpected source – the heavy-lift Proton rocket – after Zvezda was finally completed. Two successive launch failures in July and October 1999 by the normally reliable Proton led to further launch postponements.

With the Proton grounded until the causes of the failures could be ascertained and remedial actions taken, there was no choice but to leave the Service Module languishing in storage on the ground. Months went by before the necessary modifications were made to the Protons and they were cleared for flight once more. Meanwhile, so much unlaunched hardware was being stockpiled that NASA began to lose track of it all. The most embarrassing débâcle occurred in February 2000, when two gas tanks worth $750,000 were accidentally thrown away.

Until the arrival of Zvezda, all that could be done was to monitor the station's systems and continue outfitting the existing modules. In May 2000, NASA launched STS-101, a "housekeeping visit" to the orbital complex. This third Shuttle flight to the ISS – the first launch of Atlantis since September 1997 – was loaded with equipment in a Spacehab double module.

The seven-member crew (including the future Expedition Two crew, Yuri Usachev, James Voss and Susan Helms) spent five days docked to the PMA-2 on Unity. Most of their time was spent checking and upgrading the station's systems. It was a matter of some concern that the new structure was already suffering from hardware problems that required significant time to be devoted to repairs.

Among the primary tasks undertaken during a 6 hour 44 minute spacewalk on 21–22 May was the replacement of an early communication antenna and a power distribution box on Unity. Astronauts Williams and Voss also secured a US-built crane and completed assembly of the Russian Strela crane. Inside Zarya, four out of six batteries, a radio telemetry system and three fire extinguishers that had reached the end of their design lives had to be replaced. The crew also installed 10 new smoke detectors and four new cooling fans.

In an effort to solve the problems with stale air that had been reported by the previous visitors, the crew collected air samples from the station, and measured air circulation and carbon dioxide levels. They later ran air ducts through the two modules to improve air circulation.

Among the 1.5 tonnes of supplies unloaded by the crew were sewing kits, trash bags, books, note pads, can openers and an IMAX movie camera. Other

Jeffrey Williams hangs onto one of the newly installed handrails on the exterior of PMA-2/Unity during a 6 hour 44 minute spacewalk on 21–22 May 1999. (NASA photo STS101-374-007)

Susan Helms carries a treadmill from the Pressurised Mating Adapter and through Unity. (NASA photo STS101-348-018)

items left for future station occupants included an exercise treadmill, a bicycle ergometer, a resistive exercise device, clothing and four 45-litre (12-gallon) containers of water.

By the time Atlantis departed, the station's orbit had been raised some 43 km (27 miles), leaving it in the optimum path for the arrival of the Service Module (now known as Zvezda, the Russian for Star) – scheduled for July.

A key advance came on 26 June 2000, when Yuri Semenov, general designer and president of RKK Energia, and Tommy Holloway, director of NASA's International Space Station programme, signed an "international certificate" stating that the Service Module was ready to be launched to the ISS on 12 July. Assuming all went well, the launch of the first long-term expedition to the station was set for 30 October.

With no back-up module available, the Russians were taking no chances. Although Zvezda was expected to dock automatically with the Zarya–Unity complex, difficulties with such procedures had been experienced in the past with Mir. In the event that the automatic docking system failed, a Russian "zero" crew made up of cosmonauts Gennadi Padalka and Nikolai Budarin was ready to be sent to the ISS with special equipment that would enable them to dock Zvezda to the station under manual control.

Fortunately, the 12 July launch of the Proton-K rocket from Baikonur went flawlessly, although the cash-strapped nature of the Russian space industry was reinforced by the appearance of the Pizza Hut logo on the side of the launcher.

Artist's impression of Zvezda deploying its solar arrays prior to docking with the ISS. (NASA photo JSC2000-E-17849)

Russian newspapers reported that the American fast food chain paid $1 million for the unusual advertising opportunity.

Post-launch operations proceeded according to plan. Once in orbit, pre-programmed commands activated Zvezda's systems, allowing the solar arrays and communication antenna to be deployed successfully. The Service Module then became the passive "target" vehicle while the Zarya–Unity combination completed a link-up with the aid of the Russian Kurs automated rendezvous and docking system. Two nail-biting weeks after lift-off, on 26 July 2000, Zvezda pulled in at Zarya's rear docking port. The third element of the ISS had arrived.

NASA and its partners sighed with relief. The fact that Zvezda was nearly two years behind schedule – a delay that is thought to have cost the multinational project as much as $3 billion – was put to one side. "Now we can say with confidence that the International Space Station will go ahead," commented a proud Anatoli Kiselov, head of the Khrunichev company that built Zvezda.

Certainly, without the Service Module, no long-term visits to the multibillion dollar complex would be possible. The engines of Zvezda, which was to be the central hub of the ISS, were required to regulate the height of the station's orbit. It was also equipped with the main docking ports for Russian Soyuz and Progress vehicles and would act as the living quarters for the first long-stay crews. Its home comforts included a workshop, two sleeping areas, a bathroom, galley and exercise facilities, as well as 13 windows for observing the Earth and heavens.

The next step was to outfit the stripped down module in preparation for the arrival of the Expedition One crew – the station's first long-term inhabitants. On 9 August, an unmanned Progress M1 cargo craft pulled alongside – the first of its type to visit the ISS. Among the supplies for Zvezda were 1.6 tonnes of fuel, life support equipment, clothes and food. The fuel was particularly vital, since it would eventually be transferred to tanks on the Service Module and used in the station's main propulsion system. The remainder of the fuel would be consumed by the Progress to boost the station's orbit.

The success of the flight was dampened somewhat by a dire warning about future Russian involvement in the ISS programme. Yuri Semenov, head of RKK Energia, told the assembled press that the launch of Zvezda and the subsequent Progress launch were "goodwill" gestures by his organisation, and then declared that "the corporation has no other spacecraft for future launches paid for by the state".

Semenov went on to explain that Energia was owed some $40 million by the Russian state, adding that this "most serious problem should be solved jointly with the government". The corporation "cannot stake its all any longer," he said.

STS-106 – A FLYING VISIT

Preparations for the first expedition continued with the launch of STS-106 on 8 September. After docking with the mating adapter on Unity, the main objective of the crew – five Americans and two Russians – was to make Zvezda operational. The first task was to integrate its systems with the rest of the station. During a 6 hour 14 minute spacewalk, Edward Lu and Yuri Malenchenko connected nine cables between Zvezda and Zarya. These included four power lines that would later be used to bring electricity from US-built solar arrays to the Russian segment, four cables for video and data transmission and a fibre optic telemetry link for sending Russian spacesuit data to the ground during future EVAs.

During a final task, the tethered duo had to work more than 30 m (110 ft) above the Shuttle cargo bay in order to attach a magnetometer that would serve as a back-up navigation system for the ISS. Acting as a space "compass", the instrument's magnetic field measurements would be used to tell Zvezda's

Shuttle Atlantis is seen docked to the front end of the ISS in a picture taken during the spacewalk by Malenchenko and Lu. (NASA photo STS106-349-002)

computers the direction in which the station was oriented, so reducing the consumption of valuable propellant by the module's thrusters.

Entry into the enlarged station the next day was even more complicated than on previous visits, with no fewer than 12 hatches to open in sequence before the crew could reach the Progress at the far end of Zvezda. Precautionary breathing masks and eye goggles proved unnecessary as the crew entered the Russian module for the first time.

Once the ventilation system was operational, the tasks of the next few days included installation of three batteries in Zvezda (to save weight at launch it had only been outfitted with five of its full complement of eight). However, it was quickly found that one of them was not working properly. Replacement of two batteries in Zarya also proved less straightforward than expected: mission specialists Dan Burbank and Boris Morukov had to resort to a hammer and chisel to remove floor panel rivets before they could access one of the units.

Later in the mission, the waste tank and hose of the station's first toilet were unloaded, along with two Orlan pressure suits and components of Zvezda's Elektron unit, which would eventually generate oxygen from waste water. A special treadmill that allows astronauts to exercise without shaking sensitive experiments was also installed in Zvezda.

An extra day of docked operations was added in order to ease the onerous task of unloading and carefully stowing nearly 3 tonnes of cargo from Atlantis and the newly docked Progress. The inventory included more bags of water, food, office supplies, a vacuum cleaner, a food warmer, ham radio, computer and monitor – more or less completing the onboard stowage of supplies for the resident crew. Trash and redundant hardware were loaded into the Progress, for eventual incineration, or into Atlantis for a less destructive return to Earth.

After a successful week of docked operations, the Shuttle departed on 17 September. The crew took the opportunity for the usual fly around before pulling away from the 43-m (143-ft) long station.

JAPAN AND THE ISS BACKBONE

As the programme at last began to increase momentum after a long period in the doldrums, only three weeks went by before yet another Shuttle lifted off from Florida on its way to the ISS – although the 11 October launch might have ended in disaster if sharp-eyed technicians had not spotted a stray metal pin that could have been sucked into one of the main engines.

Discovery carried a crew of seven, including mission specialist Koichi Wakata, the first Japanese astronaut to visit the International Space Station.

Koichi Wakata, the first Japanese astronaut to enter the ISS, floats through the supply-laden Zarya module. (NASA photo STS092-340-033)

Also on board were a 3-tonne conical tunnel, called Pressurised Mating Adapter 3, that would provide an additional Shuttle docking port, and a 9-tonne latticework truss, that would become the "backbone" of the International Space Station. This exterior framework, called the Z1 truss, would house gyroscopes and communications equipment as well as provide enhanced voice and television capability.

"The foundation for the International Space Station has been laid and this mission begins the true station build-up in orbit," said Space Shuttle Programme Manager Ron Dittemore. "With multiple space walks planned and multiple components to attach, we're taking the level of complexity up a notch over the past few station construction flights."

Docking with the Unity module was successfully completed as they passed 385 km (240 miles) above Russia. The next day, despite a short circuit in some of his support equipment, Wakata deftly manoeuvred Discovery's robot arm to grapple and lift the Z1 truss from the cargo bay, then berthed it to Unity's exterior. Inside Unity, pilot Pam Melroy used a laptop computer to command the tightening of 16 bolts that secured the truss to the module.

The first of four spacewalks took place on 15 October when Leroy Chiao and Bill McArthur stepped outside to connect electrical and computer data cables between the truss and Unity, then relocated and deployed two communication antennas.

The next day, during a 7-hour EVA, astronauts Jeff Wisoff and Michael Lopez-Alegria assisted in the relocation of the Pressurised Mating Adapter 3 to its new home on Unity, in a position opposite the Z1 truss, then readied an attach point for future American solar arrays on the truss.

Chiao and McArthur returned to action in the third EVA, installing two converter units that would regulate voltage output from the US solar arrays. They also reconfigured power cables to send electricity from PMA-2 to PMA-3 in preparation for the arrival of Shuttle Endeavour on mission STS-97.

The fourth successive day of spacewalks saw the Wisoff–Lopez-Alegria team carry out some "wrap up" tasks. They removed a grapple fixture from the truss, tested a latch assembly for the solar array truss to be delivered in December, deployed a power supply tray that would be used by the Destiny laboratory and tested latches for the US science module. Each astronaut also tried out a nitrogen-powered SAFER backpack during a gentle 16-m (50-ft) flight.

After 27 hours and 19 minutes of outside activities, the remainder of the docked operations concentrated on completing connections for the Z1 truss and transferring equipment. The four Control Moment Gyros in the truss were given a trial run by spinning them to 100 revolutions per minute and turning on their heaters. Melroy and Wisoff gathered samples from surfaces inside Zarya

This front view of the ISS shows the newly installed Z1 truss and its antenna, together with the newly relocated PMA-3 docking port. (NASA photo STS092-712-059)

to check for any microbial growth and wiped some areas, including storage bags, with fungicide.

Transfer of supplies to Zarya took longer than expected, so the undocking was slightly delayed until the morning of 20 October. On board Discovery during its return to Earth was a protein-growth experiment that had been set up the previous September – the first microgravity science experiment to be conducted on the ISS. Now all eyes turned east to await the launch of the station's first resident crew.

EXPEDITION ONE

The historic moment came on 31 October 2000, 17 years after US President Ronald Reagan first proposed the construction of an international space station, when a Soyuz rocket blasted off from Baikonur Cosmodrome carrying the first in a long line of crews to permanently inhabit the ISS. On board were

American expedition commander William Shepherd and two cosmonauts, Soyuz commander Yuri Gidzenko and flight engineer Sergei Krikalev.

"I'd say there's a decent chance that [today] may in fact be the last day that we don't have humans in space," enthused NASA flight director John Curry.

Two days later, the Soyuz capsule pulled in at the aft docking port on the Zvezda Service Module to complete a smooth, automated link-up. A little over one hour elapsed before the hatch leading into Zvezda's living quarters was opened, signifying the start of human occupancy of the international complex. It was 4:23 a.m. Houston time. In deference to the fact that the crew had arrived in a Russian spacecraft and were about to enter a Russian-built module, Shepherd diplomatically invited Gidzenko and Krikalev to be the first to float into the station.

No time was wasted in bringing the station to life. Their first activities included checking out communications systems, switching on food warmers, charging batteries for power tools, starting up water processors, and making the toilet operational.

Three hours after they cracked open the hatch, the trio starred in their first live television show from inside the station. They also downlinked video

The Expedition One crew make themselves at home. Left to right: Soyuz commander Yuri Gidzenko, mission commander William Shepherd, and flight engineer Sergei Krikalev. (NASA photo ISS01-328-015)

footage of their entry into Zvezda and received congratulatory messages from American and Russian officials at the Mission Control Centre in Moscow.

The next few days were spent installing key life-support systems and additional communications equipment. One of their key tasks was to get the Vozdukh air regenerative system up and running so that they could stop using disposable lithium hydroxide canisters to absorb carbon dioxide. Another key Russian system, known as Elektron, was also installed, eventually taking over from the expendable canisters used to replenish the oxygen supply in the cabin atmosphere.

The air-conditioning system in Zvezda proved more of a problem. Until it became operational, the average temperature aboard the station was about 24 °C and the relative humidity was around 40–50%.

The reliance of the ISS on computer technology was emphasised by the amount of time spent on setting up the various machines. One of Shepherd's first tasks was to hook up the cables and laptop computers associated with the station's Early Communications System. Once this was operational, the crew would be able to have lengthy conversations with flight controllers, as well as send and receive electronic mail files, images and video through US relay satellites. However, teething troubles meant that there were difficulties in booting up some of the computers, as well as locating compatible cables for the various US and Russian equipment.

The crew also set up a Russian laptop computer system that would be used, among other things, to track the station's inventory of equipment and supplies, and a central post computer, a system that allowed various laptops to monitor the operation of the Zvezda systems.

Meanwhile, Gidzenko and Krikalev installed and tested the hand controllers and television monitor of the TORU manual system, in preparation for the arrival of the next Progress. The crew would turn to this back-up system in the event that the automated rendezvous system failed on the final approach of the unmanned supply ship.

With another 2 tonnes of supplies soon to be delivered, keeping tabs on all the equipment on board proved to be quite a headache. Shepherd, Gidzenko and Krikalev spent many hours conducting an inventory of the hardware, ensuring that everything was catalogued so that new items arriving on the Russian cargo ship could be properly distributed and accounted for.

Loaded with clothing, food, spare parts and gifts from their families, the second Progress spacecraft to visit the ISS was launched from Baikonur on 15 November. Two days later, the cargo ship's automatic docking system failed on final approach to the ISS, forcing Gidzenko to resort to use of the TORU manual back-up to slowly dock the Progress at the nadir, or Earth-facing, docking port on the Zarya module.

Yuri Gidzenko replacing one of the ventilation fans in Zvezda. (NASA photo ISS01-E-5187)

Over the next two weeks, the three crew members spent much of their time unloading the new arrival, filling it with rubbish and replenishing the station's oxygen supply from tanks on the Progress vehicle. They also carried out routine maintenance on the station's humidity removal system, toilet and treadmill. Repairs were also required for one of four ventilation fans in the Zvezda living quarters, which was shut down after one of the blades failed.

POWER TO THE PEOPLE

The Expedition One crew did not have long to wait until their first visitors arrived. The first of three Shuttle missions scheduled during their four-month tenure of the ISS lifted off from Florida on 30 November. As its crew of five astronauts headed for low-Earth orbit, the station's sleeping residents were 13,800 km (8,340 miles) away over the Indian Ocean.

Some 13 hours later, the trash-filled Progress pushed away from Zarya to make room for Shuttle Endeavour to dock at the nearby port on Unity. The unmanned vehicle was parked about 2,500 km (1,560 miles) away, while US and Russian officials discussed what to do with it. The Russians wanted to test a software patch as a solution to the problem in the craft's navigation system, either by attempt a redocking with the station or another rendezvous without a docking.

Meanwhile, with the occupying crew looking on, commander Brent Jett guided Endeavour to a smooth link-up with Unity, but almost a week passed before they made physical contact with the STS-97 astronauts.

Since the two crews were operating on different sleep cycles, the ISS trio went to bed shortly after the link-up, leaving their Shuttle colleagues to start the complex process of attaching the first set of US solar arrays.

With the aid of Endeavour's robotic arm, Canadian Marc Garneau grappled the 13-m (45-ft) long P6 solar array assembly, lifting it from its restraints in the payload bay and holding it aloft at an angle of 30 degrees for a prolonged "overnight" period of warming.

After completing leak checks between the two vehicles, astronauts Joe Tanner and Carlos Noriega moved through Endeavour's docking tunnel and opened the hatch to Unity so that they could leave some gifts, supplies and computer hardware just over the threshold. A failure to properly equalise the air pressure on both sides of the doorway meant that Tanner had to give it a hefty shove before the hatch would give way.

The ISS crew was delighted with the packages, which included a new laptop computer, headsets for the videoconferencing system, a new hard drive for a Russian laptop, bags of water, fresh and packaged food, plus a special "care package". Shepherd was delighted with his supply of fresh coffee and a large pair of pliers.

"It's kind of like Christmas up here going through these bags," he said.

Sunday 3 December saw the mating of the new solar array with the Z1 truss on Unity. Back at the controls of the robotic arm, Garneau moved the structure into position, then forced it into its installation point. Before its grip was slackened, spacewalkers Tanner and Noriega secured the bolts that held it in place. Pilot Mike Bloomfield then took over from his Canadian colleague to carry Noriega around the array as he connected nine power, command and data cables.

Deployment of the giant array was more problematic. A first attempt to release automatically the pins sealing the solar array blanket boxes ended in complete failure. A second try saw the pins on the starboard box opened and the panel deployed, but a single pin stubbornly remained in place on the other wing.

The eleventh ISS EVA concluded after more than 7½ hours. A few minutes later, a third attempt to disengage the pins on the baulky box apparently succeeded, but flight controllers decided not to deploy the port-side array until they could check the tension of the already functional starboard panel. The apparent slackness suggested that two of the cables that were meant to keep the array taut had slipped off their guides. Happily, the first of three photovoltaic radiators that would dissipate into space heat generated by onboard electronics was successfully deployed.

The next day the crew took a more leisurely approach to tackling the solar arrays. In an effort to prevent unwanted motion during the deployment of the port panel, a step-by-step technique was introduced. This meant that the full extension of the 30-m (100-ft) long array took some two hours instead of the 13 minutes taken by its starboard counterpart. Even then, some careful forward and backward manoeuvring had to be employed when two rows of panels stuck together near the end. In the next few hours, the 12 batteries on the power unit were fully charged, ready to send electricity to the Unity module.

The second spacewalk of the visit took place on 5 December, when Tanner and Noriega reconfigured ammonia coolant lines and cables that would be used to transfer power from the new arrays to the station. They also moved a low data rate antenna from the Z1 truss, where it had been temporarily stowed two months earlier, to the top of the solar array tower. This 6 hour 37 minute excursion was followed by the deployment of a second heat-dissipating radiator on the side of the tower.

Carlos Noriega waves to the camera during the second EVA to install the solar array structure (visible at the top). (NASA photo STS097-375-012)

The third EVA was also highly successful. Despite last-minute instructions on a procedure that would increase the tension in the starboard array, Noriega and Tanner worked so well that they were able to tackle some "get-ahead" tasks for the next scheduled spacewalks in January. The loose array was tackled by slightly retracting the wing, allowing the spacewalkers to pull the slack. The spring-loaded reels were then released while Noriega guided the cables onto the reel grooves. Their activities concluded with a "topping ceremony" when an image of an evergreen tree was placed on top of the P6 array structure alongside a probe to measure the electrical potential of plasma (electrified gas) around the station.

After 19 hours and 20 minutes of spacewalks, the Shuttle crew could at last concentrate on internal affairs. The Expedition One crew came face to face with their first visitors on 8 December – 38 days into their mission. By this time, living conditions on the station were back to normal after the repair of the broken fan on the Vozdukh carbon dioxide removal unit and the replacement of the air conditioner.

Artist's impression of the Shuttle docked to the ISS after installation of the P6 solar array assembly. (NASA photo JSC2000-E-29786)

Apart from conducting a joint news conference, the eight crew members spent their time moving supplies to the station and redundant equipment back to the Shuttle. They also carried out structural tests on the new arrays and set up a TV system that would be used to help a Shuttle crew attach the Destiny science laboratory to the ISS in January.

By the time Endeavour undocked and completed the usual pre-departure fly-around, a major step had been taken in the expansion of the station's capabilities. The US solar arrays, which measured 74 m (240 ft) tip to tip, were typically sending some 13 kW of electricity to the ISS, with a potential for up to 60 kW under optimal conditions – five times more than had been available just two weeks earlier.

CHRISTMAS IN ORBIT

To the delight of the Expedition One crew, who still had nearly three months of their tour to run, the extra power allowed them to have continuous access to the cluttered interior of Unity. It also allowed them to feed supplemental power to the Russian modules on the station.

The bad news was that the men would have to spend an extra two weeks in orbit due to the delay in the launch of their ride home. Shuttle Discovery was now targeted for launch on 1 March, in order to allow sufficient time to replace 10 of its jet thrusters. Even the January launch of Atlantis and the US Destiny laboratory was under threat after the failure of a separation bolt on the Shuttle's Solid Rocket Boosters during Endeavour's climb to orbit on 30 November.

A more immediate problem involved the future of the free-flying Progress ship. US and Russian officials eventually agreed to permit its redocking on 26 December. The first manoeuvres to raise the orbit of the Progress and close on the station took place on 20 December. Additional engine firings were commanded by Russian flight controllers on Christmas Day and the following morning, ultimately placing the 7-tonne craft about 200 m (660 ft) below Zarya's downward-facing docking port. Gidzenko then took over control of the final approach using the TORU manual navigation system. Within two hours of the redocking, Krikalev equalised pressure between the Progress and Zarya, then opened the hatches to enable the crew members to deactivate the systems on the cargo craft.

Apart from the preparations for the return of the robotic ship, the Expedition One crew spent a quiet Christmas, opening presents, having private communication sessions with their families and enjoying a dinner of rehydrated turkey. They also received a holiday greeting from NASA administrator Daniel

Goldin. As always, one of their main sources of pleasure was looking out of the windows at the Earth below.

The New Year's weekend was also marked by light duties, with further opportunities to speak to their families. US Navy Captain Bill Shepherd followed naval tradition at the turn of the New Year by writing into the ship's log a poem he had composed on board.

Between these times of celebration and relaxation, the crew concentrated on unloading ballast from the Progress, removing its faulty Kurs docking system for analysis back on Earth, performing biomedical experiments and reviewing flight plans for the January Shuttle flight to install the Destiny laboratory.

The New Year saw a continuation of the dreary, never-ending task of carrying out a thorough inventory of items as they began stowing equipment and supplies to clear passageways before the next visitors arrived. The trio also reviewed documentation for the Destiny laboratory's activation, held conferences with various technical specialists and the STS-98 crew, and a mid-tour debriefing with flight controllers.

In preparation for the relocation of the Shuttle docking port, Pressurised Mating Adapter 2, flight controllers in Houston attempted to cycle four latches on the Common Berthing Mechanism to which the PMA was attached. The first latch cycled properly, but the second was obstructed by a piece of ducting used to circulate air throughout the station while a Shuttle is docked. The problem was easily solved by pressurising PMA-2 so that the crew could enter and move the obstruction.

Unfortunately, although the ISS was ready for its next addition, the delivery vehicle was not. Shuttle managers ordered the rollback of Atlantis from the launch pad to enable inspections of cables inside its Solid Rocket Boosters. Destiny was removed from the payload bay and put in storage on the launch pad until the vehicle could be given the all clear.

The station's disappointed occupants kept themselves busy by continuing to update the station's inventory and practising emergency evacuation procedures should an air leak occur. Otherwise, it was business as usual, with the daily exercise sessions and routine housekeeping – changing filters, inspecting equipment and checking station systems.

DESTINY

The seventh Shuttle mission to the station eventually lifted off on 7 February. A video of the launch was transmitted to the ISS crew to assure them that their first visitors in two months were on their way. On board was the US Destiny module, the first of six space science laboratories that were expected to be launched to the station.

The next day – the Expedition One crew's 100th day in space – the Progress undocked from the ISS for the second and final time, clearing the way for the arrival of Atlantis. In a faster, lower orbit, the Shuttle rapidly closed the 3,200-km (2,000-mile) gap.

Less than 1 km (half a mile) from the station, commander Ken Cockrell took over manual control and edged the huge vehicle towards the downward-facing port on Unity. A little over two hours after hard dock was achieved, the two crews opened the hatches for a short-lived greeting and a rapid transfer of supplies – water bags, a spare computer for Zvezda, power cables, gifts, fresh food and DVD movies.

Destiny, the space station's first science laboratory, is moved into position by the Shuttle's robotic arm. (NASA photo STS098-331-017)

With the doors once more sealed, the air pressure in Atlantis was lowered in preparation for the next day's spacewalk. However, demonstrating that even well-trained astronauts make mistakes, the Shuttle crew found that they had left some connectors required for the Destiny laboratory on board the station. The ISS residents had to deposit them in the Unity docking compartment for their colleagues to collect.

10 February began with an early morning wake-up call from mission control and a hasty raising of the joint complex's orbit to avoid a piece of space debris that was expected to pass to close for comfort.

The all-important transfer of the Destiny laboratory from the Shuttle's cargo bay to its berth on the station went flawlessly in what NASA described as "a dazzling display of robotics finesse and spacewalking skill". The complex operation began with the removal of PMA-2 by the Shuttle's robotic arm under the control of Marsha Ivins. Once this docking port was latched in a temporary position on the station's truss, spacewalkers Tom Jones and Bob Curbeam got Destiny ready for its relocation. Gripped by the robotic arm, the 15-tonne cylinder was lifted high above the payload bay, swivelled through 180 degrees, and moved into position on the Unity module.

Once the automatic bolts tightened to hold it in place, Jones and Curbeam began to connect its cooling, electrical and data links. Unfortunately, a small leakage of ammonia crystals from a coolant line meant that Curbeam had to remain in sunlight for half an hour to vaporise the potentially hazardous chemical, while Jones brushed down his colleague's suit and equipment. Before they re-entered the orbiter, the pair performed a partial depressurisation and venting of the Shuttle airlock to flush out any residual ammonia. As a final precaution, the other three astronauts on board Atlantis donned oxygen masks for 20 minutes while the air-cleansing systems operated.

These convoluted decontamination procedures increased the duration of the EVA to 7 hours 34 minutes, more than one hour longer than planned. Inevitably, the remainder of the day's work suffered from delays, but that evening the hatches were opened once more, allowing the crews to attempt a successful powering up of Destiny.

After their tiring day, both crews were allowed an extended sleep period before they began to outfit the latest addition to the station. Apart from its scientific use, Destiny would also provide a 40% increase in living space, so the crews would have to activate air conditioners, fire extinguishers, computers, internal communications systems, electrical outlets, ventilation systems, alarms, and the installation of a carbon dioxide removal rack to augment the Russian Vozdukh unit.

The second spacewalk saw Jones and Curbeam assist Ivins as she used the Shuttle's robotic arm to move the PMA-2 from its temporary parking place on the truss to the forward end of Destiny. In this new location it would become the main docking port for all future Shuttle visits. The remainder of the 6 hour 50 minute EVA was spent preparing Destiny's exterior for future spacewalkers and attaching a base for the station's forthcoming robotic manipulator.

During the rest of the week, "control authority" tests of Destiny's computers demonstrated the ability of the laboratory's computers to take over control of the station's orientation from the computers in the Russian segment. In particular, they proved their ability to align the station through the fuel-saving use of four gyroscopes on the station's Z1 truss.

All of the laboratory's systems were reported to be working well, apart from the back-up carbon dioxide removal rack, which had to be switched off after one of its pumps failed.

The third EVA of the mission, and the 100th in US spaceflight history, took place on 14 February. The 5 hour 25 minute spacewalk proved to be fairly routine as Curbeam and Jones attached a space communications antenna to the station, checked connections between Destiny and its docking port, released a cooling radiator, inspected solar array connections and tested their ability to carry an immobilised astronaut back to the airlock.

"In 1962, astronaut Ed White made history by walking outside his Gemini 4 spacecraft for 21 minutes," said Mike Hawes, NASA's Deputy Associate

Members of the STS-98 crew join ISS commander Shepherd (rear) in the Unity node. The camera is pointing towards Zarya. (NASA photo STS098-359-021)

Administrator for Space Station. "By 2003, we will have spent more than 550 EVA hours on the construction of the space station alone."

With the excursions now completed, the hatches could be reopened between the Shuttle and the ISS. All that remained was the final transfer of supplies and equipment. Among the items stowed aboard the station were a replacement hard drive for a portable computer, the EVA suit worn by Jones, a screwdriver, tape and printer paper. The crews also squeezed in a question and answer session with school students and reporters.

By the time Atlantis departed, the ISS orbit had been raised 26 km (16 miles) and the enlarged station measured 52 m (171 ft) long, 27 m (90 ft) high and 72 m (240 ft) wide. It already boasted a larger internal volume than any space station in history.

Once more alone, Shepherd, Krikalev and Gidzenko settled down to a quiet weekend before continuing the activation of Destiny's systems and preparing to move their Soyuz capsule from the rear end of Zvezda to the Earth-facing port on Zarya. Despite being hampered by poor communications, the short excursion in the Soyuz TM-31 successfully took place on 24 February, clearing the way for another Progress supply ship. The crew amused themselves watching DVD movies during spells of inactivity. Four days later, the Progress M-44 vehicle duly docked automatically. It was loaded with 2.5 tonnes of supplies for the Expedition Two crew, whose arrival was imminent.

EXPEDITION TWO

The rapid launch activity continued unabated as the third Shuttle mission to the ISS in less than four months blasted off from Cape Canaveral on 8 March. The seven-person crew included the station's second long-term occupants – Russian commander Yuri Usachev and NASA astronauts James Voss and Susan Helms. Discovery also carried the Leonardo Multi-Purpose Logistics Module, the first in a series of Italian-built, reusable storage facilities that had been developed for the space station programme.

The 12-day mission suffered a few early hiccups when Shuttle commander Jim Wetherbee had to wait for an hour little more than 100 m (400 ft) from his destination until he received verification that one of the station's huge solar arrays was safely turned away from the orbiter's potentially harmful thruster plumes. No sooner had the two vehicles docked than the Shuttle suffered a communication malfunction, so messages from Houston had to be passed to the crew via the space station link.

Nearly two and a half hours after arrival, the hatches were opened, allowing the two crews to mingle for the first time. Leading the way into Destiny was

Space hand-over. The Expedition One crew (dark shirts) pose alongside their replacements (right) and the STS-102 crew (rear) inside Destiny. (NASA photo STS102-319-028)

Yuri Usachev, whose task would be to ensure a smooth hand-over from one expedition to the other.

Almost immediately after the greetings had ended, Usachev marked the beginning of his occupation of the station by swapping places (and personalised Soyuz seat liners) with Yuri Gidzenko. The gradual crew exchange continued on 11 March with Voss replacing Krikalev, then Shepherd finally retired to Discovery as Helms moved onto the station on 13 March. However, the Expedition One commander did not relinquish control of his "ship" until the hatches closed for the final time on 18 March.

The first full day of docked operations saw yet another EVA – the 17th devoted to assembly of the rapidly expanding station. Mission specialists Jim Voss and Susan Helms donned their pressure suits for what turned out to be the longest spacewalk ever undertaken. The unprecedented duration was partly explained by the loss of a portable foot restraint early in the spacewalk, which obliged Voss to seek out a spare from the storage location on Unity's outer skin.

The main objective of the spacewalk was to prepare the Pressurised Mating Adapter 3 for repositioning from the Earth-facing berth on Unity to the port side. After disconnecting eight cables and removing an Early Communications

System antenna, the way was clear for former Mir veteran Andy Thomas to move PMA-3 with the Shuttle robotic arm. However, before this could take place, the duo also had to attach a special cradle to Destiny in readiness for a future space station robotic arm.

The following day saw Thomas manoeuvre Leonardo into place on Unity, relying almost solely on images relayed to TV monitors on Discovery's aft flight deck. Once Shepherd connected a power cable between the cargo module and Destiny, the way was clear to start unloading more than 5 tonnes of supplies.

The second spacewalk – this time by Thomas and Paul Richards – involved installing a stowage platform for spare parts and attaching to it a back-up ammonia coolant pump. They also finished several cable connections left over from the first EVA and ascended to the top of the station's huge solar arrays to strengthen the port-side array by latching in place a fourth structural brace.

"Well, Andy, we were on top of the world there for a while," quipped Richards.

Among the items unloaded from Leonardo over the next few days were seven systems racks carrying electronics equipment, communications gear, experiments and medical facilities for Destiny. Of particular significance was the Human Research Facility, the first major science hardware to be installed on the ISS.

Leonardo did not remain empty for long. In place of the newly delivered racks, the astronauts deposited waste, redundant hardware and luggage for the returning Expedition One trio. Meanwhile, commander Wetherbee raised the orbit of the complex to ensure that it avoided the free-floating foot restraint lost during the first EVA.

Towards the end of the week, Expedition Two engineers Voss and Helms got their first real taste of things to come when they spent long hours installing a TV monitoring system in Destiny that would eventually be used by crew using the Space Station Remote Manipulator System.

When the crews awoke on 15 March, they were informed that the joint mission had been extended for a day to allow more time to properly stow everything inside Leonardo. Lead flight director John Shannon explained that ground controllers also wanted more time to analyse the positioning and weight distribution of items inside the module – factors that could affect the Shuttle's performance during re-entry and landing.

Leonardo's unberthing finally took place on 18 March, although even then there was a four-hour delay due to a leaky hose that was used to depressurise the small compartment between Unity and the Italian module. Filled with about 1 tonne of Earthbound baggage, it was gently placed inside Discovery's payload bay and securely latched inside its cradle.

The crew bade each other farewell about one hour before the hatches closed for the last time on this mission. The occasion was marked by the historic hand-over of command from Shepherd to Usachev. "We are on a true space ship now, making her way above any Earthly boundary," said the outgoing commander.

During their 136 days on board the ISS, his crew had welcomed three Space Shuttles and the space station had more than doubled in size and power with the installation of giant solar arrays and the Destiny laboratory module. Despite some minor teething troubles, the ambitious programme was now well underway.

The undocking of Discovery took place over Guyana on the evening of 18 March. A threat from bad weather abated, allowing Discovery to touch down in the Florida darkness three days later. As a precaution against an adverse reaction from the sudden onset of gravity after 141 days of weightlessness, the Expedition One crew rode home strapped into their specially designed reclining seats. After being reunited with their families, they began a long period of medical evaluation and rehabilitation.

Up above, their replacements settled gently into the busy work schedule. Awakening at midnight (Houston time) they began the daily exercise routine, then set about adapting the station toilet for its first female resident and familiarising themselves with their new surroundings. Susan Helms set up her own private sleeping space inside Destiny, in a cavity that would eventually be outfitted with a science experiment rack.

Susan Helms, the first female resident of the ISS, carries out battery maintenance beneath the "floor" of Zarya. (NASA photo STS101-390-025)

Over the next few weeks, the crew and flight control teams in Houston and Moscow had to deal with a series of minor, but irritating, breakdowns and malfunctions. Most significantly, the activation of the station's Ku-Band television, voice and high-speed data communication system was delayed by an apparent pointing error with the dish-shaped antenna. Although normal voice and low-data-rate communication could continue through the S-Band system, the only way to transmit TV images was via the laptop computer-based digital video system. This meant that no experiment data could be sent from the Human Research Facility experiment rack in the Destiny laboratory, while a press interview session on 30 March was limited to audio only.

Two of the air-conditioning systems in Destiny were also causing trouble. The condensate venting system was not working, so no moisture in the cabin air could be condensed. The back-up carbon-dioxide removal unit in Destiny had also broken down, but fortunately, with only three people on board, the Vozdukh system in Zvezda was able to cope.

The station's heavily used treadmill was also showing wear in many of the slats that were designed to prevent vibrations from affecting sensitive experiments. However, the crew was able to set up a new bicycle exercise machine called CEVIS (Cycle Ergometer with Vibration Isolation System) as a substitute.

On a lighter note, the Expedition Two crew starred in the TV broadcast of the Oscars with a pre-recorded introduction involving a life-sized likeness of the host, actor Steve Martin.

"When producers of the Oscars' ceremony approached us, we thought it was an excellent opportunity to expose a global audience to the important work being done by NASA and its international partners in orbit on the International Space Station," said NASA chief Dan Goldin.

The attached Progress craft continued to play an important role in daily activities as the crew continued to transfer equipment and water from the cargo ship. Meanwhile, an oxidiser was transferred under automatic control to the service module's storage tanks, and oxygen was fed into the station to replenish the cabin air. In another first for the programme, commands were sent from the ground via Zvezda's computers to fire the cargo ship's thrusters and adjust the orbit.

The scientific promise of the space station began to be fulfilled as Voss installed the first experiments in the Destiny laboratory. These included three experiments that measured radiation levels in the station – one sponsored by the Japanese – and the first of the laboratory's science racks. As its name suggests, this Human Research Facility was designed to house a variety of experiments related to the physiological, behavioural and chemical changes in human beings caused by spaceflight.

The crew also began their weekly photographic survey of the Earth's surface, beginning with the Parana River basin in South America, and set up a student Earth observation camera, dubbed EarthKam, which sent electronic pictures to an Internet web site on the ground.

Early April saw the Expedition Two team loading the Progress with trash and unneeded items in preparation for its undocking. Following normal procedure, the unmanned ship was eventually undocked from the station on 16 April, leaving the aft port on the Zvezda module free for the relocation of Soyuz TM-31 – the capsule that delivered the Expedition One crew to the station – two days later. This, in turn, provided the necessary clearance for the attachment to Unity of the Raffaello Multi-Purpose Logistics Module during the next Space Shuttle visit.

CANADARM 2 AND CRASHING COMPUTERS

The rapid pace continued when Endeavour and the STS-100 crew lifted off from Florida on 19 April. The multinational seven-man Shuttle crew included Canadian Chris Hadfield, Italian Umberto Guidoni – the first European Space Agency astronaut to visit the ISS – and Russian Yuri Lonchakov. Their mission was to enhance the station's in-built maintenance and assembly capability by installing an advanced robotic arm called Canadarm 2.

Space Shuttle Programme Manager Ron Dittemore was upbeat about the progress so far made in ISS construction. "The launch of Endeavour marks a significant milestone for us in that it completes a quick, safe and successful full turnaround of the Space Shuttle fleet dedicated to assembly of the station in only a few months," he said. "Once Endeavour arrives on this flight, all three Shuttles capable of docking with the station will have done so twice in the past eight months."

Endeavour eventually docked to the PMA-2 adapter on Unity as the craft passed over the southern Pacific at 8:59 a.m. Houston time on Saturday 21 April. With a spacewalk in the offing for the following day, handshakes and bear hugs were postponed until later in the mission. Instead, the Shuttle newcomers simply settled for opening a hatch into PMA-2 to retrieve a battery-powered drill that would be needed during the EVA. Before withdrawing to Endeavour, they left behind some water containers, computers, fresh food and IMAX camera film for the ISS crew.

Sunday saw pilot Jeffrey Ashby take control of the Shuttle's robotic arm to lift a Spacelab pallet, loaded with the Canadarm 2 and an Ultrahigh Frequency (UHF) antenna, onto a cradle outside Destiny. Astronauts Hadfield and Parazynski then floated into the payload bay to install and deploy the antenna –

an important step towards implementing future EVA communications via the ISS and improving station–Shuttle communications.

Stage two was the connection of power and data relay cables between Destiny and the inert Canadarm 2. After removing insulating blankets and verifying that power was reaching the 17-m (58-ft) long arm, the duo set about unfolding the hinged structure, using the pistol-grip tool to secure a series of expandable fasteners that kept the boom rigid. Problems in ensuring that the correct force had been applied to the fasteners meant that the EVA lasted for more than seven hours, but the nineteenth ISS spacewalk was eventually classed as a success. Mission control in Houston brought it to a suitable conclusion by playing the Canadian anthem in honour of the first EVA by a citizen of that country. Later that day, ISS engineers Voss and Helms successfully commanded the first motion of the station's new appendage. The way was now clear for more ambitious manoeuvres with the robotic marvel.

Ten hours of joint operations began on Monday morning as the two crews met for the first time. However, the highlights of the day involved the two robotic arms. From a work station inside Destiny, Susan Helms commanded the newly delivered Canadarm 2 to "walk off" its pallet. The free end then grasped the grapple fixture on Destiny's exterior, ready for its first lifting job the next day. Its more seasoned cousin then swung into action, latching onto the Raffaello cargo module and carrying it across to its berth on Unity. Over the remainder of the joint mission, some 3 tonnes of supplies and two experiment racks would be carried across from the second of the Italian-built multipurpose modules to the station.

With the hatches once more sealed and cabin pressure lowered inside the Shuttle, Parazynski and Hadfield prepared to continue where they had left off two days before. Their first task involved opening a panel in Destiny's exterior and connecting power, computer and video cables to the grapple fixture where the Canadarm 2 was lodged. Once the cables linking it to the pallet were disconnected, the new arm was under the control of the Robotics Work Station inside Destiny.

Not everything went according to plan. When a back-up power circuit failed to respond during a test by Helms, the spacewalkers were obliged to open up a panel and shuffle the cables to enable another connector at the base of the arm to be used. This *tour de force* brought cheers from flight controllers in Houston.

The men's efforts to remove a redundant Early Communications Antenna from Unity also ran into difficulties when a metal cover floated from Hadfield's grasp and wedged itself behind some thermal insulation on a docking port. After two unsuccessful attempts to retrieve the errant item, the men were ordered to move on to more pressing work. Their second excursion eventually concluded after 7 hours 40 minutes.

More serious problems arose on 25 April, when a scheduled operation using the Canadarm 2 to move the 1.3-tonne pallet back to Endeavour's payload bay had to be postponed. Shortly after the Expedition Two crew went to bed, ISS controllers reported that one of the station's three command and control (C&C) computers had crashed. Attempts to transfer data files to the second, and then the third C&C computer, both failed, resulting in a complete breakdown in communication and data transfer between the ISS and the ground. Fortunately, with Endeavour attached, the crew was able to speak to Houston using Shuttle communication links and the other systems on the station continued operating normally.

"Doesn't sound like you're going to get much sleep tonight," quipped Helms when informed of the situation by flight director Mark Ferring.

"No, and I think you can rest assured that anyone who knows anything about a computer is now (at Johnson Space Center) and we're all working hard on it," replied the frustrated Ferring.

The first indication that progress was being made came when the ground successfully sent a command to switch one of the lights inside Destiny on and

James Voss and Susan Helms operate the Canadarm 2 controls during the first robotic "handshake in space" on 28 April 2001. (NASA photo STS100-390-021)

off. However, it was not until the following morning that Helms was able to re-establish links between the station and the ground after conducting a series of troubleshooting routines with the ISS laptop that interfaced with the command computer. This enabled Houston to configure the ISS computers in readiness for further difficulties, and to download data that would help in analysis of the paralysing problem. However, further complications arose when two fault protection computers in Unity suddenly shut down.

Not surprisingly, mission managers decided that the Shuttle's visit would have to be extended for two extra days, though even this would only allow a truncated version of the original programme. The Russian Space Agency responded to the emergency by largely ignoring US requests to postpone the launch of the next Soyuz and its controversial crew member, American millionaire Dennis Tito.

By 27 April, specialists had determined that the hard drive had failed on command and control computer 1. With only one C&C computer operational, a back-up payload computer was called upon as a replacement. However, after an overnight operation to synchronise timers on all of the station's computers was completed successfully, controllers were sufficiently confident to allow Parazynski and Guidoni to go ahead with the retrieval of the Raffaello MPLM.

Although attempts to reload software into one of the two failed C&C computers were unsuccessful, the all-clear was given on Saturday 28 April for calibration of a visual aid for the Canadarm 2 operators, known as the Space Vision System, and the "first ever robotic-to-robotic transfer in space". Despite the lack of computer support, Helms used the Canadarm 2 to lift the redundant Spacelab pallet and hand it over to the Shuttle's remote manipulator, which was under the control of Chris Hadfield.

With the historic two-hour "hand off" now achieved, the one remaining factor preventing Endeavour's departure was the condition of the station's command computers. Fortunately, by 29 April, mission controllers were able to declare that the primary and back-up computers were operating well – although only one had full capabilities. This was sufficient progress for managers to allow the Shuttle to undock from the station later that afternoon.

Less than 15 hours passed before Soyuz TM-32 pulled in at the Earth-facing port on Zarya. The entry into Zvezda of the world's first space tourist, businessman Dennis Tito, with Kazakh commander Talgat Musabayev and Russian flight engineer Yuri Baturin, was relayed live on TV. Tito, who seemed to have overcome a bout of space sickness that struck during his Soyuz flight, smiled happily and announced, "I love space".

However, NASA and its non-Russian partners – which had opposed the commercial arrangement to send an untrained, non-professional to a space station that was still under construction – made every effort to downplay the

The ISS and its new robotic arm as seen from Shuttle Endeavour on 29 April 2001. (NASA photo STS100-710-103)

significance of the event, declaring that future TV coverage would only be available via Russian communication links. NASA boss Dan Goldin also made it clear that there was no way that the amateur astronaut would be free to venture anywhere near the American sector of the station without strict supervision.

Not wanting to upset their Russian colleagues or Tito, but also fully aware of the animosity felt by politicians and space agency officials towards the $20 million trip of a lifetime, Helms and Voss decided to compromise by greeting the space tourist with smiles rather than enthusiastic hugs. Veteran Voss became rather more animated when he and commander Usachev received their surprise Father's Day gifts – talking picture frames with photos and 10-second messages from their daughters, courtesy of US company RadioShack.

For the benefit of the newcomers, Usachev conducted an extensive safety briefing, introducing station systems and evacuation routes, then the custom-fitted seat liners were transferred from the fresh Soyuz to its elderly sister ship in case a hasty departure was required.

The first space tourist, millionaire businessman Dennis Tito. (Digital photo)

Media coverage of the "Taxi" crew's six-day sojourn on the station was inevitably dominated by 60-year-old Tito, although his activities were largely restricted to listening to opera, shooting video and photos through the porthole, helping to prepare the meals and admiring the view as the space station swept around the planet once every 90 minutes. The only problem to which he admitted was a difficulty in swallowing food. His two colleagues had their own programme to worry about, taking pictures of the Earth and conducting three experiments provided by Kazakhstan.

While Tito grabbed the headlines, NASA was at least able to gain some consolation from the fact that a second C&C computer – which had formerly acted as a payload computer – was now up and running. However, the ongoing struggle to reconfigure the systems meant that most of the experimental work and planned tests of the Canadarm 2 had to be put on hold.

The "Taxi" trio eventually set off for home on 6 May. Their televised farewell showed warm hugs between the Russian cosmonauts and Tito, contrasted with cool handshakes from the American crew.

Even after the safe landing in Kazakhstan, the controversy did not go away. Soyuz commander Musabayev told reporters, "Having entered the station, we immediately felt that the US crew members had been instructed to keep their distance from Dennis, and they followed this instruction."

Yuri Baturin, former political aide to President Yeltsin, weighed in with further criticism about the lack of hospitality, including the absence of the traditional bread and salt to greet the newcomers. "Unfortunately, some of Mir's good traditions are not observed on board the ISS," he said. "When visiting crews docked to Mir, the station main crews had always asked the visitors what they would like to have for lunch or dinner, even before the hatches separating the station from Soyuz were opened. When newcomers entered the Russian station, warm food was already waiting for them."

Musabayev was also critical of the lack of privacy aboard the ISS, notably the proximity of the toilet to the dining area. "It's like a big, 100-metre long sausage that can easily be looked through," he complained. "Unlike Mir, which had a complex configuration that provided its individual crew members with the necessary privacy, ISS gives no such privacy at all."

Meanwhile, away from all the controversy, the resident crew settled back into their normal routine, repairing the slats on the treadmill, carrying out their exercise regime and trying to ensure that their command computers were working.

On 9 May, NASA reported, "The current configuration shows C&C 2 as the primary computer, C&C 1 as a fully capable back-up, and C&C 3 available as a standby without the use of a mass storage device. C&C 3's hard drive has exhibited the same potential failure characteristics as that of C&C 1. As a result, the crew built a spare computer this week and is ready to install it as a replacement for C&C 3, if needed."

As if some malevolent spirit was at work on the station, three radiation experiments in Destiny temporarily ceased operating during a data transfer using the Human Research Facility computer, and another experiment known as the Commercial Generic Bioprocessing Apparatus also shut down during commanding from the ground.

Fortunately, things improved as the week progressed, with a successful check-out of the Canadarm 2, the activation of a protein crystal growth experiment, and the downloading to the ground of the first video from a US scientific experiment – an attempt to cultivate plants through an entire life cycle. Meanwhile, the third C&C computer was swapped with the spare that the crew had cobbled together and the water dump system was brought back on line by replacing a blocked filter.

Punctuated by two rest days, the hectic activity continued with medical tests for the crew, sampling of the cabin air, checks on the Russian Kurs rendezvous system, and activation of the Human Research Facility.

Most frustrating was the enforced interruption of a Canadarm 2 trial when a failure occurred in the back-up command system. The problem lay with the so-called Arm Controller Unit that routed computer commands from work stations inside the Destiny laboratory to the new construction crane. The unit had two so-called "strings", or pathways, over which computer commands could travel, but the back-up string failed to move the arm's wrist during one test and its shoulder joint during another. Whether the problem lay with the software or the arm itself, a solution would have to be found before it could be used to lift the Joint Airlock that was due to arrive on the next Shuttle mission.

On 20 May, a Soyuz FG rocket, equipped with improved engines and control systems, lifted off from Baikonur with the next ISS supply ship, Progress M1-6. Inside the unmanned craft were 1.2 tonnes of propellant, oxygen, scientific equipment, food and medical supplies, additional computer spare parts and parcels for the crew. While they awaited the arrival of the fourth Progress vehicle dedicated to station resupply, the trio continued to try out the Canadarm 2 and installed a vibration damping system that was designed to isolate sensitive experiments.

Usachev's expertise on the TORU manual docking system was not required as the Progress pulled in to the rear port on the Zvezda module, and the next few days saw the crew set about unloading the newly delivered cargo and struggling to keep the inventory up-to-date.

Then came another major setback. Continued communication difficulties with the Canadarm's shoulder pitch joint in its back-up mode and a problem with the brakes in the arm's wrist joint forced NASA managers to postpone the planned launch of Atlantis from 14 June until "no earlier than early July". This, in turn, meant that the subsequent launch of Discovery on STS-105 was put back to August to allow more time to evaluate whether the joint would have to be entirely replaced.

The crew took the news of their extended mission, at least publicly, with the nonchalance to be expected of experienced space travellers. "The Shuttle [schedule] always had the potential to slip to the right, so we were just mentally prepared for that," Helms said. "We expected it, and we're going to enjoy the extra time that we have."

Another disappointment came when the Payload Operations Center at NASA's Marshall Space Flight Center decided to abandon the malfunctioning Commercial Generic Bioprocessing Apparatus and return it to Earth in order to investigate what had gone wrong.

June began more promisingly, as the crew settled down to a succession of experimental work, Earth photography, TV interviews and checks on the Orlan pressure suits that would be used during an internal EVA.

The first "spacewalk" by ISS residents without a Shuttle alongside took place on 8 June, when Usachev and Voss moved into the small, spherical transfer compartment on Zvezda. With Susan Helms monitoring activities from Zarya, they removed a hatch at the bottom (Earth-facing part) of the compartment to open it to the vacuum of space.

After lashing the hatch cover to the top of the unpressurised compartment, they replaced it with a docking cone assembly that had been temporarily stowed in the transfer compartment. This cone would be used later in the year for the arrival of the Russian Pirs docking compartment. Although the 19-minute excursion finished ahead of schedule, it was later revealed that the

Russia's Mir space station photographed during the final fly-around by a US Space Shuttle on 8 June 1998. (NASA photo STS091-727-051)

Astronaut Shannon Lucid checks the progress of a wheat growth experiment on the Mir space station. (NASA photo STS79-E-5278)

The Expedition One ISS crew (left to right: Yuri Gidzenko, Sergei Krikalev and William Shepherd) practising water survival in the Black Sea during September 1997. (NASA photo S97-1-689)

Japanese astronaut Takao Doi (right) tests a crane designed for use on the ISS during Shuttle mission STS-87 in December 1997. (NASA photo STS087-341-036)

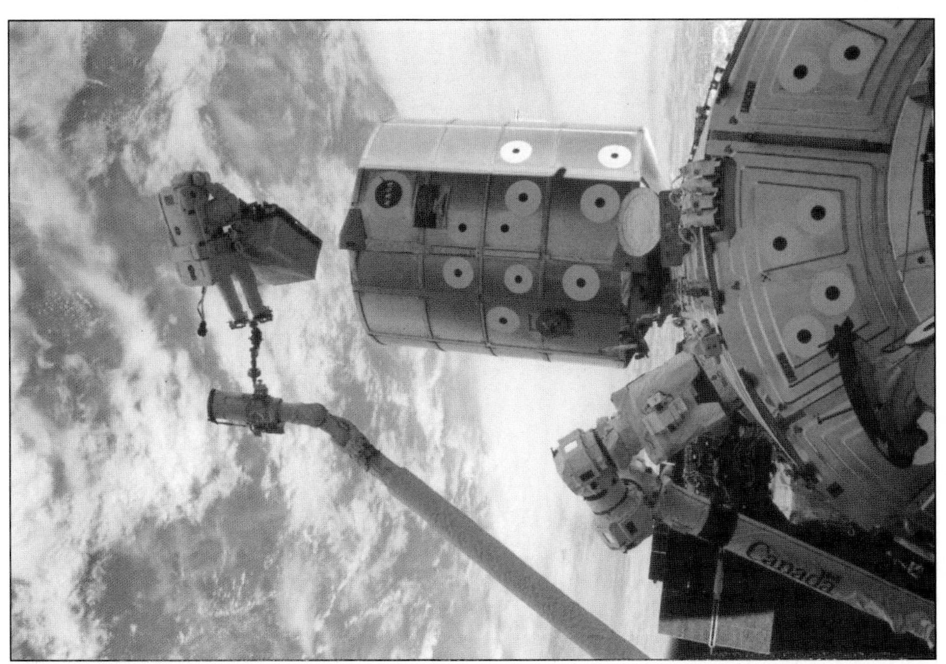

Astronaut Scott Parazynski carries a Direct Current switching unit while anchored to the end of the Shuttle's robotic arm. Also visible are the Raffaello Multi-Purpose Logistics Module (centre) and the station's Canadarm 2. (NASA photo STS100-396-019)

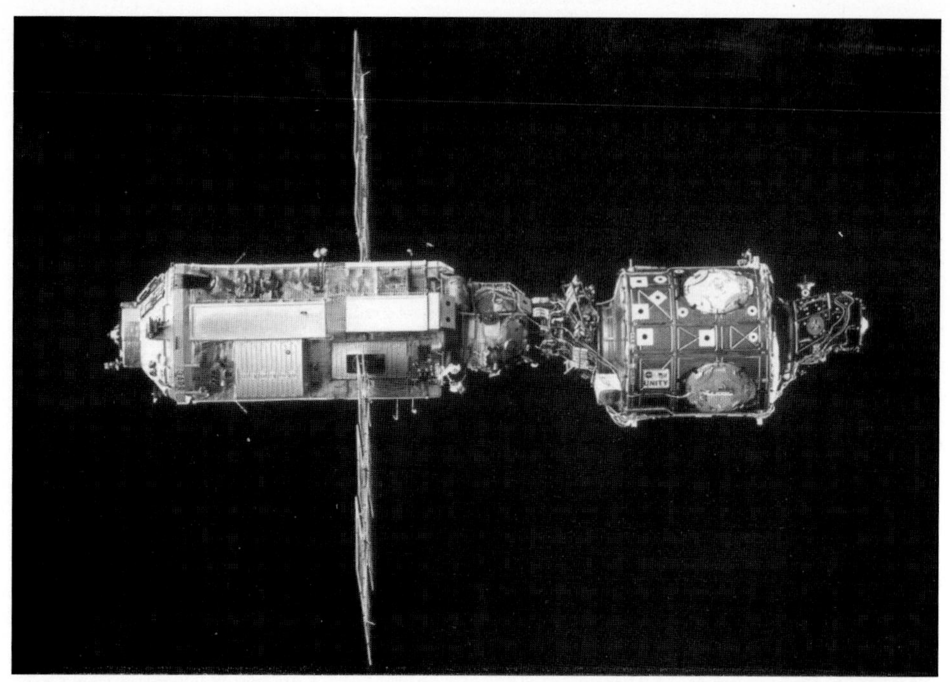

The beginning of the International Space Station. The Russian-built Zarya propulsion module (left) docked with the US Unity node in December 1998. (NASA photo)

The Progress M1-4 supply ship prepares to dock with the Zarya module on 18 November 2000. (NASA photo ISS01-324-002)

Astronaut James Voss handles the newly delivered main boom of the Russian Strela crane (21 May 2000). (NASA photo STS101-716-079)

The first robotic "handshake in space" took place on 28 April 2001 when the station's Canadarm 2 transferred its launch cradle to the robotic arm on Shuttle Endeavour. (NASA photo STS100-347-025)

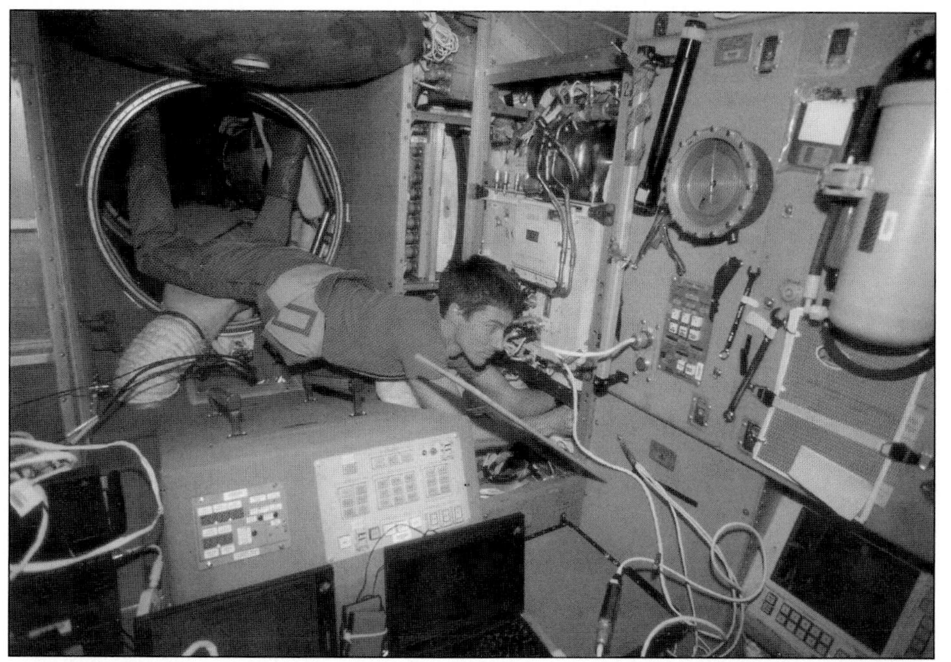

Expedition One flight engineer Sergei Krikalev works in the Zvezda module. (NASA photo ISS01-331-081)

Expedition One commander James Voss performs a task at a work station in the Destiny laboratory as Shuttle visitor Scott Horowitz enters through the hatch from the Unity module. (NASA photo STS105-304-025)

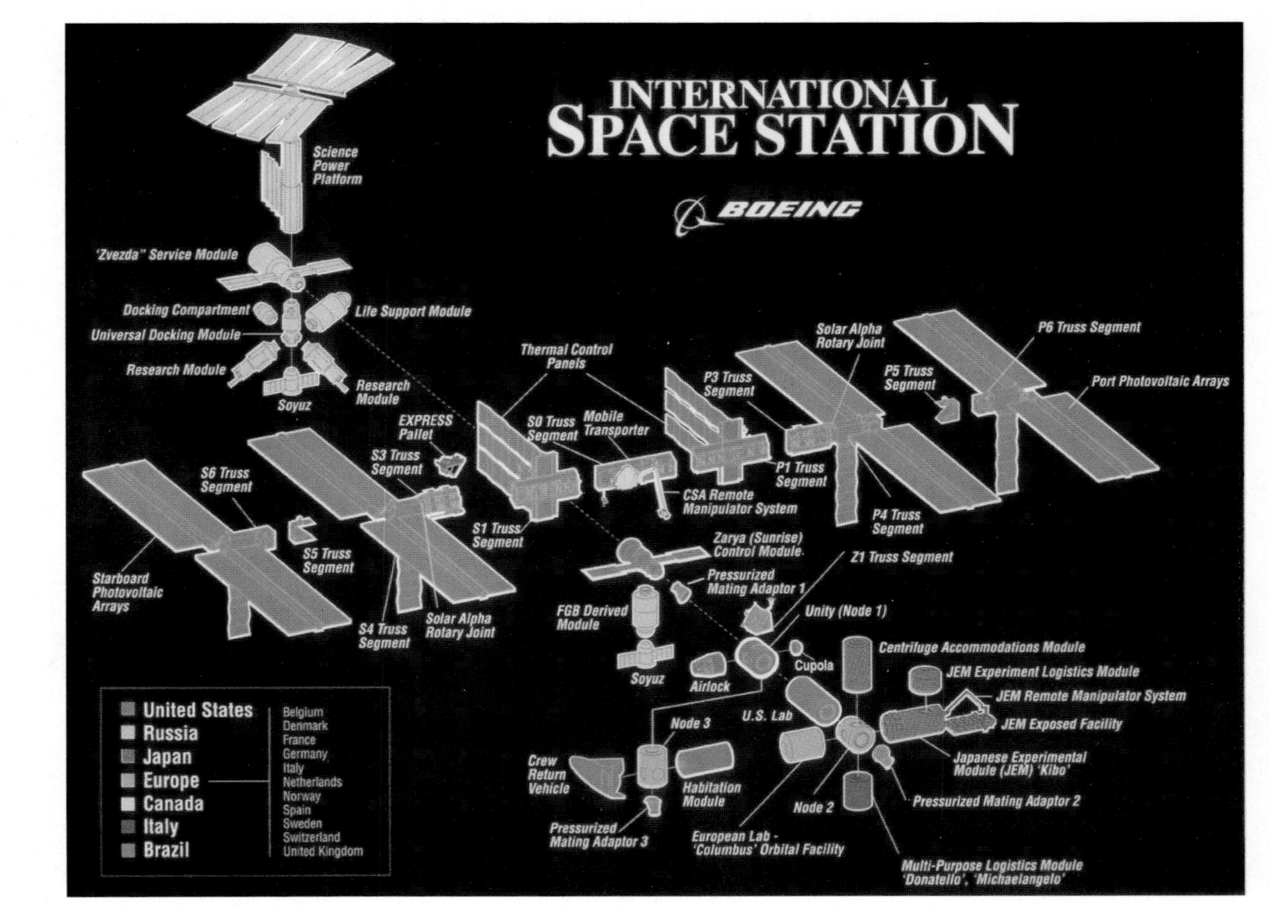

The components of the completed International Space Station, as envisaged in 2001.

venting of the air from the transfer compartment had caused the ISS to drift some 11 degrees in attitude. Normal orientation was rapidly restored.

Although the robotic arm continued to cause concern, the crew approached the benchmark 100 days in space in good spirits. The station was generally in fine shape, further experiments were brought on line, the seedlings in the Advanced Astroculture experiment began to flower, and the Earth photography programme was in full swing. The NASA status report stated, "This week (6-13 June) has been the busiest so far aboard the station for science investigations with more than 25 hours of experiment work budgeted for the crew."

On 13 June, the ISS was turned so that its main axis pointed along the orbital plane instead of the usual alignment perpendicular to the orbit. However, when the Russian control centre attempted to control the station's pitch and yaw by commanding the thrusters on the Progress to fire, the operation had to be aborted because of a high-pressure reading in a manifold. A second attempt with a different manifold was successful, but further manoeuvres were cancelled until the reason for the glitch was properly understood.

Evaluation of the latest software patches sent up to the Canadarm continued to frustrate the crew and ground teams. During a simulation of operations to lift the Joint Airlock and install it on the Unity node, the arm jumped from its grapple fixture and struck the hull of Destiny with a resounding thud. Fortunately, no damage was done to the multimillion dollar science laboratory.

The following week saw a series of intensive discussions over how to deal with the robotic arm problem. Eventually, teams of engineers from NASA, the Canadian Space Agency and its prime robotics contractor, MD Robotics, gave the all-clear to launch Atlantis in July. They concluded that a faulty computer chip was responsible for the communications error between the Canadarm 2's shoulder pitch joint and the arm's main computer commanding unit. Canadian engineers were confident that they could solve the intermittent problem by sending up a software patch that would "tell" the arm to ignore similar erroneous communications.

This certainly seemed to be the case as two dress rehearsals for the Joint Airlock installation on 21 and 28 June went without a hitch. By the time the two American crew members sent their "birthday message" to the 4th of July gala concert on the West Lawn of the US Capitol in Washington, DC, confidence was high all around. Probably the most irritating problem facing the Expedition Two team was a noisy motor on their exercise treadmill, which obliged them to operate it in an unpowered mode.

As the long-awaited launch of Shuttle Atlantis approached, the crew began to shift their daily schedule to prepare for the night-time work schedule of their colleagues on STS-104. They also continued to review the remote arm operations and held their periodic discussions with the psychological support teams. A typical list of housekeeping tasks conducted on 11 July included replacement of the toilet urine receptacle, drying out the Common Cabin Air Assembly, reconfiguring the TV system in Zvezda and transferring data from the scientific experiments.

The tenth ISS Shuttle mission lifted off from Florida on 12 July, carrying a 6-tonne airlock, now known as "Quest", and a highly experienced five-person crew. Their critical task was to attach the Joint Airlock to the station using the new Canadian-built robotic arm, opening the way for station occupants to conduct spacewalks using either US or Russian pressure suits without the presence of the Shuttle.

Commander Steve Lindsey smoothly docked the two giant craft as they passed 385 km (240 miles) above the northeast coast of South America. Some two hours later, the hatches separating the crews were opened. After the initial welcomes, the crews got down to the serious business of safety briefings and a

Canadarm 2 manoeuvres the Quest airlock onto the starboard side of Unity during the first EVA of the STS-104 mission. (NASA photo STS104-E-5068)

review of procedures for the next day's EVA. Both the Shuttle and ISS robotic arms were put through their paces during this dry run, then the hatches were sealed once more to allow a reduction in the Shuttle's cabin pressure.

The all-important spacewalk began with Mike Gernhardt removing the protective covers on the new airlock's berthing mechanism, while Jim Reilly fitted bars that would later be used to attach four high-pressure gas tanks. Heating cables that had warmed Quest during its flight to the station were then disconnected, clearing the way for Helms to grapple the mushroom-shaped structure with the Canadarm 2, lift it from the Shuttle's payload bay and move it to its berth on the right side of Unity.

After the delicate 2½-hour operation was successfully completed, the two spacewalkers set about attaching heating cables to the airlock and fitting foot restraints to assist them during the next EVA. No sooner had their six-hour session ended than the ISS crew moved into the small vestibule in the airlock's "neck" to begin its outfitting.

Expedition Two commander Usachev and Atlantis commander Lindsey celebrated Quest's successful delivery with a ribbon-cutting ceremony inside the new compartment. With the addition of the Joint Airlock, the habitable volume of the space station had increased to about 425 m^3 (15,000 ft^3), more room than a conventional three-bedroomed house.

16 July saw the crews make the first radio calls to the ground from the airlock and two American space suits. However, completing the multiple connections between the airlock and the rest of the ISS proved a problem. About half a litre (just under one pint) of coolant water was spilled as the crews installed the valves that would link Quest to the station's environmental control system.

Then a suspected leak occurred in an airlock ventilation valve, part of the system that circulates air between the airlock and the other ISS modules. Unable to pinpoint the problem, mission specialist Janet Kavandi joined James Voss to complete a temporary fix by fitting a cap over the troublesome valve. These frustrating glitches set the work schedule back about half a day, but Usachev, Lindsey and Shuttle pilot Charlie Hobaugh still found time to speak to reporters.

With the hatches once more shut and the Shuttle's air pressure reduced, Gernhardt and Reilly prepared for their second spacewalk. An unscheduled restart by the station's primary C&C computer caused an unpleasant reminder of past problems, but it was soon rebooted and operating normally. Over the next six and a half hours, the Canadarm 2 lifted the first two gas tanks from the payload bay of Atlantis and into position alongside the airlock. The duo made such good progress that they were given the go-ahead to install a third tank before it was time to call a halt.

James Reilly makes the first spacewalk from the new Quest airlock on 20 July 2001. (NASA photo STS104-E-5237)

On 19 July, Voss, Kavandi and Usachev set about replacing the leaking airlock valve by cannibalising another from the Destiny module. Then, while the ISS commander performed maintenance tasks on the Russian systems, Helms and Voss went on to assist with moving a hatch from the interface between Quest and Unity to a new position between the airlock's two rooms, the Equipment Lock (where space suits would be stored and serviced) and the Crew Lock (the exit to space). Helms also changed out the temporary C&C computer that had been assembled from a payload data computer in Destiny during the computer crisis of several weeks earlier.

The next day, the eight crew gathered inside Quest to answer questions from reporters in America and Russia. Voss reported that he had successfully sealed the slight air leak between Quest's two sections, then Lindsey sent a video tour of the fully outfitted airlock to the ground.

The third and final EVA of the joint mission took place late that evening (Houston time). Not only was it the first excursion from the gleaming new airlock, but it was also the first to be supported primarily from the Space Station Flight Control Room in Houston and the first to make use of a new technique that reduced oxygen pre-breathing time by using vigorous exercise to help purge nitrogen bubbles from the spacewalkers' bloodstreams.

Depressurisation of Quest took 40 minutes rather than the anticipated 7 minutes, but the remainder of the EVA went without a hitch. Once again working in tandem with the two robotic arms, Gernhardt and Reilly removed the fourth (nitrogen) gas tank from the payload bay and attached it to the airlock's exterior. With two oxygen and two nitrogen tanks now fitted, the hardware was in place for future ISS spacewalks. The men had time to spare to move hand-over-hand up the station's solar array truss to take a look at a swivel mechanism that was giving high current readings, but they could see nothing out of the ordinary.

Their four-hour jaunt brought the serious work of the joint mission to a successful close. With the Quest airlock now operational, Phase Two of the ambitious ISS assembly schedule had been completed.

Once the farewell speeches and ceremonies were over, the hatches were closed and the crews went their separate ways. Atlantis departed on 22 July as the spacecraft flew some 385 km (240 miles) above the coast of Newfoundland. Among the items being returned to Earth were seeds from the astroculture experiment and two of the protein crystal growth units.

The Expedition Two crew enjoyed some quiet days as they began returning to a less hectic operational schedule. They were able to sit back while Russian flight controllers fired the thrusters on the Progress, the first in a series of five orbit adjustments in preparation for forthcoming visits.

With less than a month of their five-month stay remaining, the resident trio began to prepare for their return home and the hand-over to their successors. Much of their time was devoted to removals – unpacking and stowing supplies delivered by Atlantis, loading trash into the Progress, and preparing the packages and equipment to be returned with them on their Earthbound trip. The remainder of their schedule was spent on monitoring the science experiments, physical exercises and routine maintenance.

In an attempt to help acclimatise their bodies to the forthcoming onslaught of normal gravity, the crew began to use the Russian Chibis "trousers" which pull the blood towards their legs and force the heart to pump harder. Psychological aspects of their mission were also addressed by filling in a weekly questionnaire that detailed their "interpersonal and cultural relationships".

As the Shuttle that would give them the ride home waited on the pad in Florida, all seemed set fair for their crew's final few days on the station. However, an old gremlin decided to resurface when one of the three command and control computers (C&C 1) experienced a problem reading its hard drive, or Mass Storage Device. Flight controllers attempted to reboot the computer, but with no success. Fortunately, the primary computer (C&C 3) continued to operate normally, so the back-up computer glitch had no impact on station operations. Indeed, after the earlier problems with these three items of hardware, a newly refurbished C&C computer was already scheduled for delivery by the incoming Shuttle.

STS-105

After a day's delay due to poor weather, Shuttle Discovery lifted off from the Cape in the late afternoon of 10 August. On board STS-105 were the Leonardo Multi-Purpose Logistics Module and a crew of seven that included the next long-term occupants of the ISS: Expedition Three commander Frank Culbertson, pilot Vladimir Dezhurov and flight engineer Mikhail Tyurin.

"This flight is representative of many Shuttle missions to come as station assembly and operations enter a new phase," said Space Shuttle Programme Manager Ron Dittemore "Although extremely complex and challenging assembly flights will continue, they'll be interspersed with missions dedicated to changing station crews, experiments and supplies."

Once again, the ISS crew had to adjust its daily schedule to match that of the incoming visitors. The Expedition Two team then watched in anticipation through the window on Destiny as the link-up between the two craft took place without a hitch. Two hours later, the hatches were thrown open. The science module was rather crowded, with three crews vying for room as they embraced, settled down for Usachev's obligatory safety briefing and then completed the welcoming ceremonies.

The next day saw the change in residence between the Expedition Two and Three crews, as the customised seat liners were changed in the Soyuz "lifeboat" and leak checks were completed on the Russian Sokol pressure suits. Meanwhile, mission specialist Pat Forrester moved Leonardo to the downward-facing port on Unity, preparing the way for the unloading of several tonnes of supplies. Several protein crystal growth experiments were among the first items to be transferred and activated, and the onerous task of emptying Leonardo was completed in little over one day.

Astronauts Dan Barry (upper left) and Patrick Forrester during installation of the Early Ammonia Servicer on the ISS main truss. (NASA photo STS105-725-062)

The other priority was the extensive briefing sessions as the incoming residents familiarised themselves with their new home's systems. Usachev and Dezhurov also spent time installing updated software into Zvezda's computers, in preparation for the arrival of a new Russian docking compartment. Discovery maintained control of the complex until the upgrades were operational and the gyros were able to take over.

Joint activities were temporarily terminated on 15 August, when the hatches were closed ahead of a spacewalk by Dan Barry and Forrester. During their 6 hour 16 minute excursion, the duo installed on the main truss a back-up reservoir of ammonia coolant and an experiment to expose 750 samples of different materials to the harsh space environment. The 1,000th day since the launch of Zarya was also marked with a special TV broadcast by the Expedition Three crew.

Further TV broadcasts to Russian and US audiences took place on 17 August, culminating in the official ceremony to pass command of the station to the third residential crew. Then the crews once more went their separate ways to prepare for the second Shuttle EVA – the 26th devoted to ISS assembly.

Head to head. The Expedition Two crew (dark shirts), join the Expedition Three crew

(white shirts) and the STS 105 astronauts (stripes) in the Destiny laboratory. (NASA photo STS105-717-032)

This time, Barry and Forrester spent 5½ hours attaching electrical connectors for the next section of the station's massive truss structure.

The final stages of the joint mission were marked by the return of Leonardo, packed with several tonnes of redundant hardware, trash and personal effects, to the Shuttle's cargo bay. On 20 August, the crews said their final farewells, then sealed the hatches and checked for leaks. Discovery was then backed away for the usual photo opportunity and a fly-around that gave Usachev, Voss and Helms a last view of their former home. Settled into their reclining seats, the trio eventually returned to Earth on 22 August after 167 days in space and 163 days on the ISS.

The Expedition Two crew was reported to be "in excellent shape, readapting to gravity and enjoying life back on Earth". The following day, they were flown back to Houston to continue their debriefing and recovery. It was announced that, rather than remaining in America, Usachev would return to his native Russia after 10 days or so in order to complete his recuperation.

EXPEDITION THREE

Even before the homecoming crew experienced the strange sensation of normal gravity, some significant changes had occurred up above. The launch of a fresh Progress from Baikonur was rapidly followed by the undocking of its elderly predecessor. After a two-day trip, the station's fifth Progress supply craft docked automatically to Zvezda's aft port on 23 August as the ISS passed over central Asia.

Within minutes, hooks and latches were commanded to close between the Progress and Zvezda, forming a hard mate and a tight seal between the two craft. Later in the day, Culbertson, Dezhurov and Tyurin cracked open the hatches and began unloading yet another cargo of supplies and personal effects.

Moving, unpacking and stowing more than a tonne of equipment from the cargo ship took the best part of a week. Among the fresh deliveries was the Volatile Organic Analyser (VOA) that was designed to sample the air inside the ISS on a daily basis, detecting and identifying any possible contaminants.

Other activities included the replacement of a faulty voltage converter unit associated with one of eight power-producing batteries for Zvezda and the continued monitoring of the largely automated scientific experiments, including the first cell culture experiments on the station. One of the crew's less glamorous assignments was to set up equipment for urine samples, as part of an investigation into ways of preventing kidney stones.

In early September, after completing a three-day holiday weekend of light activities that provided time to settle into their new home, the trio settled down to carry out some minor repairs. These included a check of wiring on the treadmill, tightening of a connection in a station air-conditioning system in order to stop a minute leak of freon, and the replacement of a broken videotape recorder in the Destiny laboratory. They also started the first long-term study of human lung function in space, an experiment that would measure the effects of zero gravity on exchange of gases and muscle strength in the lungs.

Even in space, the seasons have an effect – the angle of the Sun relative to the station changes with the time of year – and so, on the morning of 5 September (Houston time), flight controllers assisted the crew to alter the station's orientation slightly in order to ensure that the solar arrays received as much sunlight as possible. The Sun had previously been fairly low to the southern horizon relative to the station, and the complex was oriented so that the arrays pointed towards it. As the Sun rose higher in the sky relative to the station, it was time to move the complex back to an orientation in which the arrays were perpendicular to the station's direction of travel.

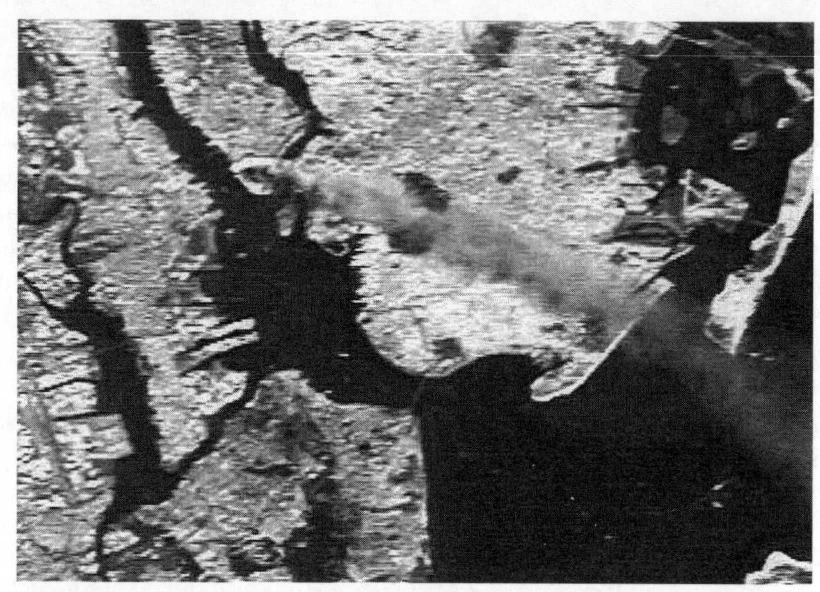

Image from video footage taken by Frank Culbertson showing a plume of smoke over New York on 11 September 2001. (NASA digital photo)

Later that day, Culbertson manoeuvred the station's robotic arm into position to allow its television cameras to focus on a dump of waste water from vents on the Destiny laboratory that was scheduled for later that week. The intention was to record what happened when the 19 litres (5 gallons) of water were dumped overboard in 10 minutes, creating a flurry of floating water crystals. Although such dumps had happened before, engineers were very keen to find out how well the jettisoned water cleared the vicinity of the station and its effects, if any, on the station's outer surface.

Then came the tragic events of 11 September, when three aeroplanes packed with passengers were hijacked and directed into the twin towers of the World Trade Center and the Pentagon. A radio conference between the crew and flight surgeons was abruptly interrupted by the terrible news. Shortly after, the station flew over New York, allowing Culbertson and his Russian colleagues to capture video footage of the plumes of smoke rising from the scenes of carnage.

The following day, NASA TV broadcast the crew's pre-dawn images of the "Big Apple" illuminated by searchlights as well as the more familiar city lights. Culbertson described the scene, "There was a large plume of smoke coming from Lower Manhattan and streaming off to the south. We were too far away to see much in the way of detail, but we could see where in the city it was, and there also was a shroud of smoke over the large part of the city. It was quite a disturbing sight ... and very heartbreaking."

ARRIVAL OF THE PIER

The next major event was the launch of a Russian docking compartment named Pirs, the Russian word for pier, on board a Soyuz rocket on 14 September. Culbertson reported that he was able to see the rocket climbing into orbit as the station passed southwest of the Caspian Sea.

The back-up manual docking system was not required as the propulsion unit and the new module, which comprised an airlock and docking port, duly linked up with the Earthward-facing docking port on Zvezda two days later. The 11-m (36-ft) long spacecraft approached the station from below and behind, then the station's thrusters moved it to the proper orientation for docking. "We really felt that," said Culbertson as Pirs linked up with its new home on Zvezda.

After docking, the crew checked to make sure there was a good seal between the station and the docking compartment, then began to equalise their air pressure before opening the hatch a few hours later. They spent the next two days unloading cargo and supporting equipment from Pirs, then removed the Kurs automated rendezvous equipment so that it could be returned to Earth for reuse on later missions.

Before the docking compartment could become operational, a number of modifications had to be made. The crew had to upgrade and test the station's Russian software to allow control computers on board Zvezda to work with Pirs' systems, install and activate the compartment's caution and warning system, and set up ventilation equipment and lighting. This was followed by activation of its communications and further systems tests.

The Pirs docking compartment and its Progress-based propulsion section approach the ISS. (NASA digital photo)

As well as working in Pirs, the crew continued their scientific work, a mixture of untended experiments and life sciences investigations which required human participation. These included an experiment to study spinal cord reflexes during long-duration spaceflight and two more that measured vibrations on the station. The crew also conducted physical examinations that are carried out periodically during the flight to gauge the effects of weightlessness.

The vibration-measuring sensors, which had recorded the arrival of Pirs, were also in operation when the docking compartment's aft instrumentation and propulsion system was jettisoned on 26 September. Mission controllers in Moscow fired pyrotechnic devices that activated spring pushrods to eject the 6-m (20-ft) long section. The stage moved away at a rate of about 4 m (13 ft) per second until it reached a point far enough away to fire its thrusters without contaminating the station. It then continued ahead of and above the station to a distance of 24 km (15 miles), when its thrusters were commanded to fire in a deorbit manoeuvre, sending it into the atmosphere to burn up upon re-entry.

The crew's satisfaction with Pirs activities was dampened somewhat by an unexpected shutdown of the Russian Elektron oxygen generation unit and an air-conditioning unit in Zvezda. Russian and American flight controllers attempted to determine the causes of the shutdowns and to ensure that there was sufficient back-up oxygen generation capability. Fortunately, in addition to the solid fuel oxygen candles, there was about a week's worth of oxygen already in the station atmosphere, and ample stores of the life-giving gas were available in the gas tanks on the Quest airlock.

Maintenance work completed by the crew in late September included the replacement of 10 smoke detectors in the Zvezda module. They also expanded their scientific investigations into a new area by studying the ability of certain chemicals to impede the formation of kidney stones. Culbertson became the first test subject, swallowing pills that contained either the active compound or a placebo, then collecting urine samples and keeping a record of his fluid and food intake. The crew also continued testing the Active Rack Isolation System through a series of tests of its ability to protect sensitive experiments from vibrations caused by everyday crew activity.

Bringing Pirs online was a major priority, and three spacewalks were scheduled to electrically mate the Docking Compartment to Zvezda and install more equipment on its exterior.

For the first EVA on 8 October – the first external spacewalk staged from the station without the presence of a visiting Space Shuttle, and the 100th spacewalk in Russian spaceflight history – Dezhurov and Tyurin closed the hatch on Zvezda's transfer compartment, then depressurised Pirs for the first time. They then floated through one of two external hatches on Pirs to connect a cable that would send telemetry and data from Russian Orlan spacesuits, then attached handrails, an access ladder, a Strela cargo boom, a docking target and an automated navigational antenna to the Docking Compartment. Watching from inside Zarya, Culbertson monitored their progress throughout the 4 hour 58 minute spacewalk and manoeuvred the Canadarm 2 so that its cameras could provide TV pictures of the cosmonauts in action.

One week later, with the Elektron once more up and running and the station's orbit boosted by the thrusters on the Progress, the way was clear for a second excursion from Pirs – the 28th spacewalk in support of the assembly of the station. This time, the cosmonauts mounted a variety of instruments outside the Zvezda service module in a spacewalk that lasted almost six hours. Moving hand-over-hand to work sites on the rear of Zvezda, they installed three sets of equipment designed to learn more about the effects of exposure to the space environment on various materials. The duo then moved on to a nearby site, where they assembled a small truss structure and attached three suitcase-sized experiment packages provided by NASDA, the Japanese Space Agency.

On their way back to the Pirs hatch, they removed a placard and exposure experiment with the image of the Russian Federation flag, and replaced it with another exposure experiment as part of a commercial agreement.

A few days later, on 19 October, the crew floated into the Soyuz TM-32 spacecraft for the short trip from the Earth-facing port on the Zarya module to the docking port on Pirs. It was the first time that a spacecraft had pulled in at the new Russian module. This brief jaunt freed the vacated port for the imminent arrival of a fresh Soyuz craft with a taxi crew of commander Viktor Afanasyev, rookie flight engineer Konstantin Kozeev and European Space Agency flight engineer Claudie Haigneré.

ANDROMÈDE

Two days after its launch from Baikonur on 21 October, Soyuz TM-33 pulled in at the station for an eight-day stay. The ISS crew watched the stubby spacecraft gradually close on the station and prepared to equalise the cabin pressures and open the hatches to greet their first visitors since they took over the complex back in August. Viewers in North America were able to watch events unfold live on NASA TV as images were relayed through the Russian TV network.

Most of the attention was inevitably grabbed by Claudie Haigneré, the first European woman to enter the ISS. Although she was a member of the European Space Agency's astronaut corps, Haigneré was actually flying under a commercial contract between CNES, the French Space Agency, and the Russian Aviation and Space Agency. During her stay, the taxi team and some

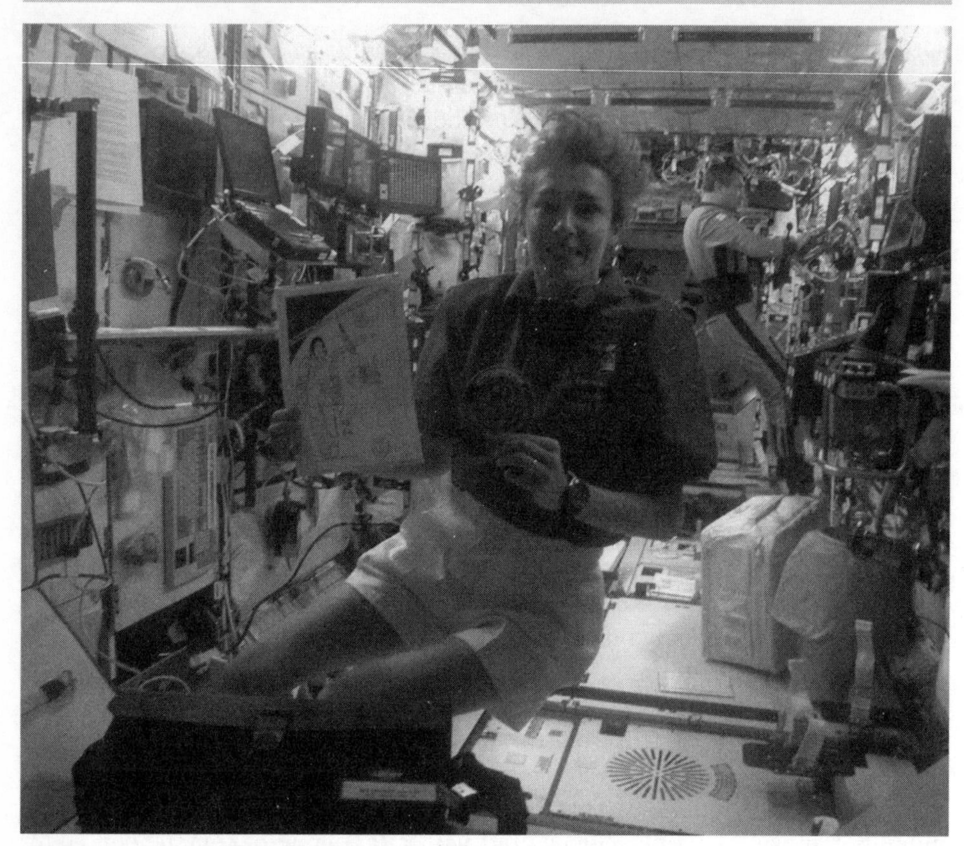

Claudie Haigneré, the first European woman on the ISS, poses for the camera in the Destiny module. (ESA/CNES photo)

members of the Expedition Three crew carried out a programme of experiments exploring life sciences, biology, materials science and Earth observation on behalf of CNES, ESA and DLR (German Aerospace Centre). On 26 October, the crew shot some documentary footage under the commercial Dreamtime contract, using a new high-definition TV camera. Housekeeping chores were also on the menu, including the replacement of the smoke detectors in Zvezda and the testing of the solid fuel oxygen-producing candles that had passed their original expiration date.

The brief occupation of the ISS by six residents came to an end on the evening of 30 October (Houston time) as the taxi crew undocked the elderly Soyuz TM-32 craft from Pirs and headed for the Kazakh steppes. On their own once again, the Expedition Three crew members headed into the final quarter of their four-month mission by continuing their scientific programme with measurements of space radiation aboard the ISS and growing protein crystals. They also prepared for a third spacewalk to complete the outfitting of Pirs.

The supposedly final Expedition Three EVA by Culbertson (who was taking part in his first spacewalk) and Dezhurov was moved from 5 to 8 November, and then 12 November, to give the men more time after the departure of the Soyuz taxi crew. Floating out of Pirs, they began by hooking up seven telemetry cables between the airlock and Zvezda, so completing the installation of the Kurs automated rendezvous system that would be used in future to guide Russian manned vehicles towards the station.

Their other tasks during the 5 hour 4 minute spacewalk included attachment of foot restraints and a survey of a solar array panel on Zvezda which failed to deploy properly following its launch in July 2000. Culbertson and Dezhurov wrapped up the spacewalk by testing the newly installed Russian Strela crane on Pirs that will be used to manoeuvre cosmonauts and cargo around the Russian modules of the ISS during future spacewalks. Using a crank, they manually extended the Strela to its full length of about 9 m (30 ft), then raised and lowered it from an operator's post at its base. Spacewalk choreographer Tyurin operated the Canadarm 2 from inside the station to provide lighting for the duo and television views for flight controllers.

With most of their major tasks out of the way, the crew took a short breather. Later in the week they cleaned, serviced and put away the Orlan spacesuits they had used on the spacewalk, and spent about 20 minutes answering questions from school students in Texas and Kansas.

PROBLEMS WITH PROGRESS

Apart from resuming their busy schedule of science work and packing all of their gear for their return to Earth in early December, the crew members had to get ready for a series of comings and goings by Progress craft. The supply craft attached to Zvezda successfully undocked on 22 November, loaded as usual with several tonnes of rubbish. Four days later, its replacement was launched from Baikonur Cosmodrome in Kazakhstan. However, on this occasion, the clockwork precision of the Russian robotic craft's manoeuvres ran into unforeseen difficulties.

Early indications were that the sixth Progress to visit the ISS successfully docked to the International Space Station on 28 November. However, the Russian control centre did not receive verification that the hooks that hard dock the supply craft to the station had closed and latched.

The station remained in free drift, with attitude control turned off for more than three hours while the Russian flight controllers evaluated the potential problem. With the solar arrays not receiving the full amount of sunlight, power generation was reduced and some non-critical items on board were temporarily switched off to conserve electricity. These included heaters that control condensation on the walls of the American modules, as well as a back-up cabin air assembly and a contaminant monitor in the Quest airlock.

Once the Russians were happy that the Progress was attached securely enough to allow the steering system to operate normally, the station returned to its normal orientation and all equipment was turned back on. However, it was clear that the food, fuel and supplies for the next residents of the orbital outpost could not be unloaded until the docking problem was fixed.

After a joint review by Russian and American flight controllers, it was agreed to carry out an unscheduled spacewalk to clear an apparent obstruction in the docking mechanism at the rear of Zvezda. As a result, the launch of Shuttle Endeavour with the Expedition Four crew was put back five days to 4 December.

In order to prepare for a probable four-hour spacewalk, the ISS crew had to adjust its sleep schedule to match the times during which the complex would be able to communicate with Russian ground sites. In fact, the EVA by Dezhurov and Tyurin – the 30th devoted to ISS assembly and maintenance – took just 2 hours and 46 minutes. After working their way from Pirs to the aft end of Zvezda, Dezhurov used a tool to cut free a rubber seal that was interfering with the docking interface. With the debris removed, Russian flight controllers commanded the Progress's docking probe to retract fully, and a hard mate between the two craft was quickly completed.

HOMECOMING

The way was now clear for the launch of Endeavour. The 12th Shuttle mission to the ISS duly lifted off from launch pad 39-B at the Kennedy Space Center on the afternoon of 5 December. The main objectives of its 11-day mission were to bring home the Expedition Three crew and to deliver the next three residents as well as several tonnes of equipment and food.

Also on board were a US flag that was recovered from the debris of the World Trade Center, a Marine Corps flag that was retrieved from the Pentagon, and an American flag from the Pennsylvania State Capitol, along with dozens of New York City police officer shields and patches, a Fire Department of New York flag and a poster with the photographs of fire-fighters who lost their lives in the 11 September attacks. In addition, 6,000 American flags were being flown as part of the "Flags for Heroes and Families" campaign. They would later be distributed to the families of the victims and survivors of the suicidal hijacks.

During the two-day rendezvous, the Expedition Four team assisted the

Shuttle crew and worked with several secondary scientific investigations. After the usual series of engine burns to close the gap, commander Dominic Gorie assumed manual control of Endeavour's approach when the Shuttle closed to within 1 km (0.6 mile) of the station. He then manoeuvred the orbiter to a point about 180 m (600 ft) directly beneath the ISS before flying a quarter of a circle around the station. Docking took place at the front port as the two craft sailed over England.

At first, the Shuttle's docking ring and the docking mechanism on the ISS did not align properly, but after allowing the two craft to dampen their relative motion, the vehicles were successfully hard mated. Once the leak checks were completed, the hatches between the spacecraft were opened for the 10 crew members to greet each other. The newcomers were only the second party of visitors the Expedition Three crew had seen in almost four months aloft.

In the first full day of joint operations, pilot Mark Kelly used the Shuttle's robotic arm to hoist the Italian-built Raffaello logistics module from Endeavour's payload bay and attach it to a Earth-facing station berthing port on Unity. Meanwhile, the formal exchange of space station crews took place as Culbertson, Dezhurov and Tyurin exchanged their customised seat liners in the Russian Soyuz spacecraft with the incoming Expedition Four crew – commander Yuri Onufrienko, and flight engineers Carl Walz and Dan Bursch.

Later that day, the three commanders on board – Shuttle commander Dom Gorie, Expedition Three commander Frank Culbertson, and Expedition Four commander Yuri Onufrienko – along with Shuttle pilot Kelly, gave interviews to several US TV stations from the Destiny laboratory. Also sandwiched into the busy week of personal briefings, handing over station responsibilities and unloading 3 tonnes of supplies from the Raffaello MPLM was a special time to pay tribute to the heroes and victims of September's terrorist attacks.

Against the backdrop of the Stars and Stripes, Culbertson declared, "All of us were affected by that day greatly. . . . We will continue, I hope, to set a good example of how people can accomplish incredible things when they have the right goals." He added: "We will continue to think of how we can improve peace around the world and how we can improve knowledge, and hopefully that will bring people together."

The other main highlight of the week was a spacewalk by STS-108 astronauts Linda Godwin and Dan Tani to install thermal blankets over two "beta gimbal assemblies" – mechanisms that enable the large US solar arrays to rotate and track the Sun. It was hoped that, by adding this insulation, temperature fluctuations would be reduced and occasional unexpected power surges in the motors would be eradicated.

After the hatches were closed between the Shuttle and the station, leaving only the Expedition Four crew on board the ISS, the cabin pressure on

Endeavour was lowered slightly. The next day, the spacewalkers exited from Endeavour's airlock and hitched a ride from the Shuttle's robotic arm about half-way up the space station's P6 truss. Carrying the blankets, they then had to inch their way, hand-over-hand, to the worksite at the top of the truss, some 25 m (80 ft) above Endeavour's cargo bay.

With the loose insulation secured, the duo attempted to secure a brace on the starboard station array, but they were unable to close the latch. On their way down from their lofty perch, they stopped at a stowage bin to retrieve a cover that had been removed from a station antenna during an earlier flight. Godwin and Tani also performed a "get-ahead" task, positioning two switches on the station's exterior ready for installation on a future spacewalk.

The successful 4 hour 12 minute excursion completed a record year for spacewalks, with 18 EVAs conducted: 12 from the Shuttle and six from the station. After the spacewalk was completed, the hatches between Endeavour and the station were reopened.

The following day, the moving of supplies to and from the Raffaello logistics module was again interrupted by a NASA TV broadcast as part of a nationwide tribute to those who lost their lives on 11 September. At the exact time of the attack three months earlier, the US and Russian national anthems were played in the Space Shuttle and ISS mission control centres at Johnson Space Center and aboard the two spacecraft.

Speaking to the crews from Houston, STS-108 Lead Flight Director Wayne Hale said, "In stark contrast to the international cooperation and unity in our effort to take mankind literally to the stars, we are reminded of our loss and sorrow due to the acts of violence and terror in an unprecedented attack on freedom, democracy and civilisation itself. More than 3,000 people perished this day three months ago, including more than 200 citizens from countries that are family members of the International Space Station programme – Canada, Italy, France, Germany, Japan and Russia."

The three commanders also shared their personal thoughts and played a special, pre-recorded taped tribute from the other astronauts in orbit. Later in the day, Yuri Onufrienko, Vladimir Dezhurov and Mikhail Tyurin spoke to Russian media in the mission control centre near Moscow, followed by a similar interview session with the American crew members and media.

These ceremonial duties and unexpected maintenance tasks on the station, including refurbishment work on a treadmill and replacement of a failed compressor in one of the air conditioners in Zvezda, contributed to the crew's falling behind schedule, so mission managers decided to extend Endeavour's flight an extra day. Mission control later reported that the rebuilding of the treadmill was completed several hours ahead of schedule, with the old treadmill parts loaded into Raffaello to be refurbished on Earth and, eventually, reused.

A third reboost of the station by Endeavour's thrusters was completed on 12 December, raising the station's orbit by a total of almost 14 km (9 miles) to an average altitude of about 385 km (241 miles). The next day, the hectic schedule had begun to ease, so the crew was allowed to sleep in for an extra hour. The wake-up song on Endeavour was "Here Comes the Sun", in memory of former Beatle George Harrison, who had recently died of cancer. On 13 December the crews gathered in Destiny for the official ceremony that marked the change of command from Culbertson to Onufrienko after an occupation lasting almost four months.

With unpacking of Raffaello and the middeck of the Shuttle completed, the crew had moved well over 2 tonnes of supplies and experiments to the station. This load included more than 386 kg (850 lb) of food, 455 kg (1,000 lb) of clothing and other crew provisions, 136 kg (300 lb) of experiments and associated equipment, 364 kg (800 lb) of space walking gear, and 273 kg (600 lb) of medical equipment.

The empty space in the logistics module was filled with more than 2 tonnes of trash, dirty laundry, experiment results, and redundant hardware, including the customised Soyuz spacesuits and seat liners. On Friday 14 December, pilot Mark Kelly used the Shuttle's robotic arm to detach the fully loaded Raffaello from Unity and lift it back into Endeavour's payload bay for the trip home. Inside the station, Dezhurov and Onufrienko worked together to replace the faulty compressor on Zvezda's air conditioner.

The only minor blot on the horizon was the failure of one of the shuttle's primary navigation units. The inertial measurement unit was immediately taken off line and its place was taken by a third IMU, but mission controllers insisted that this would have no impact on Endeavour's mission.

After a final night of preparations and farewells, the two crews went their separate ways. On the morning of 15 December, before the homeward journey of the Expedition Three crew could begin, the station's orbit had to be altered in order to ensure that plenty of leeway was left between the giant structure and a spent Russian rocket upper stage from the 1970s that was predicted to pass within 5 km (3 miles) of it. The Shuttle's thrusters were fired for about 30 minutes to raise the station's altitude by just over 1 km ($\frac{3}{4}$ mile).

This caused a slight change to the original departure plan. After undocking from the station, Endeavour flew only half of a circle around the complex instead of a 360-degree inspection. On reaching a safe distance from the station, the orbiter's engines were fired to initiate its departure from the station's vicinity. Endeavour's crew returned home to a heroes' welcome two days later, having clocked up 129 days of space time.

Another landmark in the history of the world's largest, most expensive construction project had been reached as the station's fourth resident crew settled down for a five-month haul. Onufrienko, Bursch and Walz were to remain on board the complex until May 2002. On their menu were the installation of the next truss section and up to eight spacewalks, as well as visits by Shuttle Atlantis and a Soyuz taxi crew. Meanwhile, down below, the future of their cosmic home was once more at the mercy of the politicians and accountants.

Chapter Seven

Life on the Space Station

Compared to the cramped, Spartan capsules used by the first spacefarers, the International Space Station may be regarded as a luxury home, outfitted with almost all the modern conveniences that we tend to take for granted. And yet ... how many of us would exchange our lives here on Earth to spend four to five months encased inside half a dozen metal cylinders with two foreigners who barely understand anything we say?

When completed, the pressurised living accommodation inside the numerous modules of the ISS will have a combined area equivalent to a sizeable house. This is very spacious in comparison with all previous space vehicles, including Mir, but, when occupied by six or seven crew plus all of their essential supplies, life support systems and scientific equipment, the ISS will be far from roomy. So what is it like being there for months at a time and how does one prepare for the ultimate out-of-this world experience?

TRAINING

It takes a certain kind of person to apply to become an astronaut (or cosmonaut). Today, the exacting medical requirements have been considerably relaxed since the days when the space programmes of East and West were only open to young military pilots with the highest levels of physical and mental toughness. Indeed, it is now quite common for astronauts to be in their fifties, and US senator John Glenn broke all records by entering space for a second time in 1998 at the grand old age of 77.

Nevertheless, even today's space travellers, who experience much lower stresses during lift-off and re-entry in the Shuttle, are expected to pass an intensive physical examination. There are also a number of other qualifications that anyone wanting to visit the International Space Station - apart from a wealthy private citizen who can afford to fly as a tourist - must fulfil.

The commander and pilot of a Shuttle (and the commander of a Soyuz) are invariably military pilots with considerable experience in flying high performance jet aircraft. They must also have a degree in engineering, science or mathematics. Other crew, designated as mission specialists, require an advanced degree in the same academic areas, as well as professional experience in engineering or some other space-related occupation.

Beating hundreds of other astronaut applicants in the competition for selection is only the first stage in the long process of training that is required before a candidate can set foot in a spacecraft and head for the space station.

Each of the agencies participating in the ISS programme has its own training schedule on its own premises, but, in order to ensure that all space station astronauts have similar experience and knowledge when they come together for advanced training, each agency has agreed to follow the same guidelines.

Basic training for newly selected candidates lasts for a year to 18 months. It typically covers about 230 subjects and requires some 1,600 hours of instruction. The course also includes a general introduction to space engineering and spacecraft systems; teaching of general survival techniques, scuba diving, language skills (generally learning English or Russian), physical conditioning and scientific training. Much of this programme is taught by lectures, though some hands-on practice of the more practical skills is included. Also included are spells in the low-pressure altitude chamber and EVA (spacewalk) training, while US Shuttle pilots and mission specialists learn to fly T-38 high-performance jet aircraft.

On completion of basic training, the successful candidates are certified as career astronauts. They then undergo a year or more of advanced training. This provides astronauts with the knowledge and skills related to the operation of the space station elements, payloads, transport vehicles and related communication with the ground.

Upon successful completion of advanced training, an astronaut is eligible for assignment to a mission. Based on their duty assignments, each will then receive increment-specific training, which provides an international station crew (and a back-up crew if applicable), with the knowledge and skills required to perform the particular onboard tasks for their mission. In order to develop good teamwork, the crew interact as much as possible during approximately one and a half years of joint training.

There are three types of increment-specific training: individual tasks, team training with several crew members working on multiple tasks; and multi-segment training, which takes place in the last six months and involves combining payload and systems operations for the entire station. The crew then works as a team, sometimes with ground controllers via integrated simulations.

Although most of the necessary training takes place on the ground before lift-off, it is also quite common for crews to continue their learning process in orbit. Training continues on board the space station, sometimes with the aid of special onboard simulators, in order to maintain proficiency in skills and knowledge gained on the ground, as well as to provide more specific training in medical procedures, emergency drills, maintenance and robot operations.

SIMULATING SPACE

One of the most important requirements for space station crews is familiarisation with the unworldly experience of weightlessness. This can be achieved in a number of ways.

Anyone who has been on a rollercoaster will be familiar with the sensation of "leaving their stomach behind" as the car tops each crest on the track. Similarly, astronauts can experience 25 to 30 seconds of zero gravity as they go "over the top" during parabolic training flights. In the United States, training flights take place in the KC-135A aircraft, which is large enough to allow teams of up to four astronauts and their experiments to literally fly inside the plane's cargo area. During each two- to three-hour flight over the Gulf of Mexico, the aircraft will complete about 30 parabolas, a prolonged series of steep climbs and descents that led early veterans to dub it the "vomit comet". Other aircraft are used by different agencies, but the principle is the same.

Although parabolic flights may assist in planning and rehearsing spacewalk activities, most EVA training takes place in huge water tanks at Johnson Space Center in Houston and at Star City near Moscow. Astronauts don pressure suits that are adjusted to provide a simulated weightless space environment, and, surrounded by divers who can respond in an emergency, the trainee can practise EVA techniques on full-sized replicas of the space station modules. This enables the crew and managers to assess better the possible difficulties of using the hand- and foot-holds as well as workload limitations. Practice makes perfect is the motto, so each six-hour spacewalk will typically be rehearsed 10 times in the giant pool.

A similar Neutral Buoyancy Simulator at Marshall Space Center was used during the early stages of the Freedom and ISS programmes to enable astronauts to practise space station construction techniques and for tests of the

Canadarm 2, but this tank is no longer in operation and is now designated a National Historic Landmark.

Water also plays a leading role in training for emergency splashdowns. Shuttle and ISS crews use the Neutral Buoyancy Laboratory in Houston to practise emergency evacuation from an orbiter that has ditched in the ocean. However, since the only crew rescue vehicle on the space station is currently a Soyuz spacecraft, all ISS occupants also have to learn how to fly a Soyuz during re-entry and how to deal with an unforeseen emergency return in the Russian capsule.

Included in the training schedule is some time in the Black Sea, learning how to deal with the possibility of a Soyuz spacecraft landing in water. Astronaut Michael Foale, who performed similar training for the Shuttle–Mir programme, described the procedure. "Jumping out is the key to the exercise," he wrote. "The capsule does not float level, and there is great danger of the first person rocking the capsule, so that water comes in through the top and sinks the others. We were told to simply fall, and not push off in any way with our legs."

ISS crews also have to undergo winter survival training near Star City to ensure that they can survive a landing in the frozen wastes of Siberia. This includes lighting a fire, building a shelter or igloo and setting off flares to attract search planes.

The ISS Expedition One crew undergoes winter survival training near Star City on March 1998. William Shepherd exits a mock up of a Soyuz capsule, watched by Sergei Krikalev. Also in the picture is the third crew member, Yuri Gidzenko. (NASA photo S98-04116)

Other, less strenuous methods are also employed to familiarise crews with what they can expect to encounter on their way to and from the station or when they occupy the orbital base. Mock-ups of the Shuttle flight and mid-decks, of the payload bay and space station modules are available to prepare future crew members for the essentials of life on a spacecraft, such as the layout of key systems, routine housekeeping and maintenance, waste management, communications, and EVAs.

Crews are also taught how to make a hasty exit from a Shuttle by climbing through an overhead window on the aft flight deck of a mock-up known as the crew compartment trainer and then lowering themselves to the ground with the aid of a device known as the sky-genie.

With recent advances in electronics, computer-aided instruction and flight simulation are now commonplace. Virtual reality headsets, which allow astronauts to "experience" spacewalks and experiment activities, are also being introduced.

CLAUDIE'S TRAINING DIARY

French astronaut Claudie Haigneré, who had previously spent two weeks on Mir, was the first European woman to live on the International Space Station. The French government paid the Russian Aerospace Agency some $20 million to send her on a 10-day "Taxi" flight to deliver a fresh Soyuz spacecraft to the ISS and return home in the old craft. Soyuz TM-33 blasted off from Baikonur on 21 October 2001.

Since January 2001, Haigneré had been based at Star City outside Moscow, in training as flight engineer for October's Andromède mission to the International Space Station. The first few months of Claudie's training were mainly theoretical. About 200 hours were devoted to Soyuz systems, with which she was already familiar from her previous trip to Mir. She also spent a relatively brief 150 hours familiarising herself with the Russian section of the ISS, since there would be no station maintenance assignments on her short-duration mission.

Her diary entry for 14 September 2001 read:

> I have already been through four training periods at Star City, so this time the work was much easier. I've been able to concentrate mainly on technical systems and orbital dynamics: getting into orbit, manoeuvring, approach, docking, de-orbiting and descent.
>
> Many of the Russian ISS systems are derived from the old Mir station. Two years ago, I was back-up for my husband Jean-Pierre Haigneré's long-duration flight on Mir, so I've already had some thorough training there. I was able to concentrate on new ISS equipment, especially the computer systems.

A series of exams marked the end of theory and the beginning of practical work, in which one-to-one classroom lessons were interspersed with sessions aboard Star City's Soyuz and ISS simulators. From June onward, the Andromède team began to train together as a crew. With Viktor Afanasyev as commander and Konstantin Kozeev as second engineer, Claudie spent a week at the CNES (French Space Agency) facilities in Toulouse. There, they familiarised themselves with the onboard science packages they would oversee during their mission.

Then it was back to Star City for some more hard work. The crew began with 15 four-hour sessions in the Soyuz simulator, practising flight manoeuvres and learning the best response to 'non-nominal situations' – a phrase that covers everything from a communications glitch or a sluggish attitude-control thruster to a full-blown emergency.

They spent another 15 sessions in the docking simulator, practised manual re-entry and then moved on to mock-ups of Zarya and Zvezda, the Russian ISS modules. In between sometimes gruelling simulator work, they received more training on their science payload from French and Russian instructors.

"Right from the beginning, we've also had straightforward physical training two or three times a week," said Claudie. "There are also medical examinations every three months. The first was when we started training, the second to confirm the crew choice and the final one will be a few weeks before launch."

At the end of August the Andromède crew flew to Houston to NASA's Johnson Space Center, where they spent a week in mock-ups and simulators learning their way around the American section of the ISS.

The Andromède "Taxi" crew in a Soyuz simulator at Star City. Left to right: Konstantin Kozeev, Viktor Afanasyev and Claudie Haigneré. (CNES/ESA photo 1220-35)

"Theory is never enough," said Claudie. "For real confidence, you have to see and touch. Even though we will work almost exclusively in the Russian section, we will have the run of the whole station and we have to know our way around. It was also good to get to know the ground controllers and the CapComs, the people we'll be talking to from orbit. And to meet astronauts from America, Europe, Canada and Japan: it brings home just how international the project is."

The Houston visit offered another bonus: a reunion with the ESA astronauts who were training there. "Scattered between Russia, Germany, the Netherlands and the USA, we don't often have a chance to get together. Even though we're all in the European Corps of Astronauts, our lives are very different," she said. Claudie and her colleagues were also able to spend some time with the space station's Expedition Two crew, who had returned only a few days before. "They were all in great form. They had a wonderful present for us, too: a preview of the wide-screen, three-dimensional IMAX pictures they had taken. It was almost as if we were already on board. On the way back I stopped off in Paris to pick up my little girl Carla from her grandparents. We had a wonderful reunion after our long separation, even more so since by chance her father was in Paris, too. Then it was back to Russia and for Carla, back to kindergarten at Star City. She's delighted to be with her friends and to speak Russian again."

And the training programme? "We're now in the last stages of preparation. It's quite intense; so that we are ready for the last week in September, we have an exam practically every day. For the Soyuz, there are tests on manual piloting for approach, docking and descent; for the whole crew, exams on every part of the flight and aboard the station; an exam on a typical day's work. And medical exams, too. After a series of tests in September, the GMK – the Russian medical commission – will give us our final flight clearance. We've just been trying out our spacesuits for size and checking for leaks. Everything is ready."

THE DAILY ROUTINE

On arrival at the space station, the first challenge facing the fresh crew is to adjust to the unfamiliar absence of gravity and the idiosyncrasies of their new home. No matter how realistic the mock-up on the ground may have been, the reality will inevitably be very different, with equipment and manuals scattered in places known (hopefully) only to the existing residents. The five days or so allocated to crew exchange, when advice and first-hand knowledge can be passed between crews, are an essential part of this familiarisation process.

Once the previous crew has departed, the residents have to settle down to complete the daily programme provided by mission control. Fortunately, the days of astronauts struggling to survive the unknown hazards of spaceflight have largely disappeared, but other problems have arisen to take their place. For today's space travellers, each day is much like another, with large chunks of time devoted to household chores, maintenance and scientific experiments, punctuated only by meal times, exercise and brief periods of rest and recreation.

As the first expedition got under way, a typical day for the crew began with a wake-up call at around midnight Houston time (9 a.m. Moscow time). After the first communication session with Moscow control centre, the crew would have breakfast and then participate in a second "comm pass". The rest of the morning would be taken up with sessions on the exercise machines, maintenance and installation of equipment, or loading and unloading supplies or hardware from a Progress cargo ship.

After a short lunch break, during which further discussions with the ground sometimes occurred, it was back to the mixture as before, punctuated by short ham radio sessions and occasional interviews with the media. By 16:30 it was time for another round of exercise, followed by dinner and an evening's viewing of Hollywood blockbusters. Favourites of the Expedition One crew included DVD movies of *LA Confidential*, *Apocalypse Now* and *Pulp Fiction*. The sleep period began around at 3:30 p.m. Houston time (00:30 a.m. Moscow time).

This daily routine has changed in later missions, particularly as scientific experiments have been added to the work schedule (see Table 7.1). This schedule is also shifted several hours when a Shuttle is due to arrive.

The normal work schedule for Expedition crews is based upon five-day working weeks with weekends "free" to spend time on relaxation and recreation, including opportunities to speak to families and friends.

Table 7.1 Expedition Three Timeline Review for Friday 16 November 2001

GMT	Crew* activity
06:00–06:10	Morning inspection
06:10–06:40	Personal hygiene (post-sleep)
06:40–06:50	HEMATOCRIT: measurement of hematocrit value
06:50–07:40	CDR BREAKFAST
06:50–07:00	PLT
07:00–07:10	FE-1 HEMATOCRIT: measurement of hematocrit value
07:10–07:35	PLT, FE-1 Prep for SPRUT experiment
07:35–07:50	FE-1 assist, PLT SPRUT: experiment
07:40–07:55	CDR REFLEX-N: equipment set up
07:50–08:05	PLT assist, FE-1 SPRUT: experiment

07:55–08:10	CDR REFLEX-N: set up and activation of PC
08:05–08:15	SPRUT: concluding ops
08:15–08:45	
09:00–09:10	
	FE-1, PLT BREAKFAST
08:25–08:45	
09:00–09:10	CDR Work prep
08:45–09:00	Daily planning conference
09:10–09:40	FE-1 REFLEX-N: FE-1 subject
09:10–09:30	CDR Daily status check of US payloads
09:10–09:20	SSAS deactivation
09:20–09:40	PLT collecting FMK monitors
09:30–09:55	CDR SAMS filter cleaning
09:40–10:40	FE-1, PLT Dismantling SSC network. Conference with ground specialist (S-Band)
09:55–10:25	SAMS ICU: drawer 1 relocation
10:25–10:40	CDR SAMS ICU activation
10:40–11:10	PLT REFLEX-N: PLT subject
10:40–12:10	FE-1 Physical exercise (TVIS + RED)
10:40–11:10	SSC router relocation
11:10–12:10	CDR Physical exercise (RED)
11:10–12:10	PLT Physical exercise (cycle)
12:10–12:48	CDR
12:10–13:10	FE-1, PLT LUNCH
12:48–12:53	Prep for ISS ham radio session
12:53–13:03	CDR ISS ham radio session
13:13–13:35	ISS3/ISS4 crew conference (S-Band)
13:35–14:35	UF-1 timeline review
14:35–14:55	UF-1 timeline A/G tagup (S-Band)
14:55–15:25	UF-1 timeline review
15:25–15:55	CDR Maintenance of ???
15:40–16:40	FE-1 Physical exercise (cycle)
15:40–17:10	PLT Physical exercise (TVIS+RED)
16:00–16:30	CDR REFLEX-N: CDR subject
16:30–16:45	CDR REFLEX-N: equipment stowage
16:40–17:15	FE-1 Delta file downlink prep
16:45–18:15	CDR Physical exercise (TVIS)
17:15–18:15	FE-1, PLT Connecting SmartSwitch Router (???). Conference with ground specialist (S-Band)

18:15–18:30	Review of plan for upcoming day
18:30–18:55	Report prep
18:55–19:10	Daily planning conference
19:10–19:30	Report prep
19:30–20:00	DINNER
20:00–20:30	Daily food ration prep
20:30–21:30	Personal hygiene (pre-sleep)
21:30–06:00	SLEEP

* PLT = Pilot; CDR = Commander; FE-1 = Flight engineer

UNEXPECTED CHALLENGES

As on Mir, one of the main headaches during the early stages of ISS assembly has been keeping track of the tonnes of hardware and supplies that were being delivered and stowed on the station. The Expedition One crew found that the station's computerised inventory system often didn't work at all. Many hours were spent chasing after "lost" equipment or bar-coded packages suspended in cargo nets. The log entry for 2 January 2001 exemplifies the frustrations that could occur.

> One of the bags stowed is a non-standard white bag with a paper tag on it "ECOMM HARDWARE". No bar code. No serial number on the bag, but there is a NASA part number. Shep puts new bar code on. Marks the bag on all sides with this number. Inside the bag [is] a power cable. Cable has a label, but no bar code. Search for the bag's part number in the database – not found. Search for the cable in the database to find some more clues as to what this bag is. About six like cables turn up. Serial number for this cable says it is located in ... Definitely not the same bag Shep is holding. Pull up the history of changes, and it has been moved, moved again and deleted on the same day, and then duplicated and added back to the database, all done by the ground, and all, we think, talking about the "wrong" bag and generating data that at best is confusing.

Unexpected leakages and breakdowns are also particularly frustrating, since they throw the routine out of synch and may even threaten the entire mission. The worst example so far was the simultaneous breakdown of the station's three command and control computers, which resulted in a complete disruption of communications and data transfer between the ISS and the ground. Fortunately, Shuttle Endeavour was docked with the station at the time, so the crew was able to speak to Houston using Shuttle communication links and the other systems on the station continued operating normally.

The problems of storing and keeping track of supplies are illustrated by this view of a packed Zarya module. Astronaut Michael Lopez-Alegria surveys the scene. (NASA photo STS092-372-016)

Even minor leaks of air, water and sometimes toxic fluids cannot be ignored. Liquid droplets floating around the station have to be dealt with very quickly since they could short out its electrical systems and even threaten the health of the crew. For example, in July 2001, mopping up a small water leak that occurred when the astronauts were linking the new Quest airlock to the station's Moderate Temperature Loop put the crews about half a day behind schedule. At the same time, they had to install a temporary cap to deal with a leaky air valve in an Intermodule Ventilation Assembly in the station's Unity node.

On such a large, complex structure, minor breakdowns and malfunctions inevitably occur. Although the crew can call upon expert advice from down below, there are no specialist mechanics or technicians on board, so the crew have to assume many roles for which they are not well prepared.

One of the main complaints by ISS residents has been the inadequate pre-flight testing that has left them saddled with hardware that does not work or items that do not fulfil the tasks for which they were designed. Ironically, the so-called "crew squawk" software, which was supposed to report systems that had broken down, was itself faulty. "We would like to start documenting anomalies and we believe the squawk tool will be a good way to do this if we can ever get it to behave," wrote Shepherd. "We would like to squawk the crew squawk for starters."

Under pressure to perform and meet deadlines, while also under continuous scrutiny from the ground, even the mildest mannered astronaut can snap from time to time. At one point the Expedition One commander was ordered to use American Kapton tape instead of the more user-friendly Russian version. Shepherd demanded to know the reason. When told that it was because the Russian tape left a residue, he expressed his feelings in no uncertain terms. "That's no big deal as long as it works. Grey tape would have taken five minutes, and using Kapton meant the task took 40 minutes. We need to get a handle on the anal-retentive engineering approach to everything."

AIR SUPPLY

Life in orbit would be impossible without the provision of a carefully controlled environment. With the space station surrounded by the vacuum of space, it is clearly essential to maintain the cabin air at the correct temperature, pressure, composition and humidity.

On the ISS, the pressure and composition of the atmosphere is similar to that on Earth – about 80% nitrogen and 20% oxygen, with no more than 0.4% carbon dioxide. Separate, automated life support systems in the Russian and US segments are used to produce this essentially Earth-like environment, although the more primitive passive methods used since the early days of human spaceflight are available as back-up.

Of all the consumables on board the ISS, oxygen is probably the most essential. An astronaut will die from asphyxiation if deprived of oxygen for only a few minutes. Even more serious is the threat of catastrophic depressurisation, when all of the air escapes into space, possibly as the result of a high-speed impact with a meteorite, a piece of space debris or even a collision with another spacecraft (as happened on Mir).

Even under normal conditions, oxygen is removed at a rapid rate as astronauts breathe – typically 0.84 kg (nearly 2 lb) per person is consumed each day. Up to half a kilogram per day may also be lost as the result of leakage, either through tiny holes in the hull or as the result of opening airlocks. This means that a crew of three will use almost a quarter a tonne of oxygen over a three-month period of occupation.

In order to keep the station's atmosphere breathable, levels of nitrogen and oxygen have to be monitored continually and topped up from high-pressure gas tanks. Additional supplies of air are delivered on a regular basis by visiting Shuttles or the unmanned Progress, but this is expensive and time-consuming.

The primary system for producing oxygen on the ISS is the Russian Elektron generator installed in the Zvezda module. The Elektron, which has been tried and tested for many years on board Russian stations, uses waste water from condensation, urine, etc., to produce hydrogen and oxygen – a process known as electrolysis. The waste hydrogen gas, which is potentially explosive, is dumped overboard. However, the power-hungry system was not brought on line as the primary system for oxygen generation on board the station for several weeks after the arrival of the Expedition One crew. Officials decided to wait until after the installation of US-developed solar arrays by the STS-97 Shuttle crew, after which the Elektron could run continuously.

If the Elektron breaks down or there is a power shortage – as happened frequently on Mir – the crew is obliged to resort to a less sophisticated back-up system, the Russian Solid Fuel Oxygen Generators (SFOG). These canisters generate oxygen during combustion of a chemical known as lithium perchlorate. On at least two occasions these "candles" malfunctioned, resulting in life-threatening fires on board Mir, but the Russians insist that the modified version is now safe for use on the ISS. Prior to activation of the Elektron on the International Space Station, the Expedition One crew relied on these expendable canisters to maintain the proper level of oxygen in the modules. One canister was able to supply enough oxygen for one person per day.

Oxygen and nitrogen gas stored in high-pressure tanks located outside the Quest airlock can also be distributed through pipes to the American part of the station. Empty oxygen tanks can be recharged from a visiting Shuttle or exchanged for full tanks.

Oxygen masks are also available in each module for use in emergencies or during sudden depressurisation, but their operational lifetime is no more than a few hours.

The other key constituent of the cabin air is carbon dioxide. Simply by exhaling, astronauts alter the nature of the atmosphere. If CO_2 levels are allowed to rise, astronauts may begin to suffer from headaches, drowsiness and lack of concentration. Higher concentrations may be life threatening. In order to prevent the build-up of this suffocating gas, ways have to be found to remove it from the cabin air and to prevent it collecting in pockets of still air.

One of the oldest and simplest methods of absorbing carbon dioxide is to use lithium hydroxide in refillable or disposable canisters through which air is pumped. The carbon dioxide combines with the chemical in the canister, allowing uncontaminated air to pass back into the cabin. Each canister lasts only a few days and then has to be replaced.

However, the primary CO_2 removal method on the ISS during its early assembly phase is the Russian Vozdukh system in Zvezda. This regenerative air-scrubbing system removes carbon dioxide from the module and vents it overboard. A troublesome vent valve in a similar system fitted inside the American Destiny module caused some (metaphorical) headaches for the Expedition Two crew.

Even with carbon dioxide removal systems in operation, pockets of exhaled gas can accumulate around an astronaut in weightless conditions. Fans and ventilation tubes leading between modules play a vital role in preventing the accumulation of invisible clouds of carbon dioxide and in circulating oxygen to all parts of the space station (as well as helping to cool electronic equipment). In order to avoid waking with a severe headache from the localised build-up of carbon dioxide, a space-wise astronaut will prefer to locate his sleeping bag near a fan.

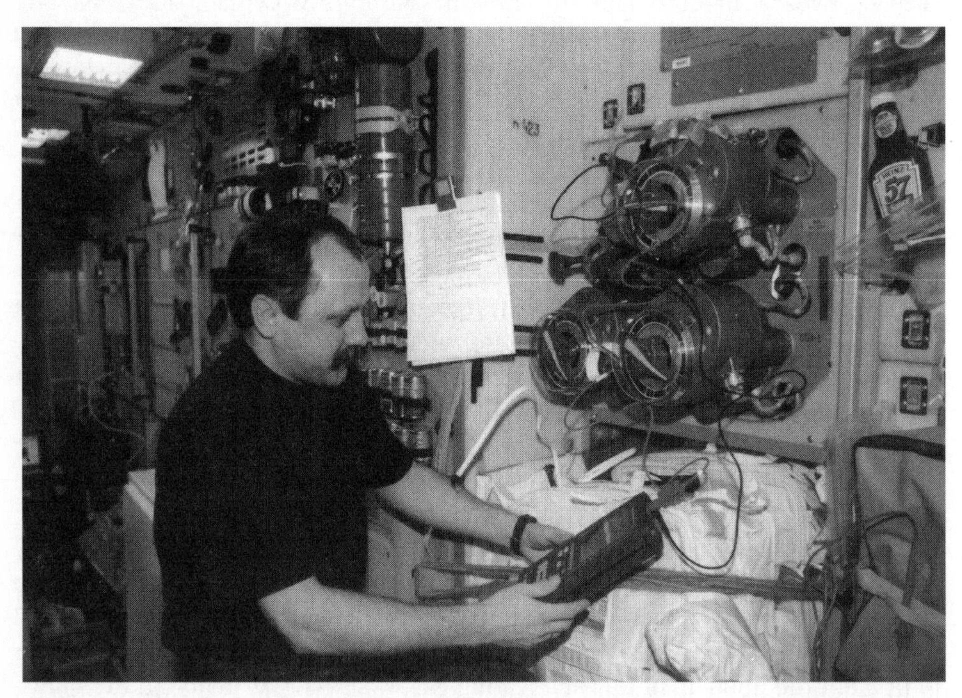

Expedition Two commander Yuri Usachev testing the Vozdukh air-purification system in Zvezda. (NASA photo ISS002-E-6111)

To maintain the comfort level, temperature is maintained at between 18 and 24 °C, with humidity levels of 30–70%. Humidity is controlled by a system that sucks in air through filters, cools the air and removes any moisture. However, this system is likely to cease operation unless it is serviced and emptied periodically. For example, on 8 November 2000, the ISS air conditioner shut itself down due to an excess amount of water in the condensate collection system. The unit was soon reactivated and returned to normal operation. One month later, an entire air-conditioning unit failed and had to be replaced with a spare brought up in a Progress craft.

A major upgrade of the station's life support was due to be provided by the American Habitation Module and the Italian-built Node 3, but these have been delayed for at least two years.

NO SMOKE WITHOUT FIRE

In an enclosed environment, it is essential to remove even small amounts of contaminants that might be harmful to health (see Case Study 7.1). The Russian Zvezda module and the Destiny laboratory contain advanced air-processing and filtering systems that are an integral part of the space station's Cabin Air Revitalisation Subsystem. A second contaminant control system should be fitted in Node 3, if it is eventually launched.

These units are designed to ensure that the levels of airborne contaminants in the US modules are much lower than in a modern office building. They can extract over 200 trace chemical contaminants generated by the astronauts or by material off-gassing in the space station atmosphere. Gases such as carbon monoxide, hydrogen, methane and ammonia are also removed by passing the air over charcoal beds and catalytic oxidisers. The filters and the oxidiser are replaced periodically as the adsorption materials and catalyst become expended.

Early detection of potentially harmful spills or leaks is essential so that crew members can immediately take action to remedy the situation. In order to avoid reliance on the astronauts' unreliable sense of smell, NASA scientists are developing a small electronic nose - about the size of a large paperback - that will be able to measure changes in humidity and monitor the most common airborne contaminants to be found on the Space Station.

"Space crews are very, very busy," said Amy Ryan, principal investigator for E-Nose at the Jet Propulsion Laboratory (JPL) in California. "Anything we can do to automate their tasks and keep the space habitat safe is highly desirable. Now we need to further develop E-Nose's capability to detect various odours and differentiate between those that signify danger and those that do not. Our

Case Study 7.1 The Case of the Mysterious Sickness

The first minor crisis with air quality on the ISS came in May 1999. Several crew members of the Space Shuttle Discovery suffered from headaches, burning and itching eyes, flushed faces and, in one case, nausea and vomiting. The symptoms, which were particularly noticeable towards the end of the eight-day visit, seemed most acute during activities on board the Russian Zarya module, though they were also present to a lesser extent in the Unity node.

Unfortunately, afraid that revelations about "collective" sickness could harm their future career prospects and determined to preserve their personal privacy, the crew omitted to admit their predicament during the flight. The truth only surfaced two weeks after their return to Earth, far too late to allow firm conclusions to be drawn.

One possible clue was the crew's report of an odour in Zarya that was "musty or like a solvent". Outgassing from the large amount of Velcro attached to Zarya's walls was suggested as a possible cause.

The crew also indicated that the problems generally occurred when two or more astronauts were working in close proximity – for example, behind Zarya's wall panels and when the panel doors were left open. Some of the team said their symptoms resembled those experienced during carbon dioxide awareness training, indicating that their difficulties were related to a build-up of exhaled gas. However, this must have been quite localised since the Shuttle's CO_2 scrubbing system was working properly and analysis of air samples in the ISS showed no elevated levels of either carbon dioxide or carbon monoxide during the mission.

The crew attributed these symptoms to poor air quality caused by inadequate ventilation and high carbon dioxide levels. Perhaps the best guess involves a disruption of air circulation between the Shuttle and the ISS. The motion of air around the Zarya depends on free flow behind the wall panels. If this flow was prevented by open panel doors and crew operations, then the air in Zarya would not have moved through the node and into Discovery.

---

current efforts are directed towards improving the sensitivity of the E-Nose, expanding the compounds we can detect from 12 to 24, and making the unit even smaller."

A major application of the E-Nose that JPL scientists are pursuing is the detection of a fire before the blaze erupts. The problem is that fires can smoulder in closed areas, such as insulation in panelling or around wires, for some time before flames actually appear. With early detection, the fire can be extinguished safely before much damage occurs, but this relies upon the telltale smoke being circulated around the cabin and picked up by a smoke sensor.

At present, each module in the Russian segment is fitted with a dozen fire detectors, while the US modules have two area smoke detectors and other sensors in each experiment rack. When they detect particles or smoke in the air, they activate a warning light and audible alarm. Temperature and humidity control equipment in the area of the fire is then shut off to reduce oxygen flow to the site that is affected. However, actually putting out a fire that has become established in the weightless environment of space is far from easy. Despite the best efforts of the Mir crew and the utilisation of three fire extinguishers, an oxygen canister fire in 1997 continued out of control for 14 minutes before it burned itself out.

Manually operated, portable fire extinguishers are available in each module and can be used to surround the flames with a "cloud" of carbon dioxide (US design) or liquid/foam (on the Russian segment) in order to starve it of oxygen.

Of course, even if the fire is successfully extinguished, a significant amount of valuable oxygen will have been consumed during the combustion process. In addition, unpleasant, possibly toxic, fumes will be generated and spread around the cabin. After the near-disastrous conflagration in 1997, black smoke filled the Mir station, forcing Linenger and his five colleagues to resort to wearing face masks for several days.

In an extreme emergency, the affected area of the station may have to be evacuated, then depressurised, in order to extinguish the fire and remove the smoke or other toxic contaminants.

WALKING IN SPACE

Around 1,400 hours of spacewalks (officially known as EVAs or Extravehicular Activities) will be required during the eight years it will take to complete ISS construction. This total far exceeds the overall astronaut time spent in the vacuum of space over the previous four decades of human space exploration. Many more extravehicular excursions will probably be needed during the 10–15 year operational lifetime of the station for repairs, maintenance and deployment or collection of exterior-mounted experiments.

Two different pressure suits and approaches to spacewalking have evolved during the past four decades, and both are employed on the ISS. The US Extravehicular Mobility Unit (EMU) is based on the successful Shuttle pressure suit, but it has been modified to meet the unusual requirements of space station operations. Enhancements include easily replaceable internal parts; reusable carbon dioxide removal cartridges; metal sizing rings that allow in-flight suit adjustments to fit different crew members; new gloves with enhanced dexterity; a new radio with more channels to allow up to five people to talk at one time; warmth enhancements such as fingertip heaters and a cooling system shut-off; and new helmet-mounted flood and spot lights.

Each spacesuit is stored in orbit and certified for up to 25 spacewalks before it must be returned to Earth for refurbishment. It can be adjusted in flight to fit different astronauts and is easily cleaned and refurbished between each excursion. Further modifications have been necessary to allow for the fact that assembly work on the station is carried out in much colder temperatures than most Space Shuttle EVAs. Unlike the Shuttle, the station cannot be turned to provide the most optimum lighting conditions to moderate temperatures during a spacewalk.

In contrast, the Russian Orlan M pressure suit has been specifically designed for use during repairs and other space station activities. Although it is outwardly similar to the US suit, there are a number of significant differences, the most obvious of which is the method of entry.

The US pressure suit has two sections, a hard upper torso and a lower torso, which connect at the waist. The astronaut first inserts his or her legs into the trousers, then pushes the arms and upper body into the upper section. Although a spacewalker can put the suit on unaided, the normal procedure is for a colleague to assist, whereas cosmonauts simply float into the Orlan through a back door which also houses the life support system. This makes it quicker and easier to put on single-handed.

The necessity for multiple layers to protect their wearers against the vacuum of space, extreme temperatures, ultraviolet radiation and micrometeorites/space debris makes both suits extremely bulky. Each weighs more than 100 kg (220 lb) on Earth and can operate outside the space station for more than seven hours. The EMU is slightly heavier because it includes more arm and waist bearings that ease mobility. The American version will also cater for a much larger variety of sizes, but the different modular pieces must be available in orbit to provide the perfect fit. Gloves are essentially custom-made. Although it accommodates a smaller range of body sizes, hidden pulleys and cables on the Orlan suit allow a much greater degree of adjustment.

The Orlan operates at a slightly higher air pressure than the EMU: 5.8 lb/in^2 compared with 4.3 lb/in^2. This higher pressure means that, in order to avoid decompression sickness (commonly known as "the bends"), cosmonauts need to breathe pure oxygen for only about 30 minutes prior to an EVA and the cabin air pressure can remain equivalent to that at sea level (1,013 mbar or 14.7 lb/in^2). One drawback of the higher pressure is a reduction in suit flexibility, particularly of the gloves.

In contrast, astronauts would typically have to spend some four hours breathing oxygen prior to an EVA if the Shuttle cabin pressure was maintained at normal levels. In order to reduce the in-suit pre-breathing time for

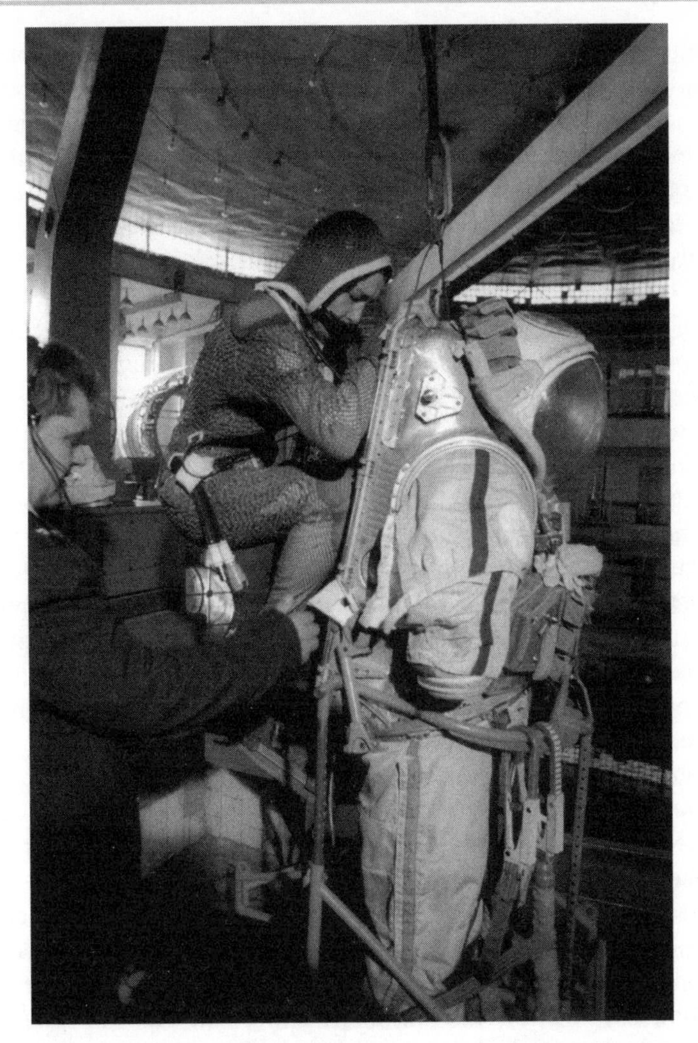

Claudie Haigneré climbs into an Orlan EVA pressure suit during training at Star City. (ESA photo)

astronauts to 40 minutes, the Shuttle's cabin pressure has to be lowered to about 10 lb/in^2.

Since the air pressures inside the Shuttle and the ISS are so different during EVAs, the hatches between a docked orbiter and the US part of the station had to be sealed before any spacewalks could take place during the early stages of ISS construction. During the first 19 months of ISS activity, all station-related spacewalks were conducted from the Shuttle, since there was no airlock system on the ISS that was suitable for US pressure suits.

The first without the presence of a Shuttle took place on 8 June 2001, when

Usachev and Voss entered the small, spherical transfer compartment at the forward end of the Zvezda Service Module. During the 19-minute internal spacewalk, they removed a hatch at the bottom (Earth-facing part) of the compartment to open it to the vacuum of space and replaced it with a docking cone assembly that would be required for the arrival of the Russian Pirs docking compartment.

However, it was the delivery of the Quest airlock in July 2001 that made a significant difference to EVA activities on the space station. This module opened the way for ISS spacewalks to be conducted using US or Russian spacesuits, without the presence of the Shuttle. It also meant that the hatches between a visiting Space Shuttle and the space station could remain open, instead of closed, during station EVAs.

Quest consists of two sections, a "crew lock" that is used to exit the station and an "equipment lock" used for storing suits and other gear. The equipment lock also enables overnight "campouts" by the crew, during which the internal air pressure is lowered while the rest of the station remains at the normal sea level atmospheric pressure. The night spent at 703 mbar (10.2 lb/in^2) in the airlock purges nitrogen from the spacewalkers' bodies and prevents decompression sickness when they go to the 296 mbar (4.3 lb/in^2) pure oxygen atmosphere of a spacesuit. This technique shortens the pure oxygen pre-breathing time to only minutes for the crew.

The airlock works by pumping most of the air it contains back into the station. The remaining gas in the airlock is vented overboard before the hatch is opened. Nitrogen and oxygen tanks on the exterior of the airlock are used to repressurise the module and resupply space suits.

The first demonstration of the ISS crew's new independence came on 21 July 2001 (see Case Study 7.2). Not only was it the first spacewalk to use the Joint Airlock, but it was also the first EVA to be supported primarily from the Space Station Flight Control Room in Houston, and the first time that a new pre-breathing technique had been employed. The astronauts were helped to purge nitrogen gas from their bloodstreams by taking part in vigorous exercise prior to the EVA. The only blot to mar this landmark occasion was a slower than expected depressurisation of the airlock – about 40 minutes instead of the anticipated 7 minutes. Although both Russia and America have amassed considerable EVA experience over the years, and no one has been injured or killed during an extravehicular excursion, the dangers are ever present and precautions are essential.

One small, but nevertheless significant, risk for spacewalkers is the possibility of becoming untethered and detached from the station. In order to prevent the possibility of an astronaut drifting helplessly in space and eventually expiring from lack of oxygen, NASA has designed special "jet

Case Study 7.2 A Walk Outside the Space Station

Astronaut Jim Reilly recounted his experiences after a series of spacewalks from the ISS in July 2001.

When you start your training, you have to start thinking about the different ways you need to manoeuvre to get to the different positions. There are certain [moves] that you don't want to try, because you'll be working against the suit.

You learn to bring yourself to a stop, and then make yourself motionless, without any momentum remaining. Then, you can do whatever needs to be done.

As we were getting ready for our first EVA, Yuri Usachev, who was commander of Expedition Two, came by. We were chatting about suit use, and things to watch out for. He just wanted to make sure that we were aware that it's very, very dangerous outside. And it is. We were always conscious of that. But he said one thing at the end: "In spite of all that, when you're working, just make sure you take a couple of seconds to just look up every once in a while, and look around."

That was the best piece of advice he gave me, because every once in a while, just for 10 seconds, I'd stop and look around, and see what part of the planet I was over, and look at the horizon ...

One of my favourite memories is hanging out with just one hand on the space station, and then swinging out so I could look across the Earth. The atmosphere is really transparent, so you can see a lot of detail on the surface below. One time when I had a chance to hang out on the bottom of the station, the sunset was coming. I left my lights off so I could watch the Sun go down. And as it went down, the stars started popping out. Of course they don't twinkle. They're all different sizes, and even different colours, in space. ... At night you can see lightning flashes from thunderstorms on the surface down below. You can get this blue light flashing, and in this case I was able to see it flashing off the bottom of the station. And as I was watching all this, we flew through the edges of the aurora, kind of green and white curtains as we flew past. It was pretty spectacular.

powered" backpacks that allow a free-floating crew member to fly back to the station in an emergency. The SAFER (Simplified Aid for Extravehicular Activity Rescue) backpack relies on bursts of nitrogen gas from thrusters to propel the wearer forward, sideways, backwards or even to turn somersaults.

First used in September 1994, the SAFER unit has since become standard issue. The only significant problem – a failed valve that prevented the jets from firing during the STS-86 mission – was solved by a redesign and checked out during the first Shuttle–ISS docking mission in December 1998 (Table 7.2).

Table 7.2 ISS Extravehicular Activities

	Date	Participants	Duration	Main tasks
1	8 Dec 1998	Ross, Newman	7 h 21 min	Connect cables between Zarya and Unity and install handrails.
2	10 Dec 1998	Ross, Newman	7 h 2 min	Fit two early communication antennae on Unity and an external video cable. Release stuck backup rendezvous antenna on Zarya.
3	12 Dec 1998	Ross, Newman	6 h 59 min	Disconnect umbilicals used during the docking with Zarya, install a handrail, photograph the station, free a jammed rendezvous antenna and test the SAFER emergency backpacks.
4	30 May 1999	Jernigan, Barry	7 h 55 min	Transfer and install two cranes, foot restraints and tool bags on exterior of Unity pressurised mating adapters.
5	21 May 2000	Voss (James), Williams (Jeff)	6 h 44 min	Secure US crane, install Strela crane, replace faulty antenna.
6	10 Sep 2000	Lu, Malenchenko	6 h 14 min	Complete connections between Zarya and Zvezda.
7	15 Oct 2000	McArthur, Chiao	6 h 28 min	Connect umbilicals on Z1 truss, relocate two antennas, install tool box.
8	16 Oct 2000	Wisoff, Lopez-Alegria	7 h 7 min	Assist in attachment of PMA-3 to Unity.
9	17 Oct 2000	McArthur, Chiao	6 h 48 min	Complete cable connections on Z1 truss and PMA-3.
10	18 Oct 2000	Wisoff, Lopez-Alegria	6 h 56 min	Test SAFER backpack, prepare for future assembly.
11	3 Dec 2000	Tanner, Noriega	7 h 33 m	Install P6 solar array structure.
12	4 Dec 2000	Tanner, Noriega	6 h 37 min	Connect cables and coolant lines on new solar array truss, prepare PMA-2 for relocation.
13	7 Dec 2000	Tanner, Noriega	5 h 10 min	Tension slack solar blanket.
14	10 Feb 2001	Jones (Thomas), Curbeam	7 h 34 min	Assist in temporary relocation of PMA-2 on truss, and docking of Destiny.
15	12 Feb 2001	Jones (Thomas), Curbeam	6 h 50 min	Assist in relocation of PMA-2 to Destiny.
16	14 Feb 2001	Jones (Thomas), Curbeam	5 h 25 min	Attach spare antenna, inspect solar array connections, practise EVA rescue.

17	10 Mar 2001	Helms, Voss (James)	8 h 56 min	Prepare PMA-3 on Unity for relocation, remove antenna.
18	12 Mar 2001	Richards, Thomas (Andrew)	6 h 21 min	Install stowage platform, attach spare coolant pump, connect cables.
19	22 Apr 2001	Hadfield, Parazynski	7 h 10 min	Install UHF antenna and Canadarm 2
20	24 Apr 2001	Hadfield, Parazynski	7 h 40 min	Complete connections on Canadarm 2, remove antenna from Unity.
21	8 Jun 2001	Usachev, Voss (James)	19 min	Install docking cone on Zvezda for Pirs (Docking Compartment 1).
22	14 July 2001	Gernhardt, Reilly	5 h 59 min	Assist installation of Quest airlock
23	18 July 2001	Gernhardt, Reilly	6 h 29 min	Attach nitrogen and oxygen tanks to Quest airlock.
24	21 July 2001	Gernhardt, Reilly	4 h 2 min	Attach nitrogen tank to Quest. The first spacewalk from the new airlock.
25	16 Aug 2001	Barry, Forrester	6 h 16 min	Install Early Ammonia Servicer and MISSE experiment.
26	18 Aug 2001	Barry, Forrester	5 h 29 min	Fit two heater cables and install handrails on Destiny.
27	8 Oct 2001	Dezhurov, Tyurin	4 h 58 min	Attach telemetry and data cables between Pirs and Zvezda, install handrails, an access ladder, a cargo crane, a docking target and an automated navigational antenna.
28	15 Oct 2001	Dezhurov, Tyurin	5 h 52 min	Install three sets of experiments designed to learn more about the space environment on the outside of Zvezda.
29	12 Nov 2001	Dezhurov, Culbertson	5 h 4 min	Connect seven telemetry cables between Pirs and Zvezda to complete the installation of the Kurs rendezvous system. Test the Strela cargo crane on Pirs.
30	3 Dec 2001	Dezhurov, Tyurin	2 h 46 min	Remove debris from Progress docking interface with Zvezda.
31	10 Dec 2001	Godwin, Tani	4 h 12 min	Install thermal blankets over US solar array drive mechanisms.
32	14 Jan 2002	Onufrienko, Walz	6 h 3 min	Relocate Strela crane from PMA-1 to Pirs. Install amateur radio antenna on Zvezda.

Total ISS EVA time to 14 January 2002: 196 hours and 19 minutes.

DINING OUT

Dining in space has progressed a long way since the days of cold paste in aluminium tubes and delicacies such as bacon and beef bites. However, just as most of us no longer grow our own food and instead make periodic excursions to the supermarket, so astronauts rely on regular deliveries of food (and other consumables) by vehicles such as the American Space Shuttle, with its Pressurised Logistics Module, or an automated Russian Progress ship.

Today, the ISS menu includes more than 100 items, plus a large variety of snacks and drinks (Table 7.3). Assisted by nutritionists, astronauts choose 28-day flight menus approximately 120 days before launch from the wide-ranging food list. Under current agreements, NASA provides half of the food and Russia the other half, regardless of the crew's composition. There are three fixed meals per day, with a fourth "meal" made up of snacks and other items that can be eaten at any time.

Most of the Russian food is ambient stowed (freeze-dried, low moisture, or thermo stabilised) rather than dehydrated. The American food is Shuttle-based and comes in six types – thermo-stabilised, rehydratable, intermediate moisture, natural form, fresh, and irradiated.

If food is dehydrated, it cannot be eaten until the astronauts add hot water to it. Many of the beverages are also in a dehydrated form. However, since electrical power for the space station is produced by solar panels instead of the fuel cells installed on the Shuttle, there is no extra water generated on board – with the exception of water that is recycled from the cabin air, but that is not enough for use in the food system. Extra supplies have to be loaded by the crew during the regular Shuttle and Progress resupply missions.

Food on board the space station comes under several categories: Daily Menu, Safe Haven (a food reserve), and Extra Vehicular Activity (EVA) food. Foods chosen for the daily menu are similar to those eaten every day on Earth. It includes frozen entrees, vegetables and desserts; refrigerated food fruits and vegetables, extended shelf-life refrigerated foods, and dairy products; and thermo-stabilised, shelf-stable natural form foods, and rehydratable beverages.

The Safe Haven food is provided to sustain crew members for 22 days under emergency conditions resulting from an onboard failure. It is independent of the daily menu food and can provide each person with at least 2000 calories per day. Important characteristics are its small volume and ability to be stored at temperatures which range from 16 to 30 °C (60 to 85 °F). The shelf life of each food item is a minimum of two years, so it mainly comprises thermo-stabilised entrees and fruits, intermediate moisture foods, and dehydrated food and beverages.

During an EVA, astronauts used to eat a fruit bar or similar food mounted in the neck area of the EMU. Today, most spacewalkers prefer to eat beforehand. Water is stored in a bag velcroed inside the suit. Spillages are prevented by a valve that is opened when the spacewalker sucks through the tube. The drink bags come in two sizes: 0.6 litre (21 oz) and 0.9 litre (32 oz) – for an eight-hour excursion. The containers are cleaned and refilled in the galley area.

Both American and Russian foods are stored and served in individual, disposable packages – usually pouches or cans – that can simply be warmed in a small food heater. Five package sizes have been designed to ensure that they are compatible with the stowage compartments, food warmer and meal tray. Fresh fruits, bread and condiments are provided in bulk packages.

In terms of volume, fresh food plays a fairly minor role in crew nutrition, but as a means of boosting crew morale its significance is hard to overestimate. As their taste buds fade and even the most appetising space food starts to become bland, crews traditionally look forward to the new tastes and smells provided by items such as fresh bread and in-season products such as apples, lemons, oranges, tomatoes, onions, garlic and

Members of three crews share a meal in the wardroom/galley area of the Zvezda module. STS-105 commander Scott Horowitz opens a can as he floats near the ceiling, while others try various packages. (NASA photo STS105-308-029)

Table 7.3 International Space Station Food List

REFRIGERATED

Dairy: cheese slices; cream cheese; sour cream; fruit yoghurt

Fruits: apple; grapefruit; kiwi; orange; plum

FROZEN

Meat and eggs

Beef: brisket; BBQ; enchilada with Spanish rice; fajita; patty; sirloin tips with mushrooms; steak bourbon; steak teriyaki; stir fried with onion; stroganoff with noodles; luncheon meat; meatloaf with mashed potatoes and gravy

Lamb: broiled

Poultry: baked chicken; chicken enchilada with Spanish rice; chicken fajita; grilled chicken; oven fried chicken; pot pie chicken; stir fried chicken with diced red pepper; chicken teriyaki with spring vegetables; roasted duck; porcupine meatball (turkey)

Pork: Canadian bacon; baked ham with candied yams; baked pork chop with potatoes au gratin; pork sausage patties; sweet and sour pork with rice

Seafood: baked fish; grilled fish; sautéed fish; broiled lobster tails; baked scallops; gumbo with rice; shrimp cocktail; tuna noodle casserole

Eggs: cheese omelette; vegetable omelette; ham omelette; sausage omelette; vegetable and ham omelette; egg omelette, vegetable and sausage, scrambled eggs with bacon; hash browns; sausage; vegetable quiche; quiche Lorraine

Pasta mixtures: lasagne; vegetable with tomato sauce; noodles; stir fry; spaghetti with meat sauce; spaghetti with tomato sauce; tortellini with tomato sauce; cheese

Other: egg rolls; enchilada; cheese with Spanish rice; cheese pizza; meat pizza; vegetable pizza; pizza supreme

Fruit: escalloped apples; sliced peaches with bananas; blueberries; peaches with bananas; grapes; strawberries; sliced strawberries

Soups: beef stew; cream of broccoli; cream of chicken; chicken noodle; cream of mushroom; won ton

Grains: biscuits; bread; cornbread; dinner roll; garlic bread; wheat/white sandwich bun; wheat/white toast; tortilla

Breakfast items: cinnamon roll; French toast; pancakes, buttermilk; apple cinnamon pancakes; waffles

Pasta: Fettuccine Alfred; macaroni and cheese; spaghetti

Rice: fried; Mexican/Spanish; white

Starchy vegetables: whole kernel corn; baked potato; escalloped potatoes; oven fried potatoes; mashed potatoes; candied yams; succotash, squash corn casserole

Vegetables: asparagus tips, green beans; green beans with mushrooms; broccoli au gratin; broccoli; carrot coins; cauliflower au gratin; Chinese vegetables; stir fry mushrooms; fried Okra; fried peas; peas with carrots; squash; acorn with apple sauce and cinnamon; Zucchini; fried spears

Desserts

Cakes: angel food cake, chocolate brownie; chocolate fudge; shortcake, yellow cake with chocolate frosting

Dairy: ice cream, chocolate, ice cream, strawberry, ice cream, vanilla, yoghurt, frozen
Pies and Pastry: chocolate cheesecake; plain cheesecake; peach cobbler; apple pie; coconut cream pie; pecan pie; pumpkin
Beverages: apple juice; grape juice; grapefruit juice; lemonade; orange juice
Condiments: margarine; grated cheese
Hot cereals: oatmeal; cream of wheat; grits

THERMO-STABILISED
Fruit: apple sauce, fruit cocktail; peaches; pears; pineapple
Salads: chicken; tuna; turkey; three bean salad; pasta salad, potato salad, German, sauerkraut
Soups: chilli, clam chowder, egg drop, Miso, Japanese, vegetable
Desserts: butterscotch, chocolate, lemon, tapioca, vanilla puddings
Condiments: barbecue sauce; catsup; chilli con queso; cocktail sauce; cranberry sauce; dill pickle chips; bean dips; onion dips; ranch dips; honey; horseradish sauce; assorted jelly; lemon juice; mayonnaise; mustard; hot Chinese mustard; orange marmalade; Peanut butter (chunky, creamy, whipped); Picante sauce; sweet and sour sauce; maple syrup; taco sauce; tartare sauce
Beverages
Fruit juices: cranberry; cranberry apple; cranberry raspberry; assorted gatorade; pineapple; pineapple grapefruit; tomato, V-8
Milk: skim, low fat, chocolate (low fat or skim), whole

NATURAL FORM
Fruit: dried apples; dried apricots; dried peach; dried pear; prunes; raisin; trail mix
Grains: animal crackers; cold cereal; chexmix; assorted crackers; baked chips; tortillas; baked potato chips; pretzels; goldfish; tortilla chips; potato chips; seasoned rye krisp
Desserts: *cookies* - butter, chocolate chip, fortune, rice krispies treat, shortbread
Snacks: beef jerky
Nuts: almonds; cashews; macadamia; peanuts
Candy: candy-coated chocolates; candy-coated peanuts; Lifesavers; gum (sugar free)
EVA food: in-suit fruit bar

REHYDRATABLE
Beverages: apple cider; cherry drink; cocoa; coffee (assorted); grape drink; grapefruit drink; instant breakfast; chocolate instant breakfast; vanilla instant breakfast; strawberry orange drink; orange mango drink; orange pineapple drink; tea (assorted); tropical punch
Irradiated meat: beef steak, smoked turkey

---

kielbasa. Fresh foods are delivered every three months or so to provide variety and prevent food boredom.

The dining area in Zvezda incorporates a wardroom table/galley equipped with food warmers, food trays and utensils. The table includes restraints for holding the weightless crew and equipment in place, and has recessed wells for food warming. The wardroom area also includes a potable water dispenser, which dispenses hot and cold water for drink and food hydration, a trash container, and two small refrigerators. A freezer is not yet on board because the limited power available could prevent its continuous use.

When the meal is over, the containers are crushed in a trash compactor and placed in a waste container for disposal in a Progress or MPLM container. The meal tray and utensils are wiped clean and stowed ready for the next meal.

WATER, WATER EVERYWHERE

Even more than food, a constant supply of clean water is essential to keep humans alive. Our bodies are 90% water, but outer space is drier than any desert on Earth. Ideally, since water is heavy and bulky to transport, a space station would generate its own supply, but, unfortunately, in the absence of water-producing fuel cells, this is not easy. As a result, although there is a Russian-built water processor on the ISS that collects humidity from the air, most of the drinking water has to be brought in by regular deliveries.

The crew transfers water manually by moving containers from the Progress and the Shuttle – a task that is made easier by the weightless environment. Surplus water generated by the Shuttle's fuel cells, which combine hydrogen and oxygen to create electricity, can be stored in contingency water containers. Each container resembles a duffle bag and holds about 40 litres (10.5 gallons) of water.

Where water is in such short supply, this precious resource must be utilised with the utmost care. Nothing can go to waste, so luxurious hot showers are out for the foreseeable future. In fact, there are no taps on the ISS. Instead, astronauts use a damp wash cloth impregnated with soap and then give themselves a wipe over. Occupants have the choice of a "wet" shave or an electric razor.

Although waste is kept to a minimum, some losses are inevitable. Water is lost by the space station in several ways: the water-recycling systems produce a small amount of unusable brine; the oxygen-generating system consumes water; air that escapes from the airlocks takes moisture with it; and the carbon dioxide removal systems extract some water from the air.

For the moment, units in the Russian Zvezda service module and the airlocks provide the only way to extract and recycle moisture from the atmosphere. The unit in Zvezda is capable of processing up to 24 kg (53 lb) of water per day. This water is purified and recycled for drinking and food preparation.

However, if all goes according to plan, advanced Environmental Control and

Yuri Usachev carries a Russian water container in the Destiny module. (NASA photo ISS002-E-5778)

Life Support Systems (ECLSS) developed by Russia and the Marshall Space Flight Center in Alabama will eventually be able "to recycle almost every drop of water on the station and support a crew of seven with minimal resupplies". This careful recycling would save the cost of transporting 18,000 litres (4,800 gallons) of water per year from Earth – the minimum required to resupply four crew members on the station.

The water purification machines on the ISS will cleanse waste water in a three-step process. First, a filter removes particles and debris, then the water passes through the "multifiltration beds" which contain substances that remove organic and inorganic impurities. Finally, the "catalytic oxidation reactor" removes volatile organic compounds and kills bacteria and viruses.

This system will reclaim waste water from urine, from oral hygiene and hand washing, and even research animals, as well as condense moisture from the air. The ECLSS equipment should eventually be housed in Node 3.

"Lab animals on the ISS breathe and urinate, too, and we plan to reclaim their waste products along with the crew's," said Layne Carter, a water-processing specialist at the MSFC. "A full complement of 72 rats would equal about one human in terms of water reclamation."

This may not sound the most hygienic way to live, but the water leaving the space station's purification machines will be better than most of our water on Earth. "The water that we generate is much cleaner than anything you'll ever get out of any tap in the United States," said Carter. "We certainly do a much more aggressive treatment process (than municipal waste water treatment plants). We have practically ultra-pure water by the time our water's finished."

Indeed, if the water recycling systems can be improved to an efficiency of greater than about 95%, then the water contained in the station's food supply would be enough to replace the lost water.

"It takes processes that are slightly more efficient than we have developed for the space station to do that," said Marybeth Edeen, deputy assistant manager of environmental control and life support at Johnson Space Center. "Those are being developed now, but they're not ready for space flight yet."

WASTE NOT, WANT NOT

It may eventually be possible to reclaim more than 95% of water used or exhaled on the ISS, but other wastes cannot be recycled as efficiently, particularly solid waste from food containers, experiments, empty fuel containers, and other onboard activities. So what happens to the trash?

Again, the Progress and Shuttle come to the rescue. Every Shuttle brings fresh supplies, packed inside a Multi-Purpose Logistics Module. And when it leaves, the MPLM is filled with a tonne or more of rubbish. Bags and containers of sealed trash are brought back to Earth, making it the world's most expensive waste haulage system.

More spectacular is the Russian method of waste disposal. Once the fresh supplies are unloaded from a visiting Progress, the trash bags are piled in and the spacecraft is sealed. After it undocks from the ISS, it manoeuvres into a lower orbit, brakes and makes a controlled re-entry during which the entire craft is incinerated over the ocean.

A similar system will eventually be available with the European Automated Transfer Vehicle and the Japanese H-II Transfer Vehicle.

Human toilet waste is generally dealt with in a similar way. The toilet, which

Yuri Gidzenko surveys the rubbish and redundant hardware loaded inside the Leonardo Multi-Purpose Logistics Module prior to its return to Earth in March 2001. (NASA photo STS102-343-008)

is compatible for both male and female residents, uses a flow of air to suck the wastes into a collection unit. Each astronaut has his or her personal attachment fitting into which they urinate. The fluid – several kilograms per day for each person – is purified and recycled. The solid faeces are collected in a container, where they are isolated and allowed to biodegrade before being stored for analysis on Earth or incinerated inside a Progress during re-entry.

Other human waste products – gases, dead skin, hair, etc. – are automatically filtered out of the cabin air before they contaminate the station. However, a vacuum cleaner is also provided, and the Expedition Three crew took advantage of its presence to try out a hairdressing technique used since the days of Skylab some 18 years earlier – capturing the floating hairs with the suction from the cleaner.

MICROSCOPIC STOWAWAYS

Wherever humans go, microbes inevitably follow. Microbes can ride in the air on particles of dust or dead skin, or in tiny clumps of bacteria or fungi. On

Earth, there might be hundreds or thousands of microbes in each cubic metre of air, and the space station is no exception. Even before the first long-term expedition boarded the ISS, viruses, bacteria and fungi had already established themselves on board. They arrived on station hardware or by courtesy of earlier assembly crews.

"Just stand and breathe, and you're releasing microbes," said Monsi Roman, chief microbiologist for the ECLSS project at Marshall Space Flight Center. "You can wash and scrub and use antiseptic soap, and you'll still have microbes on your skin. You have them everywhere: in your clothes, on your skin, in your hair, in your body – everywhere you could think of."

Each person has his or her own personal set of microbes, but there will inevitably be a transfer of microbes from one astronaut to another in the confined environment of the ISS. They also tend to mutate – change form – as time passes. Fortunately, the majority of these organisms are harmless, and several types are actually beneficial to humans, but certain microbes can pose a health threat to the crew and even attack the materials and hardware of the station itself. "When the crew goes up to the station, they'll each have their own microbial flora, and when they return back, for the most part they've exchanged that flora with each other," said Roman. Most of these exchanged microbes are fought off by the crew's immune systems and their own resident microbes, but the potential for infection is there.

The first step in protecting the health of the crew is testing each crew member for infection before launch. Only healthy crew members are allowed to fly into space, and they're quarantined before launch to prevent them from contracting harmful germs at the last moment.

Once on the space station, the air, water and surfaces with which the crew come into contact must be kept clean. Fans and ventilation systems keep the air in the modules in constant motion, and all the air passes through High Efficiency Particle Air (HEPA) filters on its way to the temperature and humidity control systems. The main purpose of the filters is to remove small particles in the air, but they are also very good at removing microbes.

Most food is heat-treated or irradiated to destroy microbes before it is packaged and sent to the station. Water is disinfected by a machine called a "catalytic oxidator", which heats the water to 130 °C (265 °F). Nearly all of the microbes are destroyed by this process, but, just to be sure, the water is then treated with iodine.

As the never-ending battle against the proliferation of mould and other fungi on the Russian Mir space station showed, microbes not only survive in the metallic world of a space station, they thrive. For the health of the crew as well as the station's hardware, microbes must also be kept from growing on surfaces and in nooks and crannies. If left to their own devices, fungi will produce acids capable of eating into the fabric of the station – rather like bathroom tiles that have been overgrown by mould.

Growth of microbes is controlled in several ways. First, all materials used in the space station are tested for resistance to fungi, and paint containing a fungus-killing chemical is also widely used. By controlling the humidity of the air to not more than 70%, microbial growth is also discouraged. And, of course, the crew clean the station the old-fashioned way, regularly wiping down surfaces with a cloth containing an antiseptic solution. However, despite the crew's best efforts, some 250 species of microbes were identified on Mir, and many of these became resistant to fungicidal wipes over the years.

NOISE

Few things are more irritating than an unwanted noise intruding into one's peaceful existence – whether it be while one is attempting to sleep or listening to a favourite piece of music. Unfortunately, life inside a metallic structure such as a space station, where sound bounces off the uncarpeted walls and machines are always operating, 24 hours a day, inevitably suffers from noise pollution.

In the case of the ISS, the Russian segment has come under particular criticism. The general specification set by NASA states that noise levels should not exceed an average of 55 decibels over a 24-hour period. Russia refused to accept this specification, and, with NASA in a hurry to get the hardware into orbit, a compromise solution was found, with the understanding that mufflers would be added later.

Not surprisingly, the first occupants of Zarya found that noise levels on the first module were between 65 and 74 decibels (dB). Even the Russians recognised that was too high, so they worked with prime contractor Boeing to produce noise reduction devices. These were installed on board Zarya in May 1999 and brought the noise down to a more acceptable 62–64 dB.

Unfortunately, the Zvezda module at the heart of the station also suffers from a noise problem. The fans that cool equipment and circulate air in Zvezda are so loud that the Expedition Two crew, who spent over five months on board, considered noise to be a major issue.

Astronaut James Voss and Russian commander Yuri Usachev slept in the Service Module, which contains the toilet, galley, treadmill and a multitude of noise-generating life support systems. Their situation was not helped by a need to remove the doors to their sleeping compartments in order to improve ventilation. "It's always there," said Voss. "It's sort of like being in maybe a factory. Not a factory where they're stamping steel, but a normal production

James Voss conducts maintenance work in the Zvezda module. Above his head is one of the troublesome fans that contributed to his temporary hearing loss. (NASA photo ISS002-E-5078)

facility where there's constant noise in the background all the time from machinery that's running."

Sergei Krikalev found the carbon dioxide scrubber in Zvezda, which tends to produce a loud blast of sound every 10 minutes, particularly annoying. "In the Mir it was located in a different module," he pointed out. "I always said that having it in the Service Module was not a good idea."

After similar experiences and reports of hearing loss involving residents of Mir, ISS crews have been advised to wear ear protection when possible, but this is unacceptable in an environment where crew members need to communicate frequently with each other and a warning alarm may go off at any time.

Even though Voss wore ear plugs every night while he slept, the astronaut suffered a partial hearing loss during his spell on the space station. In particular, his ability to detect higher frequencies decreased, although it returned almost to normal after his return to Earth. Such a deterioration in hearing could also have important safety implications, since the crew members have to repeat what they say, with the possibility of making mistakes.

"If you don't hear the numbers correctly, you put in the wrong thing, it could have bad repercussions," said NASA engineer Jerry Goodman.

Russia plans to replace its fans with quieter units, but financial problems mean that this is unlikely to happen until at least 2003. Meanwhile, engineers are working on other solutions, such as isolating and insulating the noisy equipment. Fortunately, noise pollution is not such a significant problem on the American segment of the ISS. Susan Helms, who slept in the quieter Destiny laboratory, suffered no adverse hearing effects, although even she decided to occasionally wear noise-muffling headsets to sleep.

HEALTH AND FITNESS

If you don't like exercise, then life on a space station is not for you. A significant portion of crew non-working time is taken up with exercise and physical activity in an effort to counteract the significant physical decline of the human body that occurs during long-duration space flights.

Bones, cardiovascular systems, muscle tissue and organs all change in zero-*g*, and the longer an astronaut stays aloft, the more marked the changes that take place. Physiological changes noted by long-duration crews include loss of bone mass in the form of calcium, and a weakening of the heart, which no longer has to fight against gravity to pump blood around the body. In addition, body fluids shift upwards, causing facial puffiness and ear–nose congestion, while blood volume first increases, but then experiences a drop.

First-hand information from Skylab, the Space Shuttle, and Mir clearly demonstrates the importance of exercise in maintaining physical fitness and strength, with the added advantage of easing the difficult process of re-adaptation to normal gravity. Fortunately, although the re-adaptation to life on a planet whose gravity offers resistance to every motion may take many months, the changes are reversible after return to Earth, as shown by the successful completion of several prolonged flights by Russian cosmonauts.

Each ISS crew member will typically spend up to two hours a day on the various exercise machines, except on the day of an Extravehicular Activity (EVA) or within 24 hours of a periodic fitness evaluation. Periodic fitness evaluations monitor the crew members' fitness level and determine how much their condition has deteriorated, allowing the crew surgeon to alter exercise and countermeasure protocols, if required.

The space station's exercise hardware consists of a treadmill, a resistive exercise device, and a cycle ergometer. The Treadmill with Vibration Isolation System (TVIS) is used to simulate walking and running in normal gravity. Similar to the treadmills used on Mir and several long Shuttle missions, the TVIS was installed by Atlantis's astronauts on the STS-106 mission in September 2000. It is located in a recess inside Zvezda, so its running surface is flush with the floor of the module. The display unit folds down when not in use.

Unlike its predecessors, the treadmill may be operated in an active (powered) or passive mode, when it is driven by the user's feet. Other major modifications include placing the restraints at the sides of the runner instead in front and behind and addition of a Vibration Isolation System (VIS). This minimises shaking that might affect other ISS systems or disturb sensitive microgravity experiments and does not transfer a load greater than 2.25 kg (5 lb) to surrounding connections or structures on the station.

The cycle ergometer is used for aerobic conditioning and can be used to perform independent arm and leg cycle activity. The machine is located beneath the floor of Zvezda and is used both for crew fitness and as part of an experimental research payload known as the Human Research Facility. As with the TVIS, Vibration Isolation System (VIS) isolates vibrations during use. It was used for the first time on 6 November 2000, when all three Expedition One crew members pedalled their way around the world.

The Resistive Exercise Device (RED) prevents weakening of the muscles in the legs, hips, trunk, shoulders, arms, and wrists by allowing the user to complete a series of physical exercises while restrained by elastic bungee cords. It is mounted to the treadmill to ensure isolation from other space station systems. Up to 195 kg (430 lb) of resistance can be provided in steps of 2.25 kg (5 lb).

RED information, including the set number, repetition number, and resistance load, is stored and viewed and/or downlinked to the ground via the Medical Equipment Computer (MEC). This portable computer also stores and sends back exercise data from the ergometer and TVIS, physiological data such as electrocardiogram and heart rate, and other environmental health information. The MEC also contains medical records, medical reference, and psychological support software.

Yuri Usachev works out on the cycle ergometer in Zvezda. (NASA photo ISS002-E-6136)

Essential as it is, a seemingly never-ending regime of hard physical exercise can take its toll. Individuals may start to dread the daily drudge and be tempted to skimp their full programme, with unfortunate results when they attempt to step onto land once more. Even dedicated space heroine Shannon Lucid admitted that the thought of more work-outs on the Mir treadmill filled her with dread. The weightless environment also has an unfortunate side-effect for heavily perspiring astronauts. The moisture that accumulates does not readily evaporate or drip onto the floor, but clings uncomfortably to their skin and can only be removed by wiping with a towel.

A number of other techniques are also used by the Russians to help crews to maintain muscle tone and prepare for their return to Earth. All-in-one elasticated "penguin suits" provide modest resistance and require the wearer to use his leg and back muscles to maintain an upright posture.

Less popular with fashion conscious cosmonauts are the airtight "Chibis" trousers, which are designed to simulate the effects of gravity on the lower body. They are more commonly used as a long mission draws to a close. Air is sucked from the trousers, creating a partial vacuum, with the result that blood is "pulled" into the legs.

As with so many areas of life on the goldfish bowl known as the International Space Station, every aspect of an astronaut's physical health is monitored in some way. For example, in addition to the heart rate receivers located on the treadmill and ergometer, the crew often wear personal heart rate monitors – watches that can be worn to give pulse information and allow them to control their exercise levels during daily workouts.

Astronauts are generally very fit and undergo a series of thorough medical examinations throughout their careers, so medical emergencies in orbit are rare, but not unknown. Although each crew member has paramedic-level knowledge to deal with accidents and sickness, there may be times when this is not enough. If there is genuine cause for concern, telemedicine facilities using the Internet enable specialist physicians on the ground to be contacted for a diagnosis and advice over the best form of treatment. If the condition is thought to be sufficiently serious, the patient will be ordered home at the earliest opportunity.

Among the sophisticated medical equipment on board the Destiny research laboratory is a system that can obtain ultrasound images and help to evaluate and support the general medical condition of the ISS crew. The ultrasound images obtained from the HDI 5000 are transmitted to Earth to allow scientists to study the effects of microgravity on bloodflow, the heart and other organs as well as a range of musculoskeletal and cardiovascular functions.

According to David S. Martin of NASA's Johnson Space Center, "The stakes are extremely high if there is an injury on board, so we have to be prepared for sudden ailments such as fractures, pneumothorax, abdominal trauma, musculoskeletal injuries, and appendicitis. The ultrasound system can be of help in making an accurate diagnosis should one of these occur."

Meanwhile, the more obvious problems of headaches, colds, cuts and bruises can be dealt with using the onboard medical facilities. The ISS medical kit includes the usual medications and bandages, while simple aids available include blood pressure cuff, stethoscope, intravenous infusion pump, oxygen masks and a defibrillator for a stopped heart. A medical restraint system is available for strapping down a patient who needs attention and surgical implements are kept in an emergency pack.

Not all illness in space has a surgical remedy. Disorientation in weightlessness can be a problem, particularly in the first few days of a mission. In order to establish a common sense of direction, the designers of the ISS have ensured that all of the modules have a consistent "up" orientation, including the labels on the walls.

COMMUNICATION AND RECREATION

Boredom and a sense of isolation are among the greatest enemies to be overcome during a long-duration spaceflight. For most of their time in orbit, crews are fully occupied with busy programmes of housekeeping, maintenance and scientific experiments. But, if a sense of well-being and mental alertness is to be maintained, then time for relaxation and recreation is essential.

Information gained from the experiences of the Russian Mir space station crews indicated that isolation from friends and family is one of the biggest problems a long-duration crew has to face. To prevent this on the ISS, astronauts are able to receive a weekly video telephone call from home, while daily e-mail messages also link them to family and friends. Regular news updates also keep them in touch with events back on Earth.

PR plays a major role in all ISS activities, so regular press conferences, communication sessions with school children and one-off links to events such as the 2001 Oscar ceremony are arranged for the crews.

Although the crew may be physically separated from loved ones and mission control, they are fortunate that communication links last much longer and are more reliable on the ISS than on any previous space stations. After the first few months of communication sessions limited to less than 20 minutes, the Expedition One crew began to enjoy more or less continuous contact with the ground. This is made possible by the utilisation of a fleet of NASA Tracking and Data Relay Satellites, located in geostationary orbits 36,000 km (22,000 miles) above the Earth, that can transmit huge amounts of data between the station and the ground, as well as by direct communication links with ground stations.

Many space enthusiasts on the ground are avid followers of the ISS crew's exploits, and they eagerly listen in to any transmissions they can pick up. Apart from this serendipitous approach, amateur radio enthusiasts around the globe are provided with information on frequencies which will enable them to speak to the astronauts as they pass overhead. Crew members often look forward to these fleeting "ham radio" conversations to break the monotony of the daily routine.

If boredom sets in, there are plenty of windows from which to admire the ever-changing view of planet Earth as the space station sweeps over most of the continents once every 90 minutes. Searching for familiar landmarks, watching glowing auroras or lightning flashes, and waiting for one of the 16

Yuri Lonchakov (left) and Yuri Usachev talk to amateur radio operators on the ground from a special work station in the Zarya module. (NASA photo STS100-343-020)

spectacular sunrises and sunsets that take place each day offer an endless potential for fascination.

Should the astronaut tire of Nature's wonders, there is a wide choice of leisure pursuits, since crew members are allowed to take along their personal choice of entertainment. These range from chequers or chess sets to CDs and tape players. Apart from recorded music, the more musically talented astronauts indulge themselves in playing instruments. Favourite with the Russians is the acoustic guitar, but Expedition Three commander Frank Culbertson was more adventurous, practising on a trumpet. However, while sing-alongs may be good for morale, the individual musicians must take care not to drive their colleagues to distraction with overloud, overzealous performances in the confined space of the station!

Time to relax and bond together is essential for any crew. Having become familiar with each other's tastes and peculiarities during months of training prior to launch, the team must continue to share ideas and concerns once in orbit. Meal times are generally set aside for such periods of friendly chat and discussion about everyday and work matters, and evenings often see the residents settling down to watch the latest DVD movies.

However, even the latest "home entertainment" gadgets are not always to

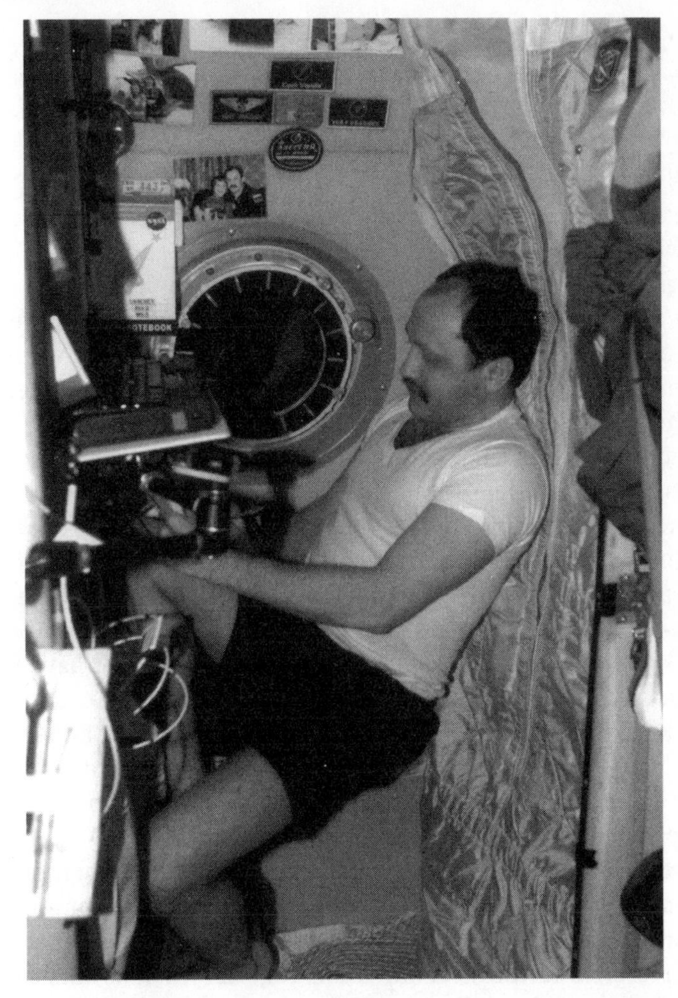

Yuri Usachev in one of the two "private" bedrooms in Zvezda. (ISS002-E-5360)

everyone's taste. The Expedition One crew decided that the DVD screen was too small and its sound output too weak – so limiting its use during exercise workouts.

Another factor limiting enjoyment of movies and other joint activities is the language problem. Even after months or years of learning a foreign language, verbal communication between Russian and American colleagues can sometimes be stilted or even lead to unfortunate misunderstandings. William Shepherd complained that the translations of some of the Russian-only labels on the medical supplies were not included in his language training. He went on to describe his struggle with the strange accents of the Russian ground controllers or "Glavnis". "I felt that, at the end of the flight, I could understand half of the Glavnis well, and for the other half I only understood 80% of what they were trying to convey. I don't feel Glavnis were making special efforts to adjust for me. Some of them needed to slow down if they wanted to be understood."

Large as it may be by spacecraft standards, lack of privacy remains a problem on the current ISS. As on Mir, there are two personal sleeping cabins in Zvezda where a cosmonaut may retire to read, listen to music or enjoy the view through the porthole. Unfortunately, these cabins are very close to the main galley/dining/work area on the station as well as the toilet area, and poor ventilation means that the door has to be left open most of the time.

Since there are three residents on the station, the third crew member has to find his or her own sleep area and personal space – usually in the less crowded Destiny laboratory. This lack of accommodation was partially solved by the addition of a temporary sleep station in Destiny for the Expedition Three crew. Even so, being cooped up with two relative strangers who speak a foreign language and have very different customs and backgrounds can lead to psychological problems and even conflicts. Other strains can erupt to the surface when ground controllers put pressure on an already stretched crew schedule. This can be a particular problem for residents who have to cope with the more hierarchical Russian system in which the crew is expected to follow "orders" without question. As Yuri Gidzenko griped when an air conditioner broke down, "They plan an activity to take one hour and we all know it will take five hours." The Russian ground controller at the sharp end of the line eventually pleaded with the men to stop swearing at him.

In an effort to avoid such situations, the crew is asked to complete a weekly questionnaire as part of an "Interactions" study dealing with interpersonal and cultural relationships among the crew. Specialists also monitor conversations between the crew and the ground, while regular sessions are scheduled for private discussions with medical staff.

One other solution may be to ensure that at least one woman is assigned to each long-term crew. As French astronaut Claudie Haigneré said, "The evidence clearly shows that mixed crews perform best – in organising their work, in decision-taking, in conflict management and in their contacts with ground control."

AVOIDING A COLLISION

The International Space Station circles the Earth at an altitude of around 395 km (247 miles). Unfortunately, this is one of the most congested regions of near-Earth space – not by other spacecraft, but by space junk. Today, the

US NORAD monitoring network keeps track of some 6,000 objects in low-Earth orbit that are larger than 10 cm across, but countless hundreds of thousands of smaller pieces of debris cannot be detected or catalogued. The good news is that Nature has its own waste removal system. As the Sun becomes more active every 11 years, the Earth's atmosphere expands, increasing the drag on the debris and causing it to re-enter. Most debris in low orbits is cleared out of the way after only a few years at the most.

An additional hazard is offered by cosmic debris – tiny particles of dust scattered through space by comets and asteroids. Since the velocities of these so-called meteoroids can be extremely high, they obviously pose a significant threat to any spacecraft in their path. Some incoming meteoroids, such as the November Leonids, may travel as fast as 70 km/s (44 mile/s) and carry a powerful punch. At such speeds, a particle only one-sixth of a millimetre in diameter will have the same impact as a rifle bullet.

Although no manned spacecraft has ever been punctured or disabled, and no spacewalk has ever had to be terminated as the result of a debris or meteoroid strike, the threat is far from negligible. After each mission, some heat-resistant tiles on the Shuttle have to be replaced as the result of impact damage. Pitted windows and peppered solar arrays are also commonplace. Indeed, the launch of the first ISS element, Zarya, was postponed until 20 November 1998, after the Leonid shower, as a safety precaution.

This decision was also partly influenced by the design of the Russian module. NASA safety studies indicate that neither Zarya nor Zvezda have adequate shielding against orbital debris. Although all of the space station partners have agreed to design their respective modules so that an aluminium sphere 1 cm (0.4 in) across cannot penetrate the station's walls and cause damage, the Russian segment was launched with a lower specification. This was partly due to financial and time constraints and partly because the additional shielding would make the modules too heavy to lift off the ground into the correct orbit.

A March 2000 report by US General Accounting Office stated that, "The Service Module was supposed to have no more than a 2.4% probability of a penetration over a 15-year period, but it has been assessed as having a 25% probability with current shielding." The shielding will have to be added in orbit at a later date, probably in 2004, and even then the level of protection will still not meet the original specification.

The report went on to complain that vital equipment in the Russian modules will fail if they become depressurised. Zarya, which was purchased as an "off-the-shelf" item by NASA, was exempted from the general ISS specifications that all equipment in pressurised sections of the station should be able to continue functioning if their cabin air escaped into space. However, the same problem arises with Zvezda, since much of its hardware requires air for cooling and will fail in a vacuum.

Windows on the Russian modules also came under criticism. With only a five-year guarantee, the authors expressed the fear that an outer pane might crack and break at some stage in the lifetime of the ISS. However, they went on to add, "The Russian Space Agency has developed metal covers that can be installed over damaged windows". In contrast, windows on the US modules are fitted with four panes of glass, each between 1.3 and 3.2 cm (0.5 and 1.25 in) thick.

Exterior shielding on the US and non-Russian modules comprises a dual-wall buffer system. A thin aluminium outer panel is covered with a 10-cm (4-in) thick insulated blanket made of Kevlar (the material used in bullet-proof vests) and other impact-resistant materials. Although a particle several millimetres across might penetrate this outer shield, its energy will be dissipated on impact so that it will not break through the inner wall. The heaviest shielding is placed on the forward-facing (US–European–Japanese) modules that are most likely to receive hits.

Models of impacts for the ISS suggest a collision with a 1 cm or smaller object every 70 years on average. However, as a last resort, the ISS can be moved out of the way if a large piece of debris is forecast to be on a near-collision course. If sufficient warning is provided by US Space Command, flight controllers can manoeuvre the station to a slightly different orbit to avoid the debris.

RADIATION

Radiation hazards come in many forms, as anyone who has been sunburnt on a beach or dazzled by reflected sunlight can testify. Space is filled with dangerous ultraviolet light and X-rays from the Sun, but these short-wavelength forms of electromagnetic radiation cannot penetrate the walls of the station.

Solar radiation will also heat up the exposed surfaces of the station to more than 100 °C (212 °F), and this threat of overheating has to be dealt with by the thermal control system. Liquid ammonia is used as a coolant to collect the heat from station components and transport it via various heat exchangers to giant radiators on the station's truss.

Spacewalkers are protected from the effects of UV light by a layer of aluminised Mylar and extreme cold by layers of insulating material. Overheating is prevented by water-cooled undergarments together with oxygen circulating inside the suit.

More serious from the point of view of astronaut health is the population of energetic charged particles that fill near-Earth space. Most of these originate in the Sun, although cosmic rays from beyond the Solar System are an additional hazard.

Solar flares may send showers of energetic protons towards our planet. Other energetic particles are trapped in the Van Allen radiation belts, and these pose a particular threat to the space station in a region known as the South Atlantic Anomaly, where the radiation belts are much closer to the surface than normal. Measurements taken on board Mir show that there were typically three to six high-energy particle events per square cm per second over most of the orbit, rising to about 60 "hits" over the South Atlantic Anomaly.

On a 51.6-degree inclination orbit, the ISS crosses the Anomaly three or four times a day, leaving just 10 hours relatively radiation free. Since spacesuits do not offer adequate protection against energetic particles, EVAs are avoided if a proton event is taking place or the station is passing through the South Atlantic Anomaly.

Although the vast majority of the incoming particles cannot penetrate the walls of the station, the most energetic of them do pose a significant health threat. For this reason, certain areas of the station, where extra "shielding" is offered by water or fuel tanks, are designated as crew storm shelters during particle events.

In the case of the ISS, a major increase in solar flare activity on 9 November 2000, which was expected to continue for the next 48 hours, caused mission control to request the crew to set up a radiation detection monitor in the Zvezda living quarters. The monitor would signal a tone if radiation levels reached higher than expected levels. As a precautionary measure, flight controllers asked Shepherd, Gidzenko and Krikalev to sleep in the aft portion of Zvezda for the next two nights near the so-called transfer compartment, where there is increased shielding.

Monitoring of radiation is taken very seriously on the ISS, partly to assess the threat to the health of the residents and partly as an area of scientific research. Radiation Area Monitors are attached with Velcro to walls and surfaces throughout the ISS – four to six per module and two to four in each node. At the end of each four to five month residency the results are recorded to determine radiation levels throughout the ISS, then the monitors are replaced.

In addition, each astronaut's exposure to radiation is checked by personal dosimeters that are worn continuously by each crew member throughout the flight. These too are analysed post-flight to determine the overall radiation exposure of the crew member.

Despite such precautions, the doses of radiation accumulated by astronauts over long-duration missions are markedly higher than those on Earth. Astronaut-physician Norman Thagard reckoned that the radiation dose he received during his 115 days in space was equivalent to about 1,000 chest X-rays. Although there were no obvious side-effects, the long-term consequences of such exposure could be genetic changes and increased risk of cancer.

While humans may be able to take refuge, the orbiting structure itself has to suffer the full punishment offered by energetic cosmic radiation. The results include damaged solar cells, which result in a reduction in power output over time, and degraded computer chips, which may cause "bitflips" and errors in the machine's memory.

Chapter Eight

Value for Money?

During his State of the Union Address to the US Congress on 25 January 1984, President Ronald Reagan said:

> We can follow our dreams to distant stars, living and working in space for peaceful, economic and scientific gain. ... A space station will permit quantum leaps in our research in science, communications, in metals, and in life-saving medicines which could be manufactured only in space. We want our friends to help us meet these challenges and share in their benefits.

Eighteen years later, the controversial International Space Station is under construction, its first science laboratory has been delivered and activated, and the first experiments have been carried out. What has been achieved so far and what are the prospects of Reagan's vision becoming a reality in the years ahead?

EARLY CONCEPTS

For more than 100 years, space enthusiasts and visionaries have dreamed of a space station that will revolutionise our lives (see Chapter 2). As it orbited the Earth, the huge structure was envisaged as a multipurpose facility to assist communications and navigation, weather forecasting and military surveillance.

By the early 1980s, when President Reagan announced his desire to build an international space station "to develop the next frontier", the endeavour was being linked to an equally extensive and ambitious set of goals. Instead of the routine, two-week jaunts on the Space Shuttle, near-Earth space would be opened up for long-term utilisation and exploitation.

The multifunctional facility would be a laboratory for research in astronomy, space plasma physics, Earth sciences, materials research and life sciences. It would also be used to develop new space technologies, such as electrical power generation, robotics and automation, life support systems, Earth observation sensors and communications.

The introduction of a permanent human presence in space would provide a unique opportunity to learn how to live and work productively in space for prolonged periods. Apart from providing comfortable living quarters for multinational crews, the station would also become a major transportation hub. Crews could be rotated every 90 days, with the Shuttle also available to deliver supplies and bring back finished products and experimental results.

A space station was also seen as an ideal space garage, from which many satellites could be checked, repaired and monitored. With the aid of an orbital manoeuvring vehicle that was expected to be operational by 1991, satellites could be deployed and gently towed to and from their low-altitude operational orbits. Maintenance and refuelling of spacecraft could be carried out during visits by astronauts or at the space station itself. Eventually, "smart" repair and servicing robots would replace astronauts for maintenance work. Among the satellites envisaged as benefiting from this attention were scientific observatories, Earth observation platforms and free-flying facilities where microgravity processing and long-term experiments would be mounted.

Looking further into the future, the station would offer advantages as a space hotel and halfway house, a "way station" to other worlds rather than an end in itself. The permanently occupied facility would pave the way for new programmes of lunar exploration and unmanned voyages to Mars and other planets, followed by eventual settlement of the Moon and the Red Planet.

One powerful argument for such a way station was the reduction in costs by avoiding the "gravity well" of the Earth. Everything carried into orbit from Earth must be lifted out of a gravitational well 6,400 km (4,000 miles) deep. In other words, to lift an object entirely free of Earth's gravitational clutches uses as much energy as if the payload was being pulled against the full force of gravity for a distance of 6,400 km. Such large amounts of energy must be expended to carry a payload into orbit that, at present, it costs more than $20,000 to launch one kilogram of payload on board a Shuttle.

Launching a rocket or spacecraft from a base in low-Earth orbit cuts the gravity well by half, so by constructing a Moon or Mars craft at a space station and then sending it on its way becomes a much cheaper option than doing so from a launch site on Earth.

So fashionable was the space station concept that a 1986 report by the U.S. National Commission on Space suggested that the future infrastructure required for the initial exploration and occupation of the inner Solar System might require three additional space stations – one in high Earth-orbit, one in lunar orbit and one around Mars.

ARGUMENTS FOR

Ever since the idea of a space station was first mooted, there has been controversy and debate, particularly among the scientific community. One of the most obvious areas of concern has been the high cost of assembling and operating the gigantic structure, and even among supporters of the programme it is generally accepted that the vast expenditure on the current International Space Station can only be justified by a broad range of potential uses for the giant structure.

As John-David Bartoe, the ISS research manager at Johnson Space Center commented, "The budget for the space station is not solely for the purpose of basic research. We are doing this for space exploration, to learn to live and work in space. And while we're there, we're going to take full advantage of this environment to do research."

Nevertheless, NASA and prime contractor Boeing have done their best to put the expenditure into perspective. Even if the overall bill of around $100 billion is taken into account, this figure looks much more modest when considered on an annual basis. As a Boeing flyer stated, the US expenditure of $2.1 billion a year works out at just 2.2 cents per person per day.

The promise of groundbreaking science has been one of the key selling points for the 16 nations that have invested in the programme and the space agencies that are building the massive amount of hardware. Until now, there have been only sporadic opportunities for researchers to fly experiments in space. Even on the permanently manned Mir, only modest laboratory equipment was provided, and the three-person crew was not large enough to devote the necessary time to onboard research.

"The space station is a unique space laboratory where we will be able to perform experiments for longer periods than ever before, in sophisticated facilities and under conditions that are more controlled," said Ron Porter, manager of the Biotechnology Programme at NASA's Marshall Center.

"One of the advantages of space station research is the flexibility to continue long-duration research over several expeditions, modifying research procedures and parameters to take advantage of intriguing results," said John Uri, lead scientist at Johnson Space Center. "We want science on the station to be as much like science in an Earth-based laboratory as possible, but of course without gravity."

Although there are currently only three long-term residents on board the

ISS, the intention has always been to increase those numbers to six or seven (three Americans, three Russians and one other) by the time its construction phase comes to an end. If the ISS crew is eventually doubled in number, the amount of time devoted to onboard experiments will dramatically increase.

At present, there is only one scientific laboratory at the station, but five more are promised, and there is certainly no question that the ISS, if completed, will provide the largest and most advanced space science facility ever built.

The argument according to NASA is as follows: "The International Space Station will establish an unprecedented state-of-the art laboratory complex in orbit, more than four times the size and with almost 60 times the electrical power for experiments – critical for research capacity – of Russia's Mir. Research in the station's six laboratories will lead to discoveries in medicine, materials and fundamental science that will benefit people all over the world. Through its research and technology, the station will also serve as an indispensable step in preparation for future human space exploration."

The key factor that makes such research so promising is the absence of gravity. The ISS is seen as the only experimental laboratory where fundamental physical, chemical and biological processes can be studied in a near-weightless environment. However, critics argue that the presence of astronauts who continually carry out exercises and bump into the walls of the station, as well as arrivals and departures by visiting spacecraft, inevitably disturb the microgravity environment, with a detrimental effect on the sensitive experiments.

In an effort to overcome this degradation, NASA has installed an Active Rack Isolation System in Destiny. This acts as a powered shock absorber to help isolate tiny potential vibrations. When sensors detect disturbances from the station, the actuators act to counter the effects by delivering an opposite force between the payload rack and the laboratory module.

Without gravity to cause heavy objects to sink and hot air or liquids to rise, materials of different densities can be mixed to create new alloys and compounds. Larger protein crystals could help pharmaceutical companies to design more effective drugs, while new insights may be gained into the factors that influence plant and animal development, the changes in human physiology due to zero gravity and basic physical processes such as combustion.

With the scientific breakthroughs, so the argument goes, there will come the development of new commercial products, with the possibility of a 21st century industrial "boom in space". One of the beauties of pure research is that one never knows what unexpected breakthroughs and revolutionary new applications or technologies will result.

One example of the advances that may come about involves research by New

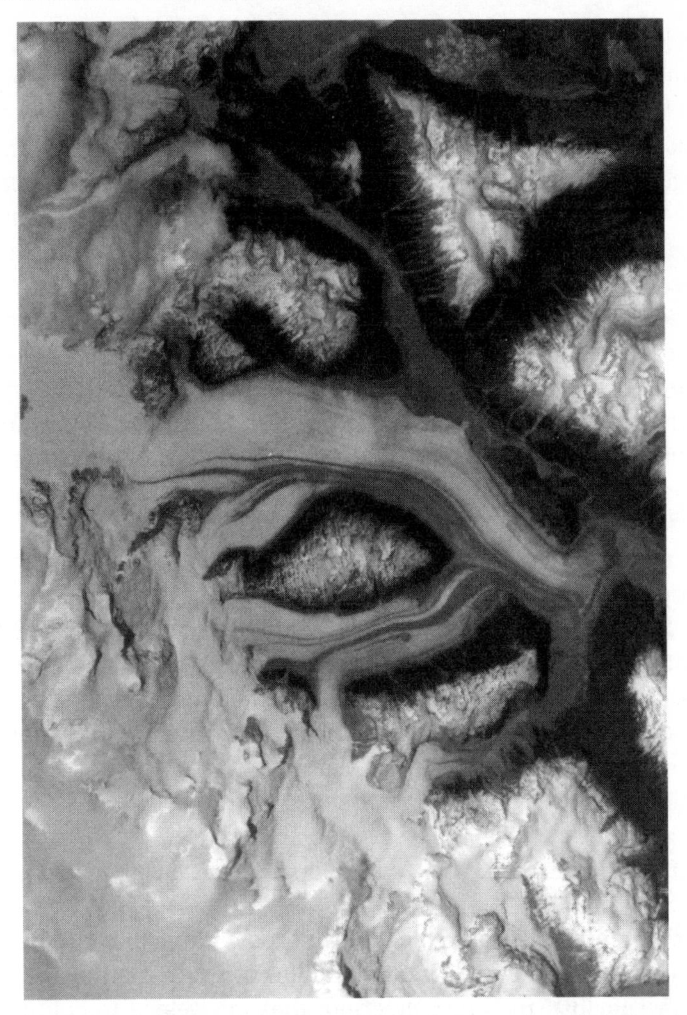

A digital camera image of glaciers in southern Chile that was downlinked from the ISS in December 2000. (NASA photo ISS01-E-5107)

York-based international pharmaceutical company, Bristol-Myers Squibb. The company has sponsored experiments into antibiotic production on the Shuttle and ISS. Initial study results indicate that microgravity stimulates microbial antibiotic production, with increases in productivity of up to about 200% compared to ground control samples.

"Our collaboration with NASA not only puts our researchers in the forefront of science, but also gives us the opportunity of being first in our field to develop major new technologies and products," said Ray Lam, senior principal scientist of the Bristol-Myers Squibb natural products department.

Materials research aboard the International Space Station may also help shape society in the 21st century according to Carolyn Griner. "You are ready to take the quantum leap into the future," said Griner, then acting director of NASA's Marshall Space Flight Center told delegates at a major space station science conference on 15 July 1998. "You have the real potential to make a difference. ... You have a responsibility to do the best possible research because I believe the future of materials is vested in this discipline."

Furthermore, supporters state that the station will be an ideal platform for Earth observation, studies of the atmosphere, high-energy astrophysics, and observations of the Sun and planets.

Despite all of these arguments, it is clear that, as in the case of the Apollo Moon programme, science is not the main driver behind the space station. Indeed, the ISS was almost scrapped on a number of occasions, and in the end it was another argument that saved the day – the opportunity to foster international relations, put an end to the Cold War, and offer employment to Russian scientists who might be tempted by offers from anti-Western governments.

Less prominent, but nevertheless significant, factors were the bolstering of national prestige, the need to maintain and justify the existing human spaceflight infrastructure, especially the Shuttle fleet, and the employment opportunities offered to aerospace workers and scientists in all the participating nations (along with their inputs to the local economies and the political votes gained for station supporters).

ARGUMENTS AGAINST

The introduction of Russian expertise and technology into the ISS programme raises an interesting conundrum. Although Soviet-Russian space stations have been in orbit for 30 years, critics argue that the desire to continue launching cosmonauts to such destinations has more to do with national prestige and commercial exploitation of the West than scientific progress. After all, the argument goes, what miracle drugs or breakthrough technologies have the Russians to show for all their investment after three decades of sending humans to orbital stations?

Particularly galling is the cost of building the huge structure that will house the experiments. NASA expenditure alone will reach $25 billion during the construction phase, with billions more each year to operate the station and send up regular Shuttle resupply flights. By the time the ISS is abandoned, its lifetime costs will probably have soared to around $100 billion. This compares with a research budget of around $3 billion a year for the US National Science Foundation.

"The cost of the space station is far beyond any justifiable scientific purpose or any justifiable practical purpose," said James Van Allen, a professor at the University of Iowa and the scientist who discovered the Earth's radiation belts.

While many scientific societies have damned the station with faint praise, others have actively opposed it, including the American Physical Society, the European Physical Society and the American Society of Cell Biology.

One of the main arguments is that the ISS is a drain on the rest of NASA's science programme, and will remain so for the foreseeable future. Apart from the magnitude of the overall expenditure, critics complain that, with two-thirds of NASA's budget allocated solely to manned spaceflight, i.e. the Shuttle and ISS, there is little left over for the agency's science programme proper. Indeed, budget overruns on the ISS programme have been blamed for funds being transferred from science to manned spaceflight. This has forced NASA to cancel scientific missions and resort to shortcuts in the form of a "faster, cheaper" approach in the development of robotic spacecraft.

It is often said that, with the development of more capable robots and teleoperation techniques, human oversight and intervention is not required for most ISS experiments. Those experiments that do require human involvement are usually related to how people adapt to life in zero gravity – a self-perpetuating area of research.

Furthermore, since robots do not require sustenance or sleep, and life support systems are unnecessary, they are far preferable to humans where orbital research is concerned. Indeed, the three-person crews on Mir and the ISS have spent at least half of their "work" time on housekeeping and repairs rather than research.

Instead of paying a fortune to send astronauts into space and provide the expensive life support systems and back-up hardware to keep them alive, these scientists argue that the same money could be better spent on dozens of unmanned, robotic missions or on improving ground-based research facilities.

As James Van Allen argued, "the overwhelming majority of scientific and utilitarian achievements in space have come from unmanned, automated and commandable spacecraft". He went on to scornfully dismiss the destruction by space station crews and visiting spacecraft of the very microgravity environment that they were supposed to be exploiting. "Some experiments one would like to carry out in space require highly stable platforms and the accurate aiming of scientific instruments, and so they must be free of vibrations and accelerations," he wrote. "An astronaut's sneeze could wreck a sensitive experiment in a microgravitational field: clouds of gas or droplets from thrusters of the spacecraft or from dumps of water or urine ruin the local vacuum and optical observing conditions, and complex magnetic and electric fields associated with manned spacecraft preclude certain kinds of radio observations."

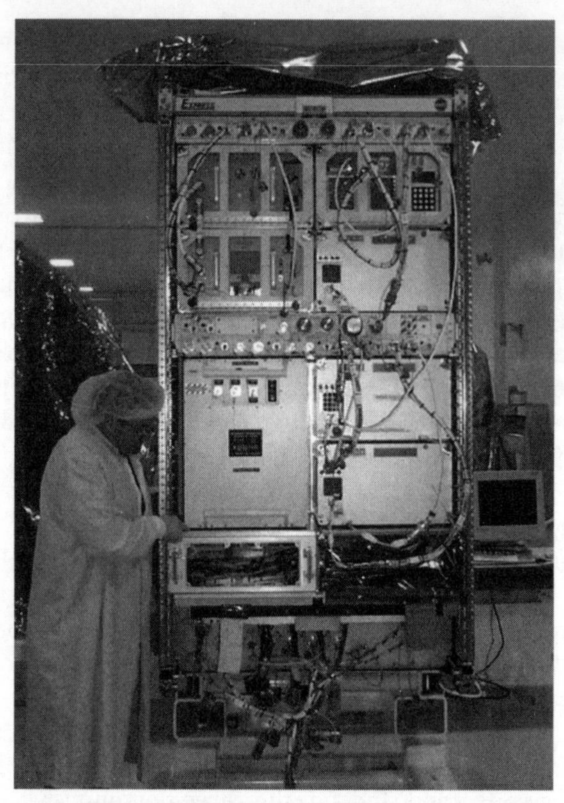

Many experiments on the ISS are automated. Standardised EXPRESS racks enable quick, simple integration of multiple payloads aboard the International Space Station. (NASA digital photo)

The quality of the scientific research that can be conducted on the station is also questioned. Many scientists are incredulous of NASA's claims that space station research may lead to miracle drugs and cures for Aids, influenza or cancer. Instead, they argue, the outcome will be a plethora of interesting but marginal science.

"It's good science, but it's not science that's going to have any profound impact," said Robert Park, director of the Washington office of the American Physical Society. Only a few scientific studies can be carried out more effectively in the absence of gravity, he added, and they "are not high priority".

As for the supposed new products and manufacturing boom that will result from orbital research, this is likely to be stillborn unless the cost of sending payloads aloft is drastically cut. As one researcher commented, "If Rumpelstiltskin took straw into space and spun it into gold, he'd still lose money."

Many of these criticisms were summarised by the Astronomer Royal, Sir Martin Rees, in an article written in 1999:

Without manned spaceflight, we're often told, there would be no sustained public support for a space program. Most people over 35 can remember Neil Armstrong's "one small step". For the middle-aged among us, the film Apollo 13 was an evocative reminder of an episode we followed anxiously at the time. But to a younger audience, the gadgetry and the "right stuff" values seemed almost as antiquated as those of a traditional Western film. The Apollo program, a spinoff from superpower rivalry in the Cold War era, wasn't a step towards any longer term goal that could inspire sustained public support.

Can the space station recapture the enthusiasm produced by the Apollo program? Will people be excited, 30 years after men walked on the Moon, by a new generation of astronauts circling Earth in greater comfort than the Russians in Mir but at far greater expense? Even if it is finished – something that seems uncertain, given the ever-rising costs, prolonged delays, and risk of accidents – the space station will be neither practical nor inspiring.

A manned station in low orbit is as unsuitable for most high-precision measurements as a ship is for ground-based astronomy. The practical case for manned spaceflight has weakened as robotic and miniaturisation techniques have advanced. It will recover only if costs can be dramatically reduced.

Unmanned space probes have yielded a crescendo of discovery. A wide public has followed this exploratory quest through pictures from the Pathfinder lander on Mars, closeups of Jupiter's moons, and the marvelous images beamed down from the Hubble Telescope. Europe has, so far, wisely eschewed manned spaceflight; a French-led plan for Hermes, a mini-space shuttle, has lost momentum. France, Germany, and Italy (though, happily, not the United Kingdom) have nonetheless made a political decision to contribute substantial funds to elements of the space station. Some countries will be rewarded by the launch of astronauts. (But how much did the launch of the Mongolian and Bulgarian cosmonauts by the Soviets benefit those countries?)

Along with many European scientists who are enthusiastic about space, I'm saddened that NASA persists with the space station. I regret even more that some European countries should bolster this misdirection of resources rather than support their aerospace industries in ways that raise the profile of European space science or lead to distinctive technological advances.

RESEARCH ON THE INTERNATIONAL SPACE STATION

Many types of research will be performed on board the station. A few experiments will be human-tended, but many payloads are automated or ground-controlled. Some of the main areas of interest are described below.

Protein Crystal Studies

Often described as one of the most valuable and important areas of microgravity research. By growing more, and larger, pure protein crystals in

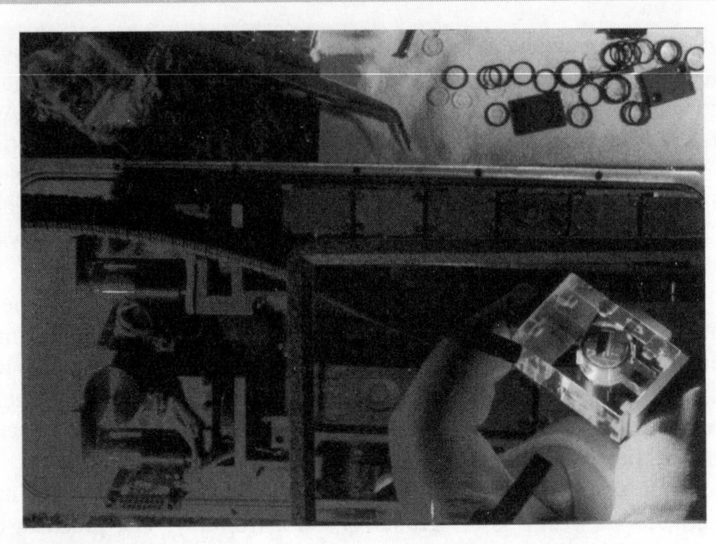

A reaction chamber of the Advanced Protein Crystallisation Facility that was installed on the ISS in 2001. (Astrium photo ASTR029)

space than on Earth, scientists hope to better understand the nature of proteins, enzymes and viruses, perhaps leading to the development of new drugs and a better understanding of the fundamental building blocks of life. Similar experiments conducted on the Space Shuttle have been limited by the short flight duration. It is often suggested that this type of research could lead to treatments for many illnesses, including cancer, diabetes, emphysema and Aids.

Tissue Culture

Another branch of biotechnology research involves growing living cells in space, where they are not distorted by gravity. NASA has previously developed a Bioreactor device that is used on Earth to simulate the effect of reduced gravity on such cultures, but experts argue that more advances will be achieved by growing cultures for long periods on board the space station. Among other uses, these tissue cultures can be used to test new treatments for cancer without risking harm to patients.

Biological Research

The effects of long-term exposure to reduced gravity on humans – for example, weakening muscles; changes in the cardiovascular system (heart, arteries and veins); and the loss of bone density – will be studied on board the ISS. Studies of these effects may lead to a better understanding of the body's systems and similar ailments on Earth. A thorough understanding of such effects and

Yuri Usachev examines a plant experiment in Zvezda. (NASA photo ISS003-E-5048)

possible methods of counteracting them is needed to prepare for future long-term human exploration of the Moon and other planets. In addition, studies of gravity's effects on plants, animals and the function of living cells will be conducted on board the station.

A Centrifuge Accommodation Module that will imitate Earth's gravity may eventually be added to the station, although this is under threat from budget cutbacks. By spinning at different speeds, the centrifuge will generate forces ranging from almost zero gravity to twice the normal gravity on Earth. This facility will be valuable for comparing processes and reactions, thus eliminating variables in experiments. It will also simulate the gravity on the Moon or Mars in order to provide information useful for future space travel.

Flames, Fluids and Metals

Flames have a more rounded shape and burn differently without gravity. Microgravity reduces convection currents, the currents that cause warm air or fluid to rise and cool air or fluid to sink on Earth. This absence of convection alters the flame shape in orbit and allows studies of the combustion process that are impossible on Earth, while molten metals or other materials can be mixed more thoroughly in orbit than on Earth. Studies in materials science may create better metal alloys and more perfect materials for applications such as computer chips. Such research may lead to developments that can add value to many industries on Earth.

Space Environment

Some experiments aboard the station will take place on the exterior of the station modules. These experiments can study the space environment and how long-term exposure to space radiation, vacuum and debris affects materials. This research can provide future spacecraft designers and scientists with a better understanding of the nature of space and enhance spacecraft design and materials.

The effects of long-term exposure to space are also of interest to the astrobiology community. By placing specimens "contaminated" with bacteria, viruses and other forms of organic material on the outside of the station, scientists will learn a great deal about the ability of micro-organisms to survive in the harsh environment. This, in turn, may provide new insights into the origins of life on Earth – for example, whether it could have been seeded by interplanetary dust or meteorites from Mars.

Some experiments will study fundamental physics, taking advantage of weightlessness to study the basic forces of nature that are weak and difficult to study when subject to gravity on Earth. Experiments in this field may help explain how the universe developed. Investigations that use lasers to cool atoms to near absolute zero ($-273\,^{\circ}$C) may help scientists to understand gravity itself. This research could lead to down-to-Earth developments such as timepieces 1,000 times more accurate than today's atomic clocks, better weather forecasting, and stronger materials.

Astrophysics

As the station evolves, opportunities will arise to deploy astronomical observatories on its exterior. Better still, there will the option to deploy and service free-flying observatories that will avoid the contamination associated with the space station's immediate environment. A number of proposals have been put forward in recent years.

Nobel prize winner Samuel Ting of the Massachusetts Institute of Technology is directing an international collaboration of some 37 universities and laboratories to use a state-of-the-art particle physics detector called the Alpha Magnetic Spectrometer (AMS). The AMS, which first flew on STS-91 in June 1998, will study the properties and origin of cosmic particles and nuclei including anti-matter and dark matter. The detector's second spaceflight is scheduled to be launched on mission UF-4 in 2003 for installation on the ISS as an external attached payload. It is expected to operate for three years before it is returned to Earth on the Shuttle.

Artist's impression of the two spacecraft of the XEUS X-ray observatory. (ESA digital image)

The European Space Agency has two candidates under consideration for operation outside its Columbus laboratory. The Extreme Universe Space Observatory would be used to detect cosmic rays. As these high energy particles smash into the atmosphere, they generate "showers" of elementary particles that may be detected in ultraviolet light by an instrument looking down from above. To reach the required sensitivity, the observatory would only operate while the station is in the darkness of the Earth's shadow.

A revolutionary type of X-ray observatory that uses optics inspired by the eye of a crustacean could be installed to study the entire sky (apart from the area obscured by the structure of the station). The Lobster telescope would provide continuous coverage in order to detect new X-ray sources – hot and exploding stars, gaseous disks circling black holes, for example – and measure the variability of known objects.

Even more ambitious is an ESA project known as the X-ray Evolving Universe Spectroscopy (XEUS) mission. This observatory would comprise two separate satellites – a mirror spacecraft, which focuses the incoming X-rays, and a detector spacecraft that carries the instruments to analyse the radiation. The first, prototype version of XEUS would be launched into a 600-km (375 miles) high orbit by an Ariane 5 rocket. After about five years of observations the two spacecraft would dock with each other so that the mirror element could be attached to the ISS, where new mirrors and other upgrades would be added.

A second detector spacecraft would then be launched, dock with the revamped mirror section, and transport it back to its operational orbit.

Other pioneering telescopes are being studied in the United States. One example is a concept known as the Dual Anamorphic Reflector Telescope (DART), which is being studied by NASA and Lockheed Martin. This low cost instrument, which would be assembled and tested at the ISS, would comprise two 15–25-m (50–83-ft) diameter "mirrors" made from an extremely light membrane. This would enable the creation of a space telescope that is 100 times lighter than the Hubble Space Telescope, but sufficiently sensitive to search for planets around distant stars.

Earth Watching

Observations of the Earth from orbit help the study of large-scale, long-term changes in the environment. Studies in this field can increase understanding of the forests, oceans and mountains. The effects of volcanoes, ancient meteorite impacts, hurricanes and typhoons can be studied. In addition, changes to the Earth caused by humans can be observed. Images taken from space can provide a global perspective unavailable from the ground, showing the effects of air pollution, such as smog over cities; deforestation – the cutting and burning of forests; and water pollution, such as oil spills.

Crew Interactions

One of the more unusual ISS experiments investigates crew psychology and interactions. It is well known that some long-term astronauts are more successful than others in adapting to the strange environment, the cramped space and the continuous close contact with their colleagues. Not only do they have to learn to adapt to the way they perform routine operations, such as eating, moving and operating equipment, but they must also learn to adjust to the internal changes that their bodies experience and to the psychological and social stresses that result from working under isolated and confined conditions. Not surprisingly, some suffer from isolation, depression or mood swings.

In the Interactions experiment, both crew members and ground control personnel in Houston and Russia complete a standard computer questionnaire once a week in an effort to identify and characterise important interpersonal and cultural factors that may impact the performance of the crew and ground support personnel during missions. The experiment will be performed on Expeditions Two to Six.

Microgravity Monitoring

For scientists sending experiments into space to take advantage of microgravity, it is clearly important to know when their instruments are disturbed by unwanted shocks or vibrations. "For instance, if there is an experiment in EXPRESS Rack 1 in the US laboratory and a crew member is exercising on the treadmill in the service module, we want to be able to tell scientists what disturbance levels they can expect from that activity," explained Kevin McPherson of NASA's Glenn Research Center in Cleveland, Ohio.

Various pieces of equipment, known as the Space Acceleration Measurement System (SAMS) and the Microgravity Acceleration Measurement Systems (MAMS), have been delivered to Destiny to monitor these disturbances. At the same time, an Active Rack Isolation System provided "powered shock absorbers" to protect experiments that were particularly sensitive to microgravity disturbances.

SAMS, in particular, has a long history. Since 1991, it has flown on 20 Space Shuttle missions and operated on the Mir space station for about four years – the longest operating US hardware on the former Russian outpost.

STUDENT EXPERIMENTS

Students from middle and high schools across America have been given opportunities to work with university and NASA scientists to prepare science experiments for the International Space Station. This hands-on education programme is sponsored by Marshall Space Flight Center in Huntsville, Alabama.

The first example of this came in February 2001, after students and teachers from 89 schools in six states mixed biological solutions and sealed them in small tubes or pipettes. The samples were frozen to -196 °C (-321 °F), then placed in a vacuum-jacketed container, known as the Enhanced Gaseous Nitrogen Dewar, for transport the station. Before thawing was complete, the STS-98 Shuttle crew moved the vacuum flask to the station where crystals slowly formed over the next few weeks.

After the samples were returned to Earth on the next Shuttle mission, scientists were able to analyse the crystals to determine the structure of their biological molecules. Some of the crystals were returned to the students so that they could compare them to crystals grown in their classrooms and at NASA workshops.

"We are pleased that students – the scientists and engineers of the future – were able to have a hands-on role in one of the first biotechnology experiments on the Space Station," said Ron Porter, manager of the Biotechnology Programme at the Marshall Spaceflight Center.

Another educational experiment involving high school students was flown during the French Andromède mission in October 2001, when various yeasts were grown in nutrients on the station in order to study how cells divide and multiply in microgravity.

EXPRESS SCIENCE

Most of the scientific research on board the station will take place in six laboratory modules. These will include the US Destiny module, the European Columbus module, the Japanese Kibo laboratory, the Japanese-built and American-operated Centrifuge Accommodation Module and two Russian laboratories.

Many experiments in these labs will be housed in EXPRESS (Expedite the Processing of Experiments to the Space Station) racks fitted within a refrigerator-size container called an International Standard Payload Rack (ISPR). Eight EXPRESS racks are being built for use on the International Space Station.

Payloads in these racks can operate independently of each other - allowing for differences in temperature, power levels and schedules - and may be controlled by the station crew or remotely from the ground by the Payload Rack Officer at the Payload Operations Center. Linked by computer to all payload racks aboard the ISS, one of these officers is on duty around the clock to oversee rack maintenance, monitor temperature and support experiments.

Each EXPRESS rack can be divided into segments as large as half of the entire ISPR or as small as a breadbox. They resemble large filing cabinets with their own electrical supply and plumbing, and include self-contained drawers for each experiment that can be easily retracted and sent back to Earth upon completion of the experiment, then replaced with another one. After being transferred from the Shuttle to the laboratory, they are tilted into position, then power, data, fluid and gas lines are connected through a utility interface panel on the front of the rack.

The EXPRESS rack system supports science payloads in several disciplines, including biology, chemistry, physics, ecology and medicine. With its standardised hardware interfaces and streamlined approach, the system enables quick, simple integration of multiple payloads aboard the station. The EXPRESS racks stay on orbit permanently, while experiments are exchanged as needed over operational periods lasting from three months to several years.

Further experiments will be mounted on "balconies" on the outside of the station, where they can be exposed to the vacuum and radiation of space for long periods of time. Research areas such as communications, space science, engineering, material processing and Earth observation can be supported on these sites.

There will be four locations on the S3 segment of the main space station truss where external payloads mounted on Brazilian-built Express pallets will be installed and removed using either the Space Station Remote Manipulator System or, in the case of a contingency, a spacewalk. Each Express pallet has six adapters with interfaces for payloads, and a control assembly that provides power and data links for the experiments.

Once they are operational, many of the experiments - both inside and outside the station - operate automatically and are only overseen by the ground, unless something goes wrong and human intervention is required. Scientists on the Earth receive much of the data in near real time so they can follow the progress of the experiment and even make adjustments if necessary.

US RESEARCH FACILITIES

The Destiny laboratory, which is the primary research module for US science payloads, was delivered to the ISS in February 2001. It is currently the only dedicated research facility on the station and will remain so for several years, although the Zvezda service module has a secondary role as the home for most Russian experiments.

The cylindrical Destiny module is 9.2 m (30.2 ft) long - including its Common Berthing Mechanism - and weighs 14.5 tonnes (32,000 lb). The laboratory consists of three cylindrical sections and two endcones with hatches that lead to other station components. A 50.9-cm (20-in) diameter window is located on one side of the central segment.

Its main purpose is to provide a pressurised, shirt-sleeve environment where astronauts can carry out a variety of scientific experiments in microgravity research, human life science, fundamental biology and ecology, as well as Earth observations, space science and commercial applications.

Computers in Destiny control power, thermal, and vacuum conditions for the experiments, as well as monitor the health and status of each payload. The module is also equipped with a number of life support systems, including cooling, air revitalisation, and temperature and humidity control. Two computers in Destiny keep the space station in its proper orientation (attitude).

The module accommodates up to 24 racks that are designed to fit against its four walls. Eleven of these racks will support space station systems, such as communications, control of the Canadarm 2 and the command and control computers, while the other 13 are specially designed to house scientific experiments. The current list of US-provided research facilities includes:

- *Human Research Facility.* The first payload rack to be brought to the station, it was installed in Destiny in March 2001. This double-rack

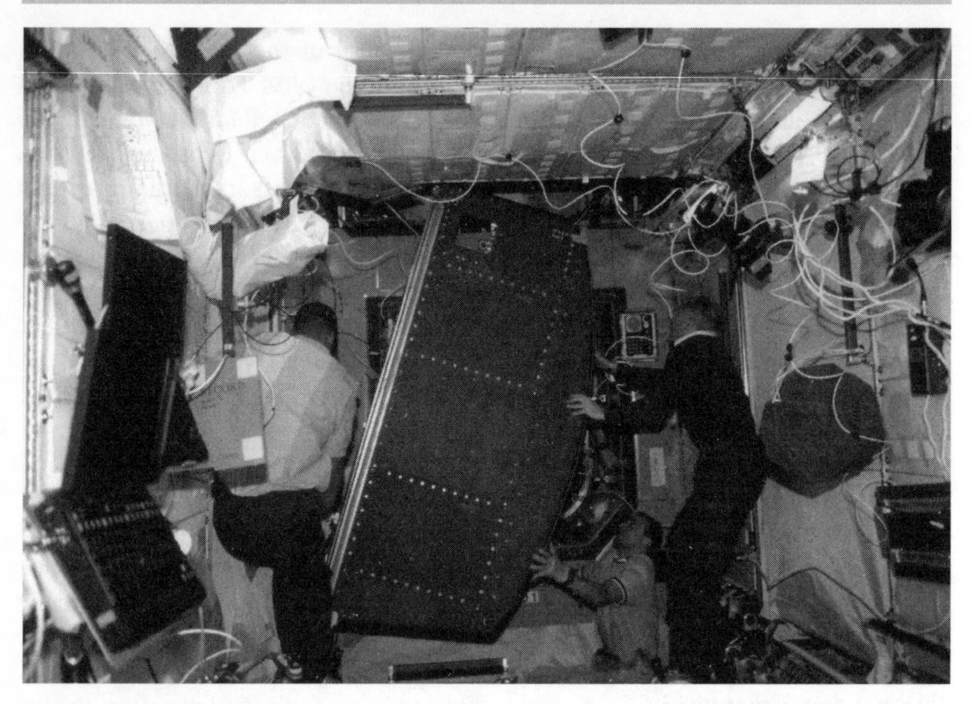

STS-98 astronauts struggle to move a rack into position in the Destiny module. (NASA photo STS098-322-001)

multi-user facility focuses on investigations to understand the physiological and psychological effects upon humans during long-term space flight. It will eventually be used in coordination with the European Physiology Modules Facility in the Columbus module.

- *Gravitational Biology Facility.* This consists of various pieces of equipment to perform investigations on a broad range of biological specimens. The centrepiece of the facility is a 2.5-m (8-ft) diameter centrifuge enabling experiments using different levels of gravity (0.01*g* to 2*g*). It also has Habitat Holding Racks to house the samples and a Life Science Glovebox to perform sample processing in a sealed environment. It is being developed by Japan but has run into development problems and is unlikely to be delivered to the ISS until 2008 or later.
- *Biotechnology Research Facility.* A single rack devoted to mammalian cell culture, tissue engineering, biochemical separations and protein crystal growth. It provides power, thermal management, video signal switching and processing, distribution of gases and bulk incubation at 37 °C. It will be installed in the Japanese Experiment Module around 2006.
- *Materials Science Research Facility.* This is dedicated to materials science investigations including solidification of metals/alloys, thermo-

physical properties, polymers, crystal growth and ceramics/glass studies. The facility consists of three autonomous racks, one of which (the Materials Science Laboratory) is provided by ESA. To be installed in Destiny 2005–6.

- *Fluids and Combustion Facility.* This is dedicated to investigations of fluid physics and combustion, and occupies three active racks and one stowage rack. The three powered racks are the Fluids Integrated Rack, the Combustion Integrated Rack and the Shared Accommodation Rack respectively. It will be installed in the Destiny module during several visits from 2005 onwards.
- *Low Temperature Microgravity Physics Facility.* This laboratory will be attached to the Exposed Facility on the exterior of the Japanese Experiment Module some time after 2006. There are two identical facilities each supporting two experiments in parallel operations.
- *Microgravity Sciences Glovebox.* This versatile, ESA-developed facility occupies a complete International Standard Payload Rack and can be used for many different investigations – materials, combustion, fluids and biotechnology. It is scheduled for installation in Destiny in May 2002, but will later be transferred to the Columbus module.
- *X-Ray Crystallography Facility.* This is a protein crystallisation facility for the analysis of large, complex crystals in a microgravity environment. The facility supports the preparation of crystals for visual evaluation and mounting, sample freezing and collection of X-ray diffraction data on selected crystals. It is scheduled for installation in Destiny in late 2005.
- *Advanced Human Support Technology Facility.* This serves as a research and development test bed for enabling technologies required to sustain human life during long-duration and deep space missions. It is scheduled for installation in the Japanese Experiment Module in 2005–6.
- *Window Observational Research Facility.* To be used for a variety of payloads for research related to geology, climatology, the atmosphere, land use, etc. It is due to be installed in Destiny in May 2002.

EUROPEAN RESEARCH

Some European facilities, such as the Materials Science Laboratory, will be located in the Destiny module. However, European researchers will eventually have access to their own 6.7-m (22-ft) long Columbus pressurised module, which is due for launch in 2005, and the Columbus External Payload Facility, where experiments can be positioned with the aid of the station's Remote Manipulator System. The Columbus laboratory can accommodate 10

Artist's impression of the European Columbus module. (ESA photo)

International Standard Payload Racks for scientific experiments. Five of these racks will be solely for use by European scientists, while the remainder will be available for NASA and other agencies that have contributed to the building of the station. There will be four European multi-user research facilities:

- *Fluid Science Laboratory.* This facility builds upon the experience gained by Europe through the Bubble, Drop and Particle Unit facility that flew on several Spacelab missions. Its purpose is to study fluid dynamic effects normally masked by gravity – convection, sedimentation, stratification or layering and fluid static pressure. Experiments may be conducted under automatic, semi-automatic or astronaut control. It is hoped that this research will feed back into manufacturing processes on Earth and improve the quality of products such as semiconductors.
- *Biolab.* This is designed to support biological experiments on micro-organisms, cells, tissue cultures, small plants and invertebrates. The major objective is to identify the role that microgravity plays at all

scales, from the effects on a single cell up to more complex organisms, including humans. The Biolab facility builds upon previous space facilities, e.g. Biorack and Biobox. It is hoped that the results will give new insights into areas such as immunology, bone demineralisation, and cellular repair. They could even have a significant influence on products in the medical, pharmacological and biotechnology fields.

- *European Physiology Module.* This also builds upon previous European experience gained through such facilities as Sled, Anthrorack and Physiolab. Human physiology experiments in microgravity are aimed primarily at increasing our knowledge of how the human body reacts to long-term weightlessness. However, this area of research also contributes to an increased understanding of problems such as ageing, osteoporosis (crumbling of bones in elderly people), balance disorders, and muscle wastage. Reference or baseline data will be collected before and after human missions to the station.
- *European Drawer Rack.* The European Drawer Rack provides a flexible experiment carrier that is not dedicated to any specific scientific discipline. One of its fundamental objectives is to support the operation of sub-rack level payloads (known as Class 2 payloads). One candidate experiment that could be accommodated in the European Drawer Rack is the Protein Crystallisation Diagnostics Facility, in which protein crystals are grown under controlled conditions.

European space science, Earth observation or technology payloads requiring accommodation on the unpressurised exterior of the ISS, will be placed on Express pallets. Up to seven ESA–NASA experiments will be exposed outside Columbus to try out new technologies. A laser-cooled atomic clock known as "Pharao" will measure time with unprecedented accuracy (about 100 times better than existing ground clocks). This will lead to more precise investigations in such areas as relativity research, atmospheric physics, geodesy (studies of the shape of the Earth), navigation and advanced telecommunications.

A solar-monitoring observatory will provide highly accurate measurements of total solar radiation and its output at different wavelengths. The fourth payload will expose various organic molecules and micro-organisms to space in order to see how they cope with such a hostile environment. Alongside will be a Sky Polarisation Observatory that will study the heavens in the unexplored frequency range of 20–90 GHz, with the aim of learning more about the cosmic microwave background – the left-over "glow" from the Big Bang that created the Universe.

JAPANESE RESEARCH

Most of the Japanese research will be conducted either inside the pressurised Japanese Experiment Module (JEM), known as "Kibo", or on an external platform. The 11.2-m (37-ft) long Kibo, which is due for launch in 2005, will be equipped with an airlock and a robotic arm, which will allow materials and experiments to be transferred to and from the Exposed Facility without having to depressurise the entire laboratory. A smaller, drum-shaped Experiment Logistics Module on "top" of Kibo will be used as a storage room for specimens, gases, equipment and consumables.

The laboratory will contain 10 research payloads, of which five will be available to Japanese users and five are allocated to NASA. Three of the five NASA racks will accommodate materials science and the other two will be devoted to life sciences. In addition, there will be a Freezer Refrigerator rack (the European-developed Minus Eighty Laboratory Freezer for ISS) and a number of stowage racks. The Japanese pressurised research facilities are:

- *Gradient Heating Facility.* This multi-user furnace will investigate the growth of crystals and semiconductors. It includes a unit for control of sample processing, control equipment and a mechanism for automatic exchange of samples.
- *Advance Furnace for Microgravity Experiments with X-ray Radiography.* A multi-user furnace that performs single-crystal growth experiments using the Floating Zone method. This provides the capability to observe

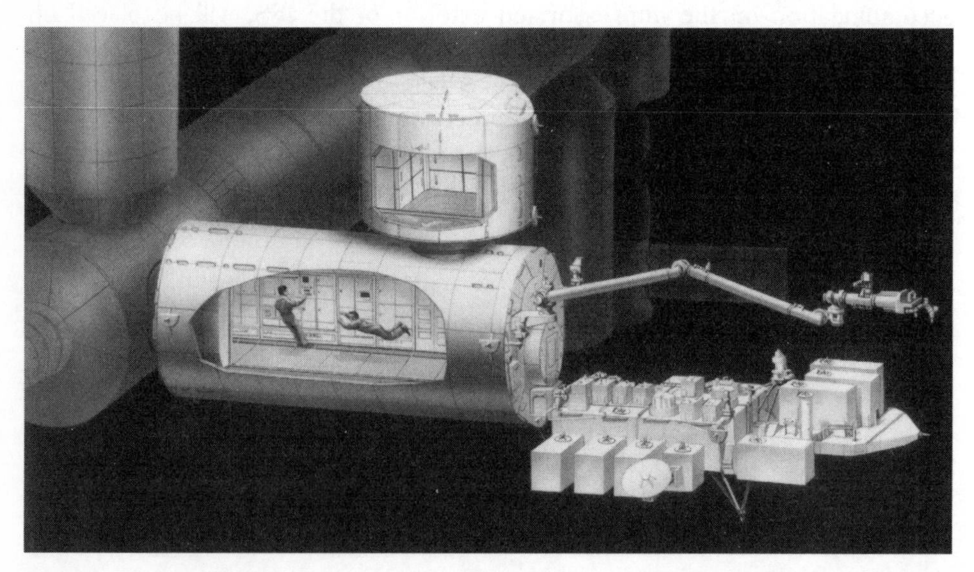

Artist's impression of the Japanese Experiment Module. (NASDA photo)

semiconductor crystallisation and the effects of convection in real time using X-ray radiography.

- *Cell Biology Experiment Facility.* This provides a controlled environment (i.e. temperature, humidity and CO_2 concentration) for the study of fundamental life sciences phenomena under microgravity, using cells, tissues, small mammals, plants and micro-organisms. It includes a centrifuge to provide variable gravity for reference experiments.
- *Clean Bench.* An enclosed working volume for processing life science and biotechnology experiment samples in a clean environment.
- *Biological Experiment Unit.* This is a multi-user system of chambers and a control unit installed on a medium-sized canister for performing experiments in the Cell Biology Experiment Facility. It can also be operated in the Clean Bench. One unit is available for plant life experiments and another for cell culture experiments.
- *Fluid Physics Experiment Facility.* A facility where fluid physics experiments can be performed in a moderate temperature environment. The primary objective is to investigate the effects of convection on experiments under microgravity.
- *Solution/Protein Crystal Growth Facility.* This focuses on fundamental studies of crystal growth of various solutions and proteins in space.
- *Image-Processing Unit.* This receives image data from various sources within the Kibo laboratory, encodes and compresses the data, and transfers it to the Kibo system. In addition, the unit can record experiment images on tape in periods when direct data downlink to Earth is not possible.

Ten payload attach points are also provided on the unpressurised pallet structure known as the JEM Exposed Facility. Five of these are available to Japanese users, with the other five for NASA payloads. The Japanese external payload facilities are:

- *Attached Payload Bus.* Developed as a simple interface to users of the Exposed Facility, it includes a power/communication interface to the ISS, a Power Distribution Unit for Attached Payloads, heater control equipment and an Extension Mechanism Assembly. Its two basic configurations are the box shape structure and pallet shape structure, both of which have "windows" that can be opened in different directions, providing flexibility for various experiments.
- *Space Environment Data Acquisition Equipment – Attached Payload.* This measures the space environment – neutrons, high-energy particles, cosmic dust, atomic oxygen, plasma, etc. – together with the effects of the space environment on materials and electronic devices.

- *Monitor of All-sky X-ray Image.* An X-ray camera with a very wide field of view to monitor high-energy astronomical objects, specifically the dynamic behaviour of distant, active galaxies.
- *Laser Communications Demonstration Equipment.* This will demonstrate extremely fast, two-way communications between the ISS and the ground. Communication experiments will be carried out at a downlink rate of 2.5 Gigabits per second and an uplink rate of 1.2 Gigabits per second, using 1.5-micron wavelength technology (fibre amplifiers and lasers).
- *Superconducting Submillimetre Wave Limb-Emission Sounder.* This will demonstrate a new method of observing the Earth's atmosphere through reception of submillimetre signals.

RUSSIAN RESEARCH

In addition to the Zvezda module, there are two other possible locations for experiments on the Russian segment: two research modules and the Science Power Platform. The research modules are expected to be located "beneath" Zvezda. Apart from these pressurised labs, there would be facilities for exposed experiments on the outside of the modules and on the central part of the Science Power Platform, on "top" of Zvezda alongside the European Robotic Arm.

Once the Russians agreed to join the ISS programme, a competition was announced in 1995 for scientific establishments, industrial companies and higher education establishments to put forward suitable experiments; 406 applications were received from more than 80 organisations in 11 major research areas.

The Long-Term ISS Russian Segment Science and Applied Research and Experiments Programme was developed and approved by Yuri Koptev, General Director of the Russian Aviation and Space Agency and Yuri Osipov, President of the Russian Academy of Sciences. The experimental programme began with the arrival of the Expedition One crew in November 2000.

Many of the current Russian experiments taking place on the ISS revolve around human adaptation to space and do not involve large amounts of complex equipment. They include an experiment to improve prediction of radiation doses in Zvezda; a test of acetominophene as a medicine; a search for genetic criteria that will provide maximum possible resistance in living organisms to the effects of prolonged space flight; studies of changes in the human cardiovascular and neurosensory systems; and the use of new computer technologies to improve the psychological condition of cosmonauts on long duration space missions. One experiment is even devoted to the problems of protecting the crew from injurious exposure to noise in the Service Module.

Earth observation has always been a primary area of interest for Russian manned missions and the ISS is no exception. One experiment is aimed at generating computer data bases of catastrophic events. Others take images of ocean colour – areas rich in algae and plankton-cloud formations, natural and artificial atmospheric pollution and other pollution effects. Another study looks at optical radiation emissions in the Earth's atmosphere and ionosphere related to thunderstorm activity and earthquakes.

Biotechnology research is largely devoted to growth and analysis of protein crystals with the aim of developing new generation medicines. Meanwhile, some effort is being put into the characterisation of microgravity disturbances and other environmental influences that affect scientific research on the ISS.

A number of other experiments provided by European scientists have been conducted in Zvezda. During the eight-day Andromède taxi mission in October 2001, Claudie Haigneré and her colleagues conducted a dozen experiments. These included Earth observation (with observations of lightning and atmospheric "sprites"), ionospheric studies, life sciences (including studies of the embryos and tadpoles of two species of amphibian), material sciences and technology. Another experiment was also devoted to ATV Control Centre operations.

TOURISM AND COMMERCIALISATION

With all of the space station partners struggling to make ends meet, any opportunity to raise funds from space station activities would seem to make sense. Curiously, the most recent converts to capitalism, the Russians, have been the most innovative in this field.

Not only did they accept $20 million from American millionaire Dennis Tito for a flight to the ISS in April 2001, despite fervent opposition from other agencies, but further tourist trips are in the pipeline. Various names have been put forward at different times, but the most likely candidate to fly on a second privately financed visit to the station seems to be South African entrepreneur Mark Shuttleworth, who was pencilled in for a Soyuz taxi flight early in 2002.

Like Tito, Shuttleworth was represented by a company called Space Adventures, which has been negotiating with the Russian Space Agency, RSC Energia, and the Yuri Gagarin Cosmonaut Training Centre since August 1999 to develop tourist flights to the ISS. According to a Space Adventures press release, "While in orbit, Shuttleworth aims to carry out a series of scientific experiments relevant to South Africa, including biomedical research on HIV/AIDS."

In recent years, sales of vacant Soyuz seats and privately financed ventures have provided almost 50% of the funding for Russian's manned space programme, so it was hardly a surprise when, in November 2001, the Amsterdam-based MirCorp announced plans to send the winners of a prime time TV game show on a Soyuz taxi mission to the ISS.

"We are looking forward to continuing this ratio for the international space station," said Alexei Krasnov, a deputy director at the Russian Aviation and Space Agency. "I get the impression from some places, particularly in the United States, that commercialising the station will compromise the image of greatness that should surround it. We are saying we should all use this as a way of showing that space can be accessible to a broader group of people."

The possibilities of using these regular Soyuz "taxi" missions to familiarise astronauts with the station and to carry out scientific research was soon recognised by the French space agency CNES, which has a long history of collaboration in manned spaceflight with the Russians. After an initiative by France's research minister, Roger-Gérard Schwartzenberg, a $17 million commercial agreement was signed between CNES, the Russian Aviation and Space Agency and RSC Energia to send Claudie Haigneré to the station.

Then, in May 2001, a broader agreement was signed under which Russia agreed to transport up to six European astronauts to the station before 2006. The European Space Agency or one of its individual member states would pay a fee for the Soyuz taxi services on a case-by-case basis. Under the agreement, ESA will have the right of first refusal for the third seat in the three-seat spacecraft for two types of missions: a week-long change-out of Soyuz craft or "increment" flights lasting up to four months. The first person to fly under the arrangement in 2002 would be Roberto Vittori, on a mission paid for by the Italian Space Agency.

High-profile manned missions and launches of space station modules have also become opportunities for Russian fund-raising. For example, the Proton launch vehicle that carried the Zvezda module into orbit was emblazoned with a Pizza Hut logo, which reportedly cost the US fast food chain $1 million. The company then paid the Russians an undisclosed sum to video the first space pizza delivery, although spinning the footage into a TV commercial was ruled out by Pizza Hut officials.

Promotional videos for *RadioShack*, *Lego* and *Popular Mechanics* magazine were also shot at the same time by Russian cosmonauts on the station. In the first of these, Russian cosmonauts Musabayev and Baturin presented colleague Yuri Usachev with a surprise gift for Father's Day – a "talking" picture frame from his daughter provided by the American electronics retail chain. NASA commander James Voss, who also received a similar present, was not allowed

Yuri Usachev with his weightless pizza (Wirepix digital photo)

to participate in the video. "We felt the first commercial filmed on the International Space Station was an extraordinary way to showcase this cool traveller's gift as an ideal Father's Day present," said *RadioShack* vice-president Jim McDonald.

In the first toy-based experiment on the ISS, Musabayev and Baturin were required to measure the mass of a *Lego* Life on Mars Red Planet Protector toy set using a special mass measurement device. The duo shot video footage of the experiment as well as a *Lego* banner and 300 toy "alien" figures that were to be given as prizes in a company competition. They also photographed a copy of *Popular Mechanics* and a banner displaying its name. The promotional package was arranged by a company called Space Media, which has been working with RSC Energia to develop and market multimedia (TV and Internet) material around the Russian space programme.

Until now, the other government-funded agencies have rejected the concept of private tourism and advertising on the station, although they are actively trying to encourage commercialisation of the ISS – with limited success so far. However, signs that NASA, at least, is beginning to revise its ideas came, in late 2001, with reports that the agency was contemplating a new strategy that would allow commercial advertising, merchandising and promotion of the US entertainment industry. Associated Press reported, "NASA might allow McDonald's to put its logo on the space station galley in exchange for McDonald's promoting space exploration to kids." In a sense, this is not too surprising, since the US Congress declared commercial utilisation to be one of the primary goals of the American space programme when it passed the 1998 Commercial Space Act and directed NASA to seek paying customers for the ISS. This led to NASA's first major commercial initiative – a partnership between the agency and Dreamtime Holdings Inc.

In return for privileged access to the station and agency resources, Dreamtime was to provide high-definition TV equipment and digitise NASA's treasure trove of video and photo archives. The intention was to provide the first HDTV coverage of astronaut activities on board the space station and make it available to schools and the general public through the company's web site. Eventually, this partnership would lead to the development of educational products and documentaries with the goal of increasing public awareness of the ISS and related programmes.

The initial step towards making this a reality came in August 2001, when the first ISS spaceflight-certified high-definition TV camera was delivered to the station. Over the next few months, crew members shot footage on a regular basis as part of their regular work routine, although Dreamtime proved rather backward in fulfilling some of its conditions, and agency oversight of the arrangement was criticised by the NASA Office of Inspector General.

A more traditional method of publicising the space station has been undertaken by the IMAX Corporation, which used 25 cosmonauts and astronauts to shoot more than 19 km (12 miles) of 65-mm film in space between December 1998 and July 2001. The resultant 3D documentary, "Space Station", narrated by film star Tom Cruise, was scheduled for release in spring 2002.

Two other commercial deals with private firms were signed with NASA in late 2000–early 2001. SkyCorp from Huntsville, Alabama, was hoping to demonstrate a novel method of building and launching small satellites from the space station.

These exceptions apart, the various agencies have largely concentrated their efforts on encouraging commercial spinoffs from ISS scientific and technological research. It is hoped that such research may benefit people on Earth, not only by providing innovative new products, but also by creating new jobs to manufacture the products.

One example has been a licensing agreement to use NASA-owned patents on a Bioreactor by Baltimore-based company StelSys. The company was hoping to develop new products, such as a machine to assist patients with liver disease and improved methods of drug production. In return, NASA would benefit through royalties and other payments.

To ensure that there will be adequate opportunities available for commercial uses, NASA (see Table 8.1) has committed to set aside approximately one-third

Table 8.1 NASA ISS Commercial Prices (liable to change)

Internal Rack Site Bundle
One 3kW ISP Rack Site: $20.8 million per site bundle
2,880 kWh energy
86 Internal Vehicular Activity (IVA) crew-hours
2.0 Terabits of data, space-to-ground

External Adapter Site Bundle
One Express Adapter Site: $20.8 million per site bundle
1,800 kWh energy
32 Internal Vehicular Activity (IVA) crew-hours
2.6 Terabits of data, space-to-ground

Space transportation is excluded from the bundles and priced separately, as premium services, allowing for privately offered space transportation services in the future.

$22,000 per kg ($10,000 per lb) each way
Space Shuttle passive pressurised cargo
$26,500 per kg ($12,000 per lb) each way
Space Shuttle passive unpressurised cargo
$15,000 per space station IVA crew-hour
$2,000 per space station kWh energy
$100 per minute TDRS transponder time

Additional services, over and above those included in the site bundles and without premium prices listed, are subject to negotiation, including services such as:

- additional space station crew training time
- space station accommodation in active rack isolation sites
- space station stowage space
- Space Shuttle transportation of active cargo

of the US share of the ISS's research capacity for private customers, ESA has reserved 30% and the Canadian Space Agency 50%. Russia (see Table 8.2) and Japan have yet to set firm policies, but both have similar plans for selling future research opportunities.

Some effort at coordination is being made, with a general acceptance that each partner has to agree to certain standards of good taste and that bidding wars should be avoided. However, at the end of the day, each partner has considerable freedom to decide how its facilities are used. Prices quoted by NASA indicate that ISS investment will only be open to large US companies for

Table. 8.2 Price limits for ISS Russian Segment Resources (liable to change)

Up payload delivery	\$10,000–20,000 per kg
Down payload delivery	\$20,000–30,000 per kg
Crew time	\$20,000–40,000 per crew-hour
Power	\$1,300–2,000 per kWh
Pressurised volume	\$800,000–1.5 million per m^3 per year
EVA	\$2 million–4 million per spacewalk
Spaceflight (guest mission)	Over \$10 million per person

the foreseeable future, whereas ESA has decided that smaller, less expensive chunks of station resources should be made available on its Columbus module.

"We think offering smaller sections of experiment space will have greater appeal to European customers," said Jochen Graf, head of the agency's space station preparation department.

The first commercial research activities on the station were begun during Expedition Two. These included experiments in the fields of biotechnology and agriculture that were developed by industry with the aid of NASA's Space Product Development Programme, which is set up to help industry through 17 Commercial Space Centres located across America. Under this arrangement, companies fund the research and pay for part of the launch costs, then are responsible for developing and marketing any new products or services that result.

Most of the centres are located on university campuses and work closely with other academic and government research institutions. The centres have agreements with almost 200 firms, including Bristol-Myers Squibb, ALCOA, Amgen, DuPont, Eli Lily and Company, Space Explorers Inc., Monsanto Company and Polaroid.

Six experiments on Expedition Four were sponsored in this way. They included investigations of treatments for bone loss, plant growth, pharmaceutical production, and petroleum refining.

THE PAYLOAD OPERATIONS CENTER

The Payload Operations Center (POC) at NASA's Marshall Space Flight Center in Huntsville, Alabama, manages all operations involving scientific research aboard the International Space Station. The centre was commissioned on 2 February 2001, shortly before the Destiny laboratory was delivered to the ISS.

The POC manages use of space station payload resources, handles science communications with the crew, and controls command and data transmissions to and from the ISS. The centre also integrates crew and ground team training and research mission timelines, plans the science missions and ensures that they are safely executed. It is also responsible for coordination of the mission planning for various international participants, all science payload deliveries and retrieval, as well as payload training and payload safety programmes for the station crew and all ground personnel.

"From this facility, we will manage fundamental scientific research that can only be done in space – research that will lead to knowledge to benefit all humanity here on Earth," declared Art Stephenson, director of the Marshall Center.

The new 1,220-m^3 (13,300-ft^3) command and control centre is located in a historic two-storey complex that previously provided engineering support for Apollo, Skylab and Space Shuttle launches, and housed the Spacelab Mission Operations Control Center from which more than 30 Shuttle-based science missions were controlled.

The centre is staffed around the clock by three shifts of between 13 to 19 flight controllers. To communicate with the ISS astronauts, POC flight controllers use the call sign "Huntsville". At the hand-over of resident crews, a fresh team of controllers takes over.

State-of-the-art computers and fibre optic cable communications feed reports

The ISS Payload Operations Center in Huntsville, Alabama. (NASA digital photo) (http://scipoc.msfc.nasa.gov/photospoc.html pocwide)

to staff manning consoles. Other computers stream information to and from the space station.

One of its major functions is to link Earth-bound researchers with their experiments and astronauts in orbit. The command centre is linked with and integrates the activities of research control centres and universities in the United States and throughout the world. These include four NASA Telescience Support Centers in Huntsville, California, Cleveland and Houston.

Four independent partner centres – in the US, Japan, Russia and Europe – prepare independent science plans for the POC. Each plan is based on requests from its participating universities, science institutes and commercial companies. Until Italy, Brazil and Canada develop partner centres of their own, the American centre deals directly with their particular requests.

EXPEDITION ONE SCIENCE OPERATIONS OVERVIEW

Since most of the crew's time was devoted to making the growing station operational, expectations for scientific research were fairly limited. The modest programme of activities focused on 10 experiments, mostly related to life sciences and human adaptation to weightlessness.

Much of the time was spent using the newly installed Treadmill Vibration Isolation System (TVIS) to study its effects on astronaut heart rate and motion, as well as its impact on the stability of surrounding structures. Another piece of exercise equipment that was also evaluated was the Interim Resistive Exercise Device (IRED) in Unity. A related experiment involved measurements of changing carbon dioxide concentrations in various parts of the station.

Materials science was covered by the Protein Crystal Growth–Enhanced Gaseous Nitrogen (PCG–EGN) experiment, which was intended to identify molecular structures that may lead to development of new drugs and improve scientists' understanding of diseases.

There were also two educational experiments. One demonstrated to schoolchildren the germination capability and growth of seeds in space. Video was downlinked to Earth for discussion. The other, a favourite from Shuttle missions, was the automated Earth Knowledge Acquired by Middle Schools (EarthKAM) experiment to take high-resolution electronic images of the Earth and transmit them to schools. The children participated by developing research projects and specifying Earth photography targets.

Earth photography was also undertaken by the crew, with a programme to photograph features of geological, meteorological or other interest.

The first hands-on experiment to be placed on the ISS was the Middeck Active Control Experiment II (MACE-II), developed by Massachusetts Institute

Yuri Gidzenko takes a still photo of a geographic target of opportunity through a window in Zvezda. (NASA photo ISS01-E-5083)

of Technology and the Air Force Research Laboratory. Launched on STS-106, MACE-II was designed to provide information on how things move and vibrate in space, and how to sense or control those motions. According to the researchers, this could lead to better telescopes, robotic arms and other devices affected by vibrations in space.

Several other experiments were flown to the ISS on board visiting Progress vehicles involving research in the fields of biotechnology, biomedical research, geophysical research and engineering. These included the station's first physical science experiment – a US–German–Russian collaboration known as the Plasma Crystal Experiment (PKE) – which was delivered to the station by a Progress ship in February 2001.

The PKE was designed to conduct basic research into the structure of matter by studying "dusty plasmas," electrically charged gases that contain tiny particles of solid matter in suspension. Instead of sinking to the bottom of the plasma under the influence of gravity, the microspheres also became electrically charged and dispersed in the plasma. They were then illuminated by a laser and imaged with a camera so that their positions and velocities could be recorded.

Also delivered by the same Progress was the Ecosystems in Space educational experiment, which contained the station's first "pets". Inside the small, self-contained cylinder that was placed above the crew's dining table in Zvezda were a number of miniature red shrimp, snails and other aquatic plants and animals. Over the next three months, video and photographic images were sent to students via the Internet, to show them how organisms adapt to microgravity.

EXPEDITION TWO SCIENCE OPERATIONS OVERVIEW

The first serious scientific research on the station began with Expedition Two and the delivery of the US Destiny laboratory in March 2001. This coincided with the start of around-the-clock operations in the International Space Station's Payload Operations Center at NASA's Marshall Space Flight Center, which linked Earth-bound researchers around the world with their experiments and astronauts aboard the space station.

Eighteen different experiments were scheduled for the four-month mission, including studies of the adaptation and reaction of the human body to the space environment, space radiation, Earth observation, crystal growth and plant growth.

The first research payload to be installed in the first ISS laboratory module was one of two racks making up the Human Research Facility. This rack contained a computer work station and portable laptop computer for crew members to command and test the equipment, store and transmit experiment data to Earth.

Some of the equipment was used for periodic assessment of crew aerobic capacity, measuring and analysing the air breathed in and out by crew members. An ultrasound imaging system provided three-dimensional images of the heart and other organs, muscles and blood vessels that could be transmitted back to Earth for study. Other experimental data handled by the Human Research Facility included measurements of radiation, changes in the nervous system and crew interactions psychological surveys.

One of the main science goals of Expedition Two was measurement of the radiation environment inside the orbital laboratory so that countermeasures may be developed for extended duration missions. Radiation exposure in high doses over long periods of time can damage human cells and cause cancer or injury to the central nervous system. In an effort to develop better shielding materials for spacecraft and space suits, three experiments – the Bonner Ball Neutron Detector, Dosimetric Mapping (DOSMAP) and the Phantom Torso – were devoted to measuring the effects of such exposure.

The Japanese Space Agency's Bonner Ball Neutron Detector comprised six spherical detectors filled with helium gas. They measured neutron radiation – uncharged atomic particles that can penetrate and damage living tissue, including the blood-forming bone marrow of humans and animals.

As its name suggests, the Phantom Torso (nicknamed Fred) comprised a dummy of a male's upper torso, sliced into 34 sections. Covered with Nomex "skin" and fitted with "organs", it appeared identical to the real thing in X-ray images. The first attempt to measure the effects of space radiation on internal organs, it contained both passive and active radiation detectors. Data from the active sensors was downlinked to Earth every 10 days.

The German-developed DOSMAP experiment comprised an electronic reader and various kinds of radiation counters scattered around the station. One type of detector in Destiny used silicon wafers to measure the flux.

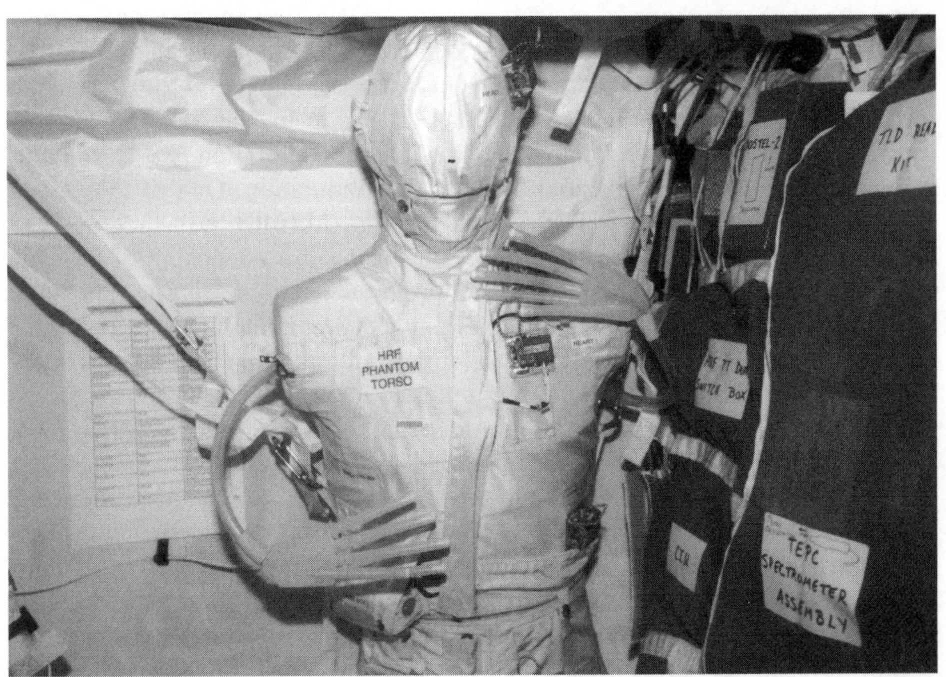

The Phantom Torso was one of three Expedition Two experiments to measure the effects of radiation on organs inside the body. (NASA photo ISS002-E-5952)

Another small, mobile type provided by the Hungarian Space Office used calcium sulphate crystals to absorb energy from ionising radiation (protons, neutrons, electrons, heavy charged particles, gamma and X-rays). Forty-eight of these were placed permanently on the station. Astronauts could measure the overall exposure by heating the crystals so that they emitted a glow of light and the brightness of the glow indicated the radiation count. This measurement could be checked back on Earth by analysing another type of detector, 12 plastic sheets which became pitted with tiny craters when exposed to heavy charged ions.

Although the experiments worked well, an unplanned shutdown during a data transfer occurred on 9–10 May. The problem involved the Human Research Facility personal computer that controlled the three radiation experiments. Start-up difficulties with the station's high data rate Ku-Band antenna and main computer had prevented the data from being transferred earlier from the recorder, which stores information on experiment results and hardware operating conditions, to the ground.

Another human life sciences investigation – Hoffman-Reflex – studied changes to the human nervous system during long-duration space flights. This Canadian experiment was begun while the Expedition Two crew was still on board the Shuttle and repeated some months later to see how their responses to a small electric shock administered to the back of the knee had changed.

Other research facilities launched to the station in late April 2001 included the first two EXPRESS racks. EXPRESS Rack 1 in the Destiny module contained experiments related to plant growth, vibration measurement, commercial protein crystallisation and drug fermentation. One of these, the Microgravity Acceleration Measurement System (MAMS), was first used to measure disturbances on board the station during the undocking of a Soyuz taxi ship and the subsequent docking of a Progress resupply craft in May 2001. Such studies of the vibration environment over the coming months would give important background information to scientists operating delicate microgravity experiments on board.

EXPRESS Rack 2 carried a vibration damping device called the Active Rack Isolation System (ARIS) that was intended to isolate sensitive experiments from outside influences, and the Physics of Colloids in Space experiment for studying the behaviour of particles suspended in a fluid. Sensors of the Space Acceleration Measurement System II (SAMS-II) were located in both EXPRESS racks. There was also a Payload Equipment Restraint System, which was designed to help astronauts access and carry equipment and tools.

Several commercial activities began during Expedition Two. Advanced Astroculture was the first plant growth facility installed in the ISS. It was used to grow plants of *Arabidopsis* (a member of the cabbage and radish family) over a complete cycle – seed germination, plant growth and seed development – and also assess the impact of spaceflight on their genes. Over a period of six to seven weeks, the crew sampled nutrients, gases and plant transpiration or "breathing" while changes were recorded on video and downloaded to Earth. The first plants and seeds produced by the experiment were originally scheduled to return on STS-105, but the plants finished their growth cycle by mid-July, so they were brought back a month early on the STS-104 mission.

The Commercial Protein Crystal Growth–High Density experiment was successfully used to grow larger and more perfect crystals from more than 1,000 biological solutions. The growth cylinders were returned to Earth on the July Shuttle mission. However, the Commercial Generic Bioprocessing Apparatus, which was meant to study why antibiotic production by microbes is enhanced in microgravity, shut down on 9 May during commanding from the ground, and was later returned to Earth for an inquiry into what went wrong.

The Physics of Colloids in Space experiment, which was also delivered in April 2001, investigated how fine particles suspended in fluids behave. Although examples of colloids such as paint, milk and ink are common in everyday life, scientists were hoping to understand better the physical properties of such substances, with the aim of eventually producing new materials and products. The crew ran into difficulties activating the experiment, but eventually got it up and running in late June.

The last experiment to be launched to the station, on its third visit to the orbiting laboratory, was a student crystal growth investigation – the Enhanced Gaseous Nitrogen Dewar – that lifted off on STS-104 in July 2001. It was returned to Earth on the next Shuttle the following month.

As usual, the Crew Earth Observation programme provided many happy hours of Earth-watching for the crew. The targets were many and varied. For example, between 6 and 9 June, scientists requested images related to global warming, agriculture and urban development. The landscapes imaged by the crew included the Red Basin in China, the Yellow River delta near Beijing, the Ganges river basin, the rift valley junction in Ethiopia, the Suez canal and agriculture east of the Nile, the Philippines, Tanzania, the Yangtse river, urban industrial centres in southeast Africa, and the glacier on Mt Kilimanjaro.

Meanwhile, more than 250 photographs of the Earth were taken with the automated EarthKAM, which operated until the cameras were deactivated and stowed on 18 May. This gave time for students to analyse the pictures via an interactive web site before the experiment resumed in the autumn.

Expedition Two payloads returning to Earth with the STS-105 Shuttle mission included the Middeck Active Control Experiment for studying the behaviour of structures in space; a biological materials experiment called the

Commercial Protein Crystal Growth–High Density; and the Dosimetric Mapping and Phantom Torso radiation experiments.

As the five-month expedition drew to a close, NASA officials reflected on the expedition and declared it a successful beginning to continuous science operations on board the orbiting research outpost. "The expedition Two crew has blazed the trail for Expedition Three by taking the first set of payloads and operating them on the station," commented Ray Echols, lead Payload Operations Director at Marshall Space Flight Center.

"We are well on our way to building and operating a world-class facility in orbit," said John Uri, lead increment scientist for Expedition Two at Johnson Space Center. "We accomplished virtually all of the goals we set out to accomplish. The crew took on the research programme as their own, giving up personal time to catch up or get ahead on tasks. The mission set the tone for the rest of the expeditions on the station."

Lybrease Woodard, lead Expedition Two Payload Operations Director at Marshall Space Flight Center, praised the first payload control team for successfully interacting with the station crew, the operations centre in Huntsville and telescience centres around the world. "It was like leading a new orchestra," she said. "Everybody learned how to play their parts and then came together like a symphony."

During the five-month mission, more than 15,000 commands were sent up to experiment payloads and support equipment on the station. Another landmark came on 1 May when researchers at a Colorado telescience centre became the first to send commands remotely to their commercial bioprocessing experiment.

Apart from the failure of one of the commercial experiments, the only serious problem encountered during the mission was the shut down on 15 May of the Medium-rate Communications Outage Recorder (MCOR) – the main storage device for science data during periods when the station was not in contact with satellites or ground stations.

EXPEDITION THREE SCIENCE OPERATIONS

The research begun by the two previous crews expanded considerably during the Expedition Three mission, with 18 American experiments undertaken. Beginning with the launch of Shuttle Discovery (STS-105) in August 2001, new research facilities launched to the station included two more EXPRESS racks (numbers 4 and 5) and the Cellular Biotechnology Operations Support System (CBOSS).

The additional EXPRESS racks brought the total number of research racks (including the Human Research Facility) in Destiny to five. CBOSS, installed in Rack 4, was the station's first experiment devoted to biological cell cultures. It was designed to augment cell growth and research aboard the space station, providing preservation, temperature regulation and proper stowage of specimens during delivery, experimentation and return to Earth. Bioreactor cell growth in microgravity permits cultivation of tissue cultures of sizes and quantities not possible on Earth, allowing research in areas pertinent to human diseases, including cancer, diabetes, heart disease and AIDS.

"This will be our first opportunity to use a sophisticated bioreactor to grow cells in low gravity created as the station orbits Earth," said John Uri. "Cells appear to grow more three-dimensional, like they do in living tissues, when they are cultured in space."

New experiments and payloads devoted to protein crystal research were the Dynamically Controlled Protein Crystal Growth C and V and the Advanced Protein Crystallisation Facility. These would be used to grow biological materials that may lead to insights in the fields of medicine, agriculture and other areas.

Apart from the continuation of the studies into muscle weakening and loss of bone mass during extended spaceflights, several other investigations were related to crew health. The Renal Stone experiment studied the reasons for an increased risk of kidney stone development during and immediately after a space flight; Xenon 1 looked into blood-pressure problems and fainting that may occur when astronauts return to Earth; and Pulmonary Function in Flight focused on measuring changes in gas exchange in the lungs and on detecting changes in respiratory muscle strength. Also taking place for the first time on the station was the Materials International Space Station Experiment to test the durability of hundreds of samples ranging from lubricants to solar cell technologies.

NASA's first commercial enterprise – Dreamtime – began on Expedition Three with the delivery of a high-definition television camcorder to the station as part of a public/private partnership to upgrade NASA's equipment to next generation HDTV technology.

Ten other experiments continued from earlier missions. They included the Space Acceleration Measurement System and Microgravity Acceleration Measurement System; the Active Rack Isolation System; a colloids/fluids science investigation; the EarthKAM experiment allowing students to select targets for an automated camera on the station; the Hoffman-Reflex experiment to study changes in nervous system; the Bonner Ball Neutron Detector; the Interactions crew–ground support personnel questionnaire; and Crew Earth Observations.

Towards the end of the mission, two cosmonauts mounted a variety of scientific instruments outside Zvezda. One was a Russian experiment called

Kromka, which was designed to collect contamination caused by Zvezda's steering jets in order to help the design of better thrusters for future spacecraft. Dezhurov and Tyurin also assembled a small truss structure and attached three suitcase-sized experiment packages provided by NASDA, the Japanese space agency.

The Micro-Particles Capturer used aerogel and foam substances to collect naturally occurring micrometeoroids and human-made particles of orbital debris. A companion Space Environment Exposure Device exposed a variety of materials such as paint, insulation and solid lubricants to the harsh environment of space. Another exposure experiment was replaced as part of a commercial agreement.

Meanwhile, a further dozen French–Russian experiments were conducted in association with the Andromède Taxi Mission (see below).

Summarising the status of the ISS science operations as Expedition Three drew to a close in November 2001, John Uri, the Expedition Four science mission manager, commented, "Since our first payload reached the Space Station in September 2000, we have launched more than 4.6 tons (4,200 kg) of research hardware and experiments, and returned more than a thousand pounds (500 kg) of hardware, samples and other data to Earth. ... The (Destiny) laboratory has five research racks, and we have accomplished the goals of 28 research payloads, supporting 41 investigations from government, industry and academia in the United States, as well as Japan, Canada, Germany and Italy."

By the time the Expedition Three crew handed over to their replacements, the overall ISS crew time dedicated to the research programme had reached nearly 500 hours, with more than 50,000 hours spent on experiments – including those operated by controllers at the Payload Operations Center.

ISS Experiments on the France–ESA Andromède Taxi Mission (October 2001)

- AQUARIUS: Three biological experiments, two to learn more about development of amphibian embryos and tadpoles, and another to study yeast. The biological fixation of the embryo containers was performed every day and filming of the African toad tadpoles also went normally. Yeasts were activated and a control experiment was performed simultaneously at the Korolev flight control centre.
- SPICA: An experiment to measure the radiation levels inside the station and its impact on electronic components. It began operation on 20 September.
- COGNI: An experiment to understand how the brain deals with human perception of three-dimensions and orientation in weightlessness, where

Claudie Haigneré using the COGNI apparatus to test her perception of space and orientation. (ESA/CNES photo)

there is no up or down. The numeric keypad required for the 3D navigation experiment developed a fault on the second day of the experiment schedule but the three astronauts successfully completed the visual orientation experiment on the same day. After checks by ground teams, the navigation experiment was performed on the fourth day using the central computer's keyboard. A second series of navigation and orientation readings was successfully taken on the sixth day.

- Cardioscience: A study of blood pressure changes during spaceflight and return to Earth. Claudie Haigneré and flight commander Viktor Afanasyev took readings on the second day. One of the two batteries discharged, and a problem was encountered transferring data to the laptop. After investigations on the ground, modified software was uplinked to the ISS and data was then transferred successfully. A second series of readings took place later in the visit.
- EAC: Claudie Haigneré answered a questionnaire provided by the European Astronaut Centre on her computer. Questions concerned sleep, moods and Mirsupio. A problem was encountered when backing up data. The experiment's scientist was on hand in Toulouse to analyse the problem and uplink a troubleshooting and recovery procedure.
- LSO: An experiment using two microcameras to study sprites and other light emissions in the atmosphere. This experiment was performed by the resident ISS crew before the taxi flight's arrival, when the ISS was oriented towards Earth. Five filming sessions were completed (one more than planned), three lasting 24 hours on the horizon and two of 24 hours at the nadir. Claudie Haigneré then checked the films recorded on the computer. All five were successful.
- IMEDIAS: A programme to photograph unusual features, such as forest fires, pollution, aerosols and cloud formations with digital and film cameras. The experiment was started on 1 October by the resident crew. A dozen sites were photographed (the Niger, Nile, Ganges, Volga and Amazon deltas, New York, Cairo, etc.). IMEDIAS continued after the departure of the Andromède crew, as soon as the ISS was pointing towards Earth again.
- GCF: A protein crystallisation experiment that was started as soon as it arrived on the ISS on 23 August 2001.
- PKE: A study of plasma crystallisation in the absence of gravity. It started on the fifth day of the visit.
- MIRSUPIO: A test of a multi-purpose bag with elastic pockets.

EXPEDITION FOUR SCIENCE OVERVIEW

December 2001 saw another change-over in orbit and on the ground. The Advanced Protein Crystallisation Facility and the Dreamtime High Definition TV camera were returned to Earth by STS-108. At the same time, a new suite of experiments was delivered to the station in the Raffaello logistics module on board Shuttle Endeavour, and a fresh team took over science operations in the

Payload Operations Center. Further scientific equipment was to follow on the STS-110 mission in April 2002.

The three new station residents were scheduled to devote about 500 hours to research during their stay on the ISS. During Expedition Four, the programme of scientific research was extended as the number of US payloads increased from 18 to 26 – seven of them new to the space station science programme and several with multiple experiments. Of the 26, two were in fundamental biology, seven in human life sciences, six in microgravity science, six in the development of space products and five sponsored by the Office of Space Flight.

Twelve experiments were managed by Marshall – the main NASA centre for microgravity research. Six of these were partially funded by industry and sponsored by NASA's Microgravity Research Programme.

"We are going to accomplish more science on Expedition Four than we attempted on any of the previous three expeditions," said John Uri, lead scientist at Johnson Space Center. "We are increasing the scope and sophistication of the science we are doing on the space station by building on what we have learned during the earlier expeditions. This month (December 2001) marks nine months of continuous research and an extraordinary increase in research capabilities aboard the station."

New experiments during Expedition Four were expected to lead to insights into bone disorder treatments, petroleum production, antibiotic production, cancer cell formation, plant growth, embryo development, biotechnology, and long-term effects of spaceflight on humans.

Future space greenhouses were pioneered by a Biomass Production System designed to validate equipment necessary to grow wheat and Brassica in microgravity and eventually develop a dedicated plant research unit for the space station. The secondary objective was to study the effects of microgravity on wheat photosynthesis and metabolism.

A variety of protein crystal growth experiments took place during the five-month mission. A new experiment competing for the prize for longest name was the Protein Crystal Growth Single Thermal Enclosure System (PCG-STES) Diffusion – controlled Crystallization Apparatus for Microgravity (DCAM), which provided a controlled temperature environment to grow large, well-ordered protein crystals in microgravity using vapour diffusion. By examining the crystals' molecular structure on Earth, scientists hoped to learn more about key biological functions.

A number of other previously flown experiments focused on developing new or improved drugs. The second flight of the Commercial Protein Crystal Growth experiment would once again enable production of crystals whose molecular structures could be studied on Earth, while the Commercial Generic

Bioprocessing Apparatus hardware, which broke down in May, was given a second chance to investigate bacterial fermentation and antibiotic production in space.

The Protein Crystal Growth Enhanced Gaseous Nitrogen Dewar, making its fourth space station flight, was once again put through its paces to demonstrate a low-cost platform for growing biological materials and studying optimum growth conditions.

A newcomer to the station's materials science investigations was the Zeolite Crystal Growth Furnace, a commercial experiment to grow larger crystals with possible applications in the chemical and electronic industries. Zeolites are crystals that can absorb and hold liquids or gases. Since they are used in the oil industry, better understanding of their structure and how they work could lead to more efficient petroleum production.

In the pharmaceutical field, a commercial experiment called the Microencapsulation Electrostatic Processing System was aimed at developing a process for producing large quantities of multilayered microcapsules of drugs that could be placed in the human body. Unfortunately, the sample containers failed a vacuum test shortly before launch in December 2001, so the experiment was not transferred to the station.

The only US radiation experiment scheduled for Expedition Four involved three active dosimeter badges to be sewn into pressure suits worn by spacewalking astronauts to determine the levels of radiation received to the skin, eyes and blood-forming organs.

Two acceleration measurements started during Expedition Four – the Space Acceleration Measurement System II and the Microgravity Acceleration Measurement System. These were to continue throughout the life of the space station, monitoring disturbances that could affect science experiments. During Expedition Four, researchers also continued to work with the Active Rack Isolation System to isolate disturbances from vibrations.

The experiment of the physics of colloids that began during Expedition Two was to be concluded during Expedition Four. The Cell Biotechnology Operations Support Systems experiment continued from Expedition Three, carrying new cell lines, as researchers tried to cultivate the cells into healthy tissue that kept the form and function of natural living tissue. This time, the experiment would attempt to grow cells of human kidneys, blood and tonsils.

With four EXPRESS Racks already installed and operational, Rack 3 was scheduled to arrive at the end of Expedition Four. The Human Research Facility Rack 1 was also to be joined by a second rack in 2002, providing expanded opportunities for biomedical research.

Human Life Sciences experiments continuing from earlier expeditions included monitoring for kidney stones; measuring lung function; measuring bone loss and recovery; monitoring nerve reflexes; measuring blood circulation; and surveys of interpersonal and cultural factors that may influence crew performance.

The Materials International Space Station Experiment – a collaborative effort by NASA, the US Air Force and private industry – located outside the station continued to test the durability of hundreds of samples, ranging from lubricants to solar cell technologies.

Further images of the Earth were to be obtained with the automated digital camera of the EarthKAM experiment and the other cameras of the Crew Earth Observations programme.

Some Expedition Two experiments sponsored by commercial companies also flew again during Expedition Four. These included a slightly modified version of the Advanced Astroculture experiment to grow *Arabidopsis* from seed to maturity. A new feature on this flight was the capability to collect samples of the growing plants while they were growing in space. Among the seeds used were some harvested from plants grown during Expedition Two.

In addition, 35 Russian experiments – most of which were already on board – were scheduled for Expedition Four. The majority of these were passive and needed very little intervention by the crew. Indeed, only 72 hours of crew time was allocated for these experiments, with the Russian commander responsible for almost all of the active operations. Four of the experiments were continuations of previous commercial contracts, with two provided by the Japanese Space Agency, one by Kodak and another experiment to verify a global timing system.

However, even as the research programme on the station continued to expand and evolve, other forces that once more threatened to turn the space station into a multi-billion dollar white elephant were gathering in the marble halls of Washington, DC.

Chapter Nine

Future Uncertain

2002

During 2001, the International Space Station had grown from a 70-tonne, apartment-sized foothold in orbit to a 150-tonne complex with more internal space than a three-bedroomed house. After a year of unprecedented achievement in assembling the largest man-made object ever to orbit the planet, 2002 was expected to be a year of consolidation. Four Shuttle flights, two Soyuz replacement missions and four Progress resupply missions were scheduled to dock with the space station.

Flights by Atlantis and Endeavour were scheduled to deliver more than 50 tonnes of components to the ISS, along with more than three dozen new experiments and two new laboratory racks. It was also to be the year of the spacewalk, breaking the record set in 2001 for the most EVAs ever conducted in a single year. Fifteen spacewalks were planned from the Space Shuttles alone, with another seven by crews from the International Space Station.

Apart from swapping the Expedition Five crew in May and the Expedition Six trio in September, the key task assigned to the Shuttle fleet was the continued construction of the station's main truss structure. This was scheduled to begin with the April delivery of the S0 truss, a base section of the framework required to connect more solar arrays to the station. Spacewalking astronauts would attach the 12-m (40-ft) long truss section to the Destiny laboratory and prepare it for the eventual addition of a further four truss segments on each side of the space station's spine.

The following month, the Shuttle would deliver the Canadian-built Mobile Base System. Two spacewalks were scheduled to install the system that would enable the Canadarm 2 to travel up and down the truss, helping with station maintenance and assembly tasks.

The next truss section to arrive would be the first starboard segment (S1) together with its cooling radiators, although these were to remain stowed until the truss could be extended still further the following year. This first piece on the right-side of the truss would be attached to the central S0 segment. A Crew and Equipment Translation Aid (CETA) cart would also be provided for use by spacewalkers as they move along truss with equipment.

Later in the year, the first port (left-side) truss segment (P1) and its radiators would be brought by Shuttle Endeavour, along with a second CETA cart. The P1 would also be attached to the central truss segment during a series of spacewalks. These activities would pave the way for the installation of additional solar arrays that would boost the station's power output in 2003.

THE BUSH BLUEPRINT

Despite the rapid progress in assembling the space station, it has been clear for some time that there is a huge cost to bear. Even before the launch of the Expedition One crew, there were signs that all was not well.

In October 2000, NASA administrator Dan Goldin and Jacob Lew, director of the US Office of Management and Budget, began to prepare the case for the defence by declaring that "NASA has worked hard to control station costs". They went on to state that "total development is 90% complete, more than half of the hardware is in orbit or at the launch site and the space station budget is beginning its spending tail-off to a planned steady level for operations and research".

However, despite the technological and operational successes, all was not well with the ISS programme. The roof was about to fall in on the agency's inefficient management and shambolic accounting.

The first problem arose at the end of October, when President Clinton signed into law an act prohibiting NASA from spending more than $25 billion on space station assembly. Based upon the agency's previous budget predictions, this would still have left a $1.2 billion safety buffer for unexpected overspend, but it soon became clear that even this amount would not be sufficient.

An early sign of things to come was NASA's decision to stop work on the $200 million Interim Control Module by the Naval Research Laboratory. The partially completed ICM was to be mothballed indefinitely. The agency justified the decision by insisting that the ISS had plenty of fuel and declaring its full confidence in Russia's continuing ability to build the Progress craft that would resupply the station and regularly boost its orbit.

By February 2001, NASA officials informed Congress that they expected the cost of building the station to increase by as much as $3 billion over the next four years. NASA field centres were put on notice that tough budget choices were looming and some projects would have to be axed. "We have come to realise that finding the level of manpower required to get hardware on orbit and operating properly is going to be very challenging," said NASA spokeswoman Kirsten Larson. "It's going to require more manpower than we initially expected."

It was subsequently revealed that about 2,300 NASA staff were assigned to the station programme, in addition to 3,400 employed by prime contractor Boeing – about 1,000 less than in 1999. Then, on 28 February 2001, President George W. Bush released his 2002 spending outline. The document, "A Blueprint for New Beginnings", stressed the Administration's commitment to "permanent human presence in space, world-class research in space and accommodation of the international partner elements". This would be achieved by providing $14.5 billion for NASA in the coming financial year – a 2% increase over 2001. The document also acknowledged that space station costs would grow by approximately $1 billion in 2001 and 2002, and by $4 billion over the next five years. However, there would a price to pay in the long term.

The increased funding would have to be largely offset through budget reductions in space station hardware and other human spaceflight programmes. In addition, space station programme management would shift from Johnson Space Center, Houston, to NASA Headquarters in Washington DC, a change that was made all the easier by the 23 February removal of JSC's influential director, George Abbey.

NASA's financial incompetence and its implications for the future of the ISS programme stunned Congressional leaders. "The revelation that the programme is $4 billion in the red only proves how poorly NASA managed the programme under the Clinton Administration," declared Dana Rohrabacher, chairman of the House Science, Space and Aeronautics Subcommittee.

FROM SEVEN TO THREE

The President's "strategy of constraining space station cost growth" received an immediate response from NASA. In a press conference called the same day, the agency's space station chief Michael Hawes told reporters that NASA would scale back or stop work altogether on its Habitation Module, Crew Return Vehicle (CRV) and Propulsion Module in order to keep the total spending on station assembly below the $25 billion spending cap imposed by Congress. Spending on scientific research over the next five years would be scaled back by $981 million, while US funding for the Centrifuge Accommodation Module – a major scientific facility being built by Japan – was also under threat.

It was immediately obvious that, without the US Habitation Module, there would be insufficient accommodation for the full complement of six or seven crew. Even more significant was the loss of the CRV and its ability to return the full number of ISS residents to Earth during an emergency, despite Hawes's assurance that, once NASA found a way of solving its budget problems, it would look for ways to return these elements to the programme.

Bush's solution "represents a strong commitment to fly off the hardware that we have built and are testing", he said. "It reflects a commitment to fly our international partners' modules. And it asks NASA to solve within its budget resources any enhancements to that configuration that will help us increase the crew size and add more capability."

This announcement inevitably led to a furore both inside the United States and beyond. In the absence of a crew complement of seven, the station's science potential was unlikely to be fulfilled, since the crew would never have sufficient time to devote to extensive research programmes. "A three-strong crew was not sometimes enough even at Mir, which is much smaller and required less operations efforts," commented a Russian space official.

Congressman Tim Roemer scathingly denounced the "incredible shrinking science programme". Even Hawes admitted that science would suffer, although he tried to minimise the consequences. "It's going to be a different kind of science programme perhaps ... But we will still deliver a science programme that is of high value to the tax payer," he said. Hawes went on to admit that the downsizing would result in a less ambitious final version of the orbital outpost. Under the first sketchy outline plan, it was suggested that NASA would end its hardware contribution to the station with the delivery of Node 2 – the docking unit to which the European Columbus and Japanese Experiment Module would be attached – which was scheduled to arrive in November 2003.

Alarmed by the implications of this downsizing, the Italian Space Agency (ASI) rapidly intervened by offering to work with NASA on construction of the Habitation Module. In exchange for building the module, ASI suggested several areas in which NASA could provide cost-equivalent services, including the launch of four Cosmo Skymed radar satellites, flights of Italian astronauts to the station and access to NASA's station facilities.

On 19 April, NASA boss Dan Goldin and ASI President Sergio DeJulio signed an outline agreement to build the station's living quarters. Joseph Rothenberg, NASA associate administrator for spaceflight, praised ASI's proposal, saying that it was "a hard offer to refuse". He went on to state that such an arrangement would be seen favourably by NASA officials, since it required NASA to put up little if any cash. However, discussions were complicated by a number of commercial proposals, including one with a large Russian component, and a subsequent investigation into NASA finances, making it impossible to decide on the future of the Habitation Module.

RUSSIA'S ENTERPRISE

On the announcement of NASA's proposed cutbacks to the ISS programme, one senior Russian official solemnly stated, "Unfortunately, NASA is encountering the same kind of problems that we have had and that is lack of financing. We have managed to cope with this problem and honour our obligations and NASA should do the same."

Anyone listening to this statement would have assumed that the Russian manned programme had successfully overcome its budget constraints and was confidently looking forward to full development and exploitation of the International Space Station. In fact, the reality could hardly have been further from the truth.

Lacking the necessary investment from government sources, Russian space enterprises had been forced into seeking cooperative commercial ventures with American aerospace companies in order to develop new ISS modules. The first of these began in late 1999, when RSC Energia and Spacehab announced a joint venture to build the first privately financed and operated module for the ISS.

Known as Enterprise, the module was originally marketed as a multimedia studio suitable for live TV and Internet broadcasts, as well as private research and possible space tourism. The entire project met with a cool reception from NASA and then hit the buffers with the dramatic decline of the dot.com multimedia industry.

Not to be disheartened, the Energia-Spacehab team saw another market opportunity appear in February 2001 when NASA caused a furore by cancelling its Habitation Module. Within weeks, Russian Space Agency officials informed their ISS partners that Enterprise had been selected as a substitute for the previously planned Stowage and Docking Module. By locating it at the Earth-facing port on the front end of Zarya, Enterprise had been reinvented as a "space hotel" which would be available for rent to customers such as NASA and the European Space Agency. Company officials eagerly offered to build more sleeping space and life support systems into the 50 m^3 (1,800 ft^3) offered by Enterprise. John Lounge, senior vice-president of Spacehab, even went so far as to predict that leasing of crew accommodation and storage space would be expected to provide about half of the $100 million annual revenue from the module during its first five years of operation.

Meanwhile, a second US–Russian partnership was also working on a commercial module for the station, using a back-up version of the FGB/Zarya module that was originally under construction. Once the first ISS section was successfully launched in November 1998, the Russian Khrunichev enterprise envisaged several uses for the spare FGB, including its operation as a giant supply ship, but in 2000, ISS prime contractor Boeing announced that it would work with the Russians to commercialise the FGB-2. This led to a rather undignified scrap for pole position with Spacehab-Energia over which project would win the right to the coveted docking slot on Zarya.

Matters came to a head in early August 2001, when the Russian Space Agency managed to broker a compromise between the opposing groups. Under the agreement, both enterprises were to work together on optimising the commercialisation of the station. All of their proceeds from such programmes were to be fed back into the national programme for reinvestment in the construction and maintenance of the Russian segment.

Since it was nearest completion, priority was to be given to the FGB-2, which would be available as a Simplified Universal Docking Module attached to the Earth-facing port on the front of the Zvezda service module. With a launch set for March 2003, commercial opportunities would be available through sponsorship and rent of excess stowage capacity.

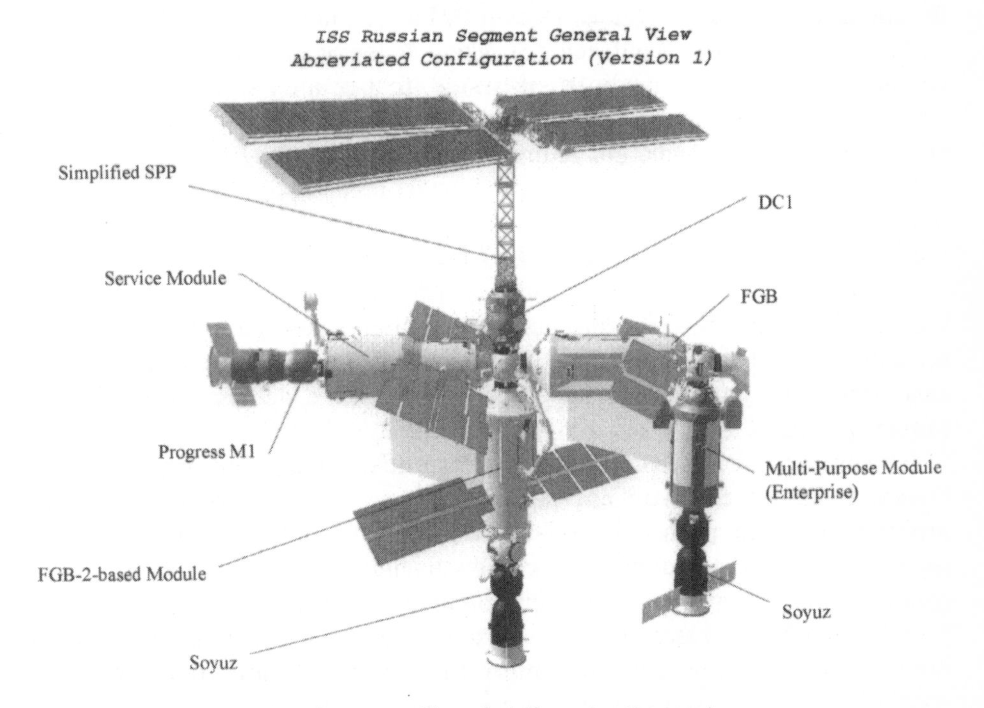

The Reconfigured Russian Segment. (Spacehab-Energia diagram)

At the same time, the Enterprise multipurpose module was assigned the favoured docking port on Zarya, near the centre of the station, and launch was expected in 2004.

Meanwhile, in an admission that government funding would be inadequate for the foreseeable future, it was decided to cut back on the size and complexity of the structure that was to provide the bulk of the Russian segment's power supply. The new plan was to move the Pirs docking compartment to the upper port on Zarya, then attach a Simplified Science Power Platform, without a pressurised compartment or attitude control thrusters. The SPP would be launched on the Shuttle (STS-122) in April 2004, one and a half years later than originally scheduled.

One of the major design changes involved the removal of a pressurised core compartment, which was to have been accessible to cosmonauts and used for research payloads. Four solar panels, instead of the original eight, would be added later, effectively halving the available energy supply. These money-saving modifications are likely to have serious implications for future research and expansion of the Russian part of the space station.

The rest of the Russian ISS programme remains in a state of flux, with little visible progress for the two (or more) Science and Application Modules intended for the ISS. In the baseline plan, these research laboratories were to be docked with the Universal Docking Module (now FGB-2). If they ever materialise, the prime contractor is likely to be RSC Energia. Although a Ukrainian-built module has been suggested, that country's finances are in an even more parlous state than those of its giant neighbour, so its development must remain pure speculation at this stage.

SAVING THE CRV

While the arguments and discussions continue to rage over the future accommodation facilities on the ISS, NASA and the European Space Agency have been struggling to find a way of continuing with development of the Crew Return Vehicle.

In November 2000, European governments, led by Germany and Italy, agreed to invest $127 million in the CRV – more than one-tenth of the projected cost. In return, European companies would be able to work on advanced technologies to be used in the vehicle's nose cone, body, wings and landing gear. In this way, ESA would be able to offset much of the money it would owe NASA in return for transport of European experiments to and from the station and for use of American data-relay satellites for direct communication with the station.

After the February 2001 announcement, NASA officials suggested that an even greater European contribution to the CRV programme might be a solution to the impasse. At first, ESA space station director Jorg Feustel-Buechl considered this to be most unlikely. "We have about come to the limit of what we can do," he declared. "We would love to help our American friends, but money does not just appear out of nowhere."

Six months later, ESA officials indicated that they might be able to increase their contribution to 40% by cancelling the launch of one of the nine Automated Transfer Vehicles that were intended to supply the ISS. However, the money was only on offer if the US agreed to continue the CRV programme.

Meanwhile, the test programme for the CRV prototype still continued. On 13 December 2001, the X-38 successfully completed its eighth large-scale flight test – the highest, fastest and longest flight to date at NASA's Dryden Flight Research Center. The subscale spaceplane was released from NASA's B-52 aircraft at an altitude of 13,600 m (45,000 ft) and descended under remote control using ESA-developed software to guide the parafoil and steer it to a safe landing.

"The X-38 tests involve innovative technologies that will be useful for many future human spacecraft as well as a crew rescue vehicle," said X-38/CRV programme manager John Muratore. "Although the production of the crew rescue vehicle for the station is deferred, we are continuing to test and mature these technologies to reduce the technical and cost risk of a future CRV production programme."

If a way cannot be found to continue with the CRV, one possible alternative that would allow six crew to live on the station is the use of multiple Soyuz spacecraft as lifeboats. However, since the seating capacity of the Soyuz only enables three persons to squeeze inside, two of the craft would have to be docked to the station at all times.

This, too, has its drawbacks. The 180-day lifetime of the Soyuz means that Russia would have to build and launch at least four spacecraft per year, compared with two under existing arrangements. This would require significant investment both by NASA and by the Russian space sector, since the Soyuz production line would have to be expanded. Each spacecraft takes about two years to build, so advance orders would be essential, and NASA would have to pay an additional $80 million for each additional Soyuz. This $800 million bill over the 10-year operational lifetime of the station would not be acceptable to many.

"Aside from the cost issues, there is no way to assure the stability of the Soyuz supply, and there is certainly no technology to be gained for us in purchasing additional Soyuz vehicles," said one European government official. "And without Soyuz, the station becomes a high-class museum in orbit. No one wants that."

THE TASK FORCE

With the agency under continuous fire from all quarters and the future of the manned programme increasingly uncertain, NASA administrator Dan Goldin had little choice but to seek outside guidance. In late July 2001, he appointed an independent task force to examine the budget and management issues facing the US space station programme. "We must ensure it is carried out in a more efficient and effective manner," he admitted.

The ISS Management and Cost Evaluation Task Force (IMCE), which included two Nobel laureates and a world famous heart surgeon, was asked to "assess the quality of the ISS cost estimates as well as programme assumptions and requirements, and identify high-risk budget areas and potential risk mitigation strategies".

While the panel deliberated, 61-year-old Goldin began to contemplate his future. As the longest-serving NASA administrator, he had already served under three Presidents and gained a reputation for a controversial management style by embracing change instead of fighting cuts in the agency's budget allocations. As it became increasingly obvious that he had allowed the ISS programme to evolve into a runaway juggernaut that consumed cash at an inordinate rate, the beleaguered NASA boss decided it was time to call it a day. On 17 October, he announced his decision to quit in one month's time.

Two weeks later, the IMCE sent a damning report to the NASA Advisory Council, recommending that both the programme and the agency be put on probation for the next two years and their performance closely monitored.

While acknowledging the outstanding technical accomplishments so far achieved, the 20-strong task force found that the existing plan to eliminate up to $5 billion of station cost overruns in the period 2002–2006 was not credible.

When the station was redesigned in 1993, NASA's original cost estimate was $17.4 billion, with completion of assembly by June 2002. In reality, the station was now four years behind schedule with the cost of assembly complete now put at more than $30 billion. Blaming most of the increase on the prolonged delays and a general inability to produce reliable cost estimates, the task force found evidence of "multiple budgeting techniques and multiple reporting techniques" involving both NASA staff and contractors.

According to the report, "Financial forecasting and strategic planning suffer from insufficient 'forward' analysis and planning due to division of financial authority and responsibility, lack of experienced financial personnel and modern tools, diverse and often incompatible accounting systems, and uneven and non-standard cost reporting capabilities."

An additional contributing factor was the imposition of yearly financial limits, which caused NASA officials to focus on meeting annual budgets rather than on total cost management. The $2.1 billion annual cost caps imposed in 1994 were condemned as "counterproductive to controlling total programme cost".

This evidence of lack of cost control and inadequate accounting was reinforced when the task force identified new cost increases totalling $366 million on top of those already notified by NASA. As a result, the panel decided that the agency's declared cost of $8.3 billion to build the so-called US core of the station (without the Habitation Module, Advanced Life Support System, Node 3 or Crew Return Vehicle) was lacking in credibility. Their conclusion was that no one knows how much the station will actually cost to build.

Even NASA's rapid switch to a skeleton "core complete" programme had not been adequately managed, with the result that necessary cutbacks in the expenditure on the research budget had not yet been implemented. As a result, the agency was still behaving as if the payloads would be operated by six or seven persons, the scientific community was confused and the international partners belligerent.

As the report stated, "the only way management will actually manage the programme, and thereby get its costs under control, is through being forced to live with less".

The task force went on to make a number of specific recommendations in an effort to "enhance the probability that a credible ISS core complete programme can be established". However, the panel also expressed the hope that its "responsible" plan left open the possibility of moving beyond core complete to a fully capable ISS, if justified by NASA performance. Meanwhile, they called for minimum funding to be provided, within existing budgets, so that the so-called "enhancements" to the core complete station could be reintroduced relatively easily at a later date.

In an effort to improve space station management, the report suggested creating a new NASA post of Associate Administrator for the ISS, with oversight of all aspects of the programme. The existing multitude of station contracts should be consolidated, with clearly defined cost reporting requirements, while new management systems to control planning and finance were urgently needed.

The task force also recommended several cost-cutting alternatives. One method to reduce expenditure would be to increase the typical on-orbit time for each expedition to six months, while at the same time reducing the Space Shuttle flight rate to four per year. This would result in the completion of the US core being shunted back to April 2004, and the international partners' elements being delayed by up to one year, but provide savings of $688 million.

Already identified options for slimming down the agency that could be achieved through facility and lab closures, together with staff reductions, might produce further savings of $350–450 million. Numbers of technical and operations staff could be trimmed by requiring them to be "on-call" rather than immediately "on-tap".

As a carrot to scientists and international partners, the group suggested the development of a credible road map with achievable performance targets, so opening the way to expanding research potential if NASA demonstrated its ability to execute a pre-planned programme and meet its commitments.

The task force unanimously agreed that the highest research priority should be given to the study of problems associated with long-duration human spaceflight, including the engineering required to support people in space. The other top science priority was the speedy completion of the Centrifuge Accommodation Module, which was considered to be essential for biological research – though no solution of how to bring it forward from 2008 was offered.

In the absence of a six- or seven-person crew, the panel proposed the use of extended duration Shuttle flights and the overlap of Soyuz taxi missions. Although most Shuttle–ISS visits lasted about one week, this time could be doubled by using an orbiter equipped with enhanced power and life support systems. This would allow Shuttle crews to give more assistance to the ISS residents, particularly with tasks involving spacewalks and maintenance.

If a replacement Soyuz was launched one month early, then its three-person crew would be on board the station for an extra 30 days, providing an opportunity for six people to carry out research and essential maintenance on the station. However, it was clear that there were significant cost implications to be clarified before either of these options could be introduced.

FALLOUT

Within a matter of days, the IMCE report received endorsement from the White House Office of Management and Budget (OMB), from most members of the independent NASA Advisory Council, and from Congressman Sherwood Boehlert, the chairman of the influential House Science Committee.

One of the most notable supporters of the task force's recommendations was Sean O'Keefe, deputy director of the OMB and the person nominated to replace Goldin as the next head of NASA. "I think, quite bluntly, the administration can use these recommendations and do something with them other than use it as a doorstop," he said.

Others, however, were not so easily impressed. Lawmakers from Texas and Florida, two of the states most likely to suffer from any layoffs at NASA and in the aerospace industry in general, were highly critical. "I just feel that, if we follow this path we are setting ourselves up for failure," said Texas

Congressman Nick Lampson. "We need to seriously consider the financial impacts on some of the communities."

Others questioned the impact on the ISS science programme. "My concern is that it started out 20 years ago as a very ambitious scientific project," said Michigan Congressman Vernon Ehlers. "I worry whether there is going to be any science left at all."

While expressing similar concerns, the international partners were particularly concerned about the effects on their multibillion dollar programmes, and what they saw as a breach of an intergovernmental treaty.

In Moscow, the Russian Aerospace Agency demanded an explanation of NASA's plans, while officials declared that the proposal for a Soyuz exchange every five months would require some form of compensation. Furthermore, a decrease in Shuttle visits would inevitably lead to more Progress supply flights – an added expense that Russia was unlikely to bear.

Canada formally requested a "government-level multilateral consultation of ISS Partners at the earliest practical time" to discuss the issue. The European Space Agency also protested to the US State Department, and the Japanese government was preparing to follow suit. "You will understand that, at a time when the European partner is preparing to make significant financial commitments towards ISS operation and utilisation on the basis of the partnership's existing international obligations ... it is a requisite to give the European partner states assurances that the ISS remains the one defined in the ISS agreements," wrote Herbert Diehl, chairman of ESA's ISS coordinating committee.

"A three-member crew means a first-class museum in orbit, not a first-class research institute," commented Jorg Feustel-Buechl, ESA space station director. "You need two and a half people to operate the station. That leaves half an astronaut's time for research. For our laboratory, that would mean 100 minutes of research per week. That is incompatible with the basic requirement that the station be a research tool. If we go to a crew of three, our laboratory and Japan's laboratory are both useless."

Following on from these comments, the ministers of ESA's member states were most circumspect at their meeting in Edinburgh on 14–15 November. Although they agreed to allocate $757 million to the space station between 2002 and 2006 – 12% less than ESA had requested – most of these funds were frozen for the coming year while the member states waited to see whether NASA would accommodate their concerns. "We will block 60% of these funds until September (2002) to see what decisions the US will make," said German research minister Edelgard Buhlman.

The outgoing NASA administrator, Dan Goldin, was less than sympathetic. In an appearance before the NASA Advisory Council, Goldin noted that the US has given its station partners slack on agreements when they had financial trouble, and expressed the hope they would reciprocate now that NASA was suffering in the same way.

On 19 December, the NASA Advisory Council (NAC) publicly pronounced its verdict on the task force report. In a letter to Acting NASA Administrator Daniel Mulville, the committee stated, "The NAC unanimously and completely endorses the findings and specific recommendations of the IMCE Task Force report." It went on, "The deficiencies in NASA's management and financial control of the ISS programme identified in the IMCE report cannot be excused and must not be ignored. Resolving these deficiencies will require major restructuring of the management, budget, infrastructure, and staffing of NASA's human space flight enterprise."

The committee agreed with the widely held view that the US core complete configuration, which supports only a three-person crew, is far from optimal as an end-state for the ISS. Nevertheless, they recommended that no commitments to augment the US core complete configuration should be made until NASA's restructuring and consolidation efforts restored confidence within the NAC, the Administration, and Congress in the management and financial controls of the ISS programme. "This 'period of consolidation' can last no more than two years," it said. The letter went on,

> Throughout the period of consolidation, NASA should consult with the ISS International Partners to ensure ongoing understanding of the progress in meeting the milestones for the US Core Complete configuration as well as the associated reestablishment of credibility.
>
> The Goals of the US International Space Station Program are not well-defined. The NAC notes that neither short-term nor long-term priorities for the United States' uses of the ISS have been firmly and clearly established. The relative priorities of exploration, international cooperation, research, education, and commerce have not been clarified. Given this lack of clarity, it is not surprising that there is little public understanding of why the United States is building the ISS.
>
> The focus of the ISS program must now change from development and construction to operation and utilisation. Indeed, this change is overdue ... The NAC recommends that NASA announce a small number of ordered priorities for US scientific usage of the ISS immediately ... The IMCE recommended that science be represented in ISS management as the Deputy Programme Manager, and the NAC concurs with this recommendation.

As for the international partners, the NAC recommended that NASA seek clarification of the US government's position concerning the possibility of an eventual enhancement of the capabilities of the ISS beyond US core complete and investigate ways in which such enhancements might be achieved.

Meanwhile, NASA should continue to inform the partners of its progress and discuss cooperative strategies to augment the core complete configuration.

Meanwhile, even as the briefings to the Advisory Council on the task force's recommendations continued, House and Senate Appropriations committees were trimming $95 million from the Administration's Fiscal 2002 request for the space station.

SCIENCE IN THE DOLDRUMS

As both the IMCE task force and the NASA Advisory Council recognised, space station science was now at the top of the agenda. With NASA-funded research on the International Space Station facing a cut of almost $1 billion over the next five years, and lengthy delays to European and Japanese modules, the scientific capability of the ISS was a major concern to politicians, researchers and international partners alike.

"We find core complete as an end state to be unacceptable," said Doug Bassett, deputy programme manager at the Canadian Space Agency. "Everybody agrees that a three-person crew will not be able to conduct science that is consistent with the level of investment."

This viewpoint was supported by the findings of the IMCE task force, which stated that "three-person crew accomplishments have not met expectations ... Crews required an additional one to two hours per day to accomplish the daily workload. It is anticipated that as more station hardware is integrated on orbit (especially the additional European, Japanese and Russian laboratories) that this problem will be exacerbated."

Scientists on the NASA Advisory Council generally accepted the task force recommendations, accepting that there must be clearly defined science milestones in order to judge whether NASA should be given more funding latitude after 2003. Even then, members agreed, trying to get the scientific community at large to go along with this scheme would be far from easy. "We'll just have to take it as a challenge," said John McElroy, chairman of the National Academy of Sciences' Space Studies Board.

A measure of the size of the challenge was provided by members of the Biological and Physical Research Advisory Committee. In recommendations to the NASA Advisory Council, the committee wrote:

> Unless alternative strategies are taken ... NASA can no longer justify to the citizens of the United States ... (a) the completion of space station build out, and (b) the maintenance of such a structure and its present operational facilities on its current "house of cards" strategy.
>
> The Biological and Physical Research Advisory Committee views this

> budgeting strategy as being totally unrealistic, and, in essence, blatantly undermines the primary rationale for constructing [the space station].

The panel went to add that the cuts bring biological and physical research on the station "to the brink of extinction".

Apart from the possible year-long delay to the start of research on Europe's Columbus module and the inadequate size of a three-person crew to conduct the experiments, the proposed reduction in flight opportunities and the US emphasis on human life sciences was a major cause of concern to the international partners. Jorg Feustel-Buechl went on record to state that ESA's research priorities were dedicated to improving life on Earth through new materials or drugs, while other nations might put commercialisation at the top of their list. The US should not set priorities for others, he said.

While the arguments raged, the development of future scientific hardware was also causing concern. Most urgent was the US-financed Centrifuge that Japan was developing as partial compensation for the Shuttle launch of the Kibo Module. On-orbit delivery of the Centrifuge, which was considered to be essential for research into the effects of microgravity on animals and plants by the IMCE task force, had already been delayed until 2006. However, technical problems had caused this date to slip by a further two years – a delay which the task force had dubbed "unacceptable". Apart from the technical and scheduling concerns, NASA and Japanese officials were also worried about the possible cost overruns facing the key project.

Financial concerns also arose over Brazil's contribution to the ISS. With costs soaring to 50% higher than anticipated, the Brazilian aerospace company, Embraer, was dropped as prime contractor for the country's space station hardware. The shift to a new industrial consortium threatened to delay the delivery of the first Express science pallet, which had been scheduled for launch in 2003.

FUTURE STATIONS?

With so many question marks hanging over the future of the $100-billion International Space Station, what is the likelihood that smaller, cheaper competition may arise in the coming decade?

One speculative venture announced by MirCorp and RSC Energia in September 2001 showed that this unlikely scenario may just become reality. Under an agreement with the Russian Space Agency, the consortium declared its intention to develop and operate the world's first private space station. Slated for launch in 2004, Mini Station 1 would comprise a Salyut-sized module with a multiple docking adapter for visiting Soyuz spacecraft.

Artist's impression of Mini Station 1 with two Soyuz spacecraft docked. (MirCorp digital photo)

The new station would be equipped to accommodate up to three visitors – including two paying customers – for 20-day stays. Over its orbital lifetime of more than 15 years, the station would be serviced by both Soyuz manned transports and unmanned Progress cargo resupply ships.

MirCorp was said to be holding extensive discussions with a range of commercial customers, while working with NASA, the European Space Agency and the other International Space Station partners to send users to the ISS until the mini-station is operational.

"MirCorp understands that the International Space Station is dedicated to world-class science and belongs to multiple governments," said Gert Weyers, company senior vice president. "We have shown there is a market for a different type of customer, whether a tourist, a commercial scientist, a film maker or anyone who is healthy and has a dream of space travel. MirCorp's mini station answers this market need."

Under the planned scenario, MirCorp Soyuz vehicles will visit both Mini Station 1 and the International Space Station. On a typical flight, the Soyuz would go first to Mini Station 1, where it would remain docked for the two-week commercial mission. It would then fly to the ISS, where the Soyuz crew would transfer to the older Soyuz already attached to the international station. The crew would return in this spacecraft, leaving a newer model as the lifeboat for the space station's residents.

MirCorp President Jeffrey Manber believed that such commercial activities would help the Russian Federation to fulfil its commitment to support the International Space Station. "This is a great agreement for ISS, its partners and everyone who dreams of flying to space," he said.

An even more ambitious rival to the ISS may be waiting in the wings. Unlike its more established rivals, China is planning to double its civilian space budget over the next five years, according to Luan Enjie, director of the China National Space Administration. A large slice of this multibillion-dollar effort will be devoted to making the world's most populous nation the third in history to send people into space.

Evidence of its ambitions is not hard to find. In recent years, the Chinese have developed the man-rated Long March 2F booster, successfully launched several unmanned Shenzhou spacecraft and constructed a large Vehicle Assembly Building at an isolated launch site in the Gobi Desert. Other facilities under construction or already completed include a new manned mission control centre, training facilities for a dozen astronauts and Shenzhou spacecraft assembly and checkout facilities in or near Beijing. China's space tracking network, including spacecraft monitoring ships, has also been modernised to handle manned flight operations.

All that remains is the first launch of men into orbit, and a group of astronauts is now in training to achieve this breakthrough, possibly as early as 2002–2003. Once the system has been successfully tried and tested, the programme will gradually progress to more ambitious flights involving extravehicular activity (EVA) and rendezvous and docking techniques, much like the US Gemini programme of the 1960s.

In the long term, an indigenous space station will be developed and launched by a new heavy-lift booster. Studies of a large Chinese space station, known as Project 921-2, have been under way for more than a decade, but this structure still only exists on paper.

The original 921-2 space station module had a total mass of 20 tonnes, a length of 15 m (50 ft), and a diameter of 4.2 m (14 ft). It was equipped with a

Computer-generated picture of the unmanned Shenzhou 2 spacecraft. (Simon Zajc)

Mir-like five-port docking section at the forward end. A scaled down version of this station was eventually given the green light in February 1999.

Although Western visitors have reported seeing illustrations of a Chinese space shuttle docked to a multimodular space station, these technologies remain far beyond current capabilities. On the other hand, it is not beyond the bounds of possibility that the Chinese could soon develop a Shenzhou craft capable of docking with the ISS, followed by a Mir-class space station some time after 2010.

Meanwhile, possible participation in the International Space Station has not been ruled out. "We want to open our manned programme," Luan said. "We want more frequent communication with entrepreneurs, engineers and enterprises in the US so that we can have extensive cooperation in the future." Without China's participation, the ISS "is not a true international programme", he added.

Before such hesitant overtures can be translated into international agreements, a number of barriers have to be overcome, not least the domination of the Chinese manned effort by the Chinese People's Liberation Army (PLA) and the barrier of secrecy that surrounds the space programme. Other stumbling blocks, which have hindered US–Chinese relations in recent years involve issues of technology transfer, human rights and weapons sales.

"The space station's supposed to stand for something better," said Congressman Dana Rohrabacher, chairman of the House Space and Aeronautics Subcommittee, referring to China's poor human rights record.

However, such attitudes may soften in the not-too distant future, especially as budget cutbacks begin to bite among the ISS partners. Chinese scientists have already begun to collaborate with their colleagues in Europe, including work on an experiment known as the Alpha Magnetic Spectrometer, an antimatter detector due to be installed aboard the ISS in 2003.

A FIERY END

It may seem premature to discuss the end of the International Space Station before it has been completed and while its future is still obscured in financial turmoil and political wrangling. However, as the fiery demise of the Mir station demonstrated, it pays to have some advance planning when deciding how to dispose of an orbital space station with a mass of several hundred tonnes.

NASA has already performed several assessments to determine the best and safest way of decommissioning and deorbiting the ISS. Not surprisingly, the risk studies determined that an uncontrolled re-entry would offer an unacceptable threat to human life, with a 1 in 20 chance that lives would be lost. This figure is far higher than the agency's stated safety target of a 1 in 10,000 chance of a single casualty.

Other alternatives include boosting the ISS to a much higher orbit to prolong its on-orbit lifetime; disassembling the pieces and returning them to Earth in the Shuttle; or attempting a controlled re-entry with a splashdown in the Pacific.

NASA concluded that boosting the ISS to a higher orbit is not a suitable option, since it would require more fuel than was available. Furthermore, this manoeuvre would only postpone the inevitable, since the raised orbit would still deteriorate over time.

Although disassembly has been favoured as an option by the European Space Agency, NASA studies consider that – since the space station was not designed to be taken apart – this would be too difficult and expensive.

This leaves controlled deorbiting as the sole remaining option, assuming that there is a sufficient and reliable propulsion capability to do the job. In one analysis, the station's termination would begin with the undocking of the Soyuz vehicles, followed by one or more deorbit burns from one or more fully fuelled Progress vehicles and the Zvezda module. Another possibility that was investigated involved the US Propulsion Module, but this system has now been cancelled and so will probably not be available when the station's 10- to 15-year operational life reaches its end.

The success of the Mir re-entry in 2001 demonstrated that such a manoeuvre is possible, but it is an inescapable fact that the ISS is much larger and even more irregular in shape than its illustrious predecessor, so predicting its behaviour upon entry to the atmosphere will be a major challenge. Clearly, in death as in life, the International Space Station will be grabbing the headlines and creating a great deal of international controversy.

Appendix 1

ISS Assembly November 1998 to January 2002

No.	Launch date (local time)	Launch vehicle	Flight number*	Payload/Mission
1	20 Nov 1998	Proton Rocket	1A/R	Control Module (Zarya)
2	4 Dec 1998	Space Shuttle (STS-88)	2A	Node 1 (Unity)
3	27 May 1999	Space Shuttle (STS-96)	2A.1	Spacehab/Supplies
4	19 May 2000	Space Shuttle (STS-101)	2A.2a	Spacehab/Maintenance
5	12 Jul 2000	Proton Rocket	1R	Service Module (Zvezda)
6	6 Aug 2000	Soyuz Rocket	1P	Progress M1-3
7	8 Sep 2000	Space Shuttle (STS-106)	2A.2b	Spacehab/Logistics
8	11 Oct 2000	Space Shuttle (STS-92)	3A	Z1 truss, PMA-3
9	31 Oct 2000	Soyuz Rocket	2R	Expedition 1 Crew
10	16 Nov 2000	Soyuz Rocket	2P	Progress M1-4
11	30 Nov 2000	Space Shuttle (STS-97)	4A	P6 Truss & Radiators
12	7 Feb 2001	Space Shuttle (STS-98)	5A	Destiny Laboratory
13	26 Feb 2001	Soyuz Rocket	3P	Progress M-44
14	8 Mar 2001	Space Shuttle (STS-102)	5A.1	Expedition Two Crew/MPLM
15	20 Apr 2001	Space Shuttle (STS-100)	6A	SSRMS (Canadarm 2)/MPLM
16	28 Apr 2001	Soyuz Rocket	2S	Soyuz Taxi Mission
17	21 May 2001	Soyuz Rocket	4P	Progress M1-6
18	12 Jul 2001	Space Shuttle (STS-104)	7A	Joint Airlock (Quest)
19	10 Aug 2001	Space Shuttle (STS-105)	7A.1	Expedition Three Crew/MPLM
20	21 Aug 2001	Soyuz Rocket	5P	Progress M-45
21	15 Sep 2001	Soyuz Rocket	4R	Docking Compartment 1 (Pirs)
22	21 Oct 2001	Soyuz Rocket	3S	Soyuz Taxi Mission
23	26 Nov 2001	Soyuz Rocket	6P	Progress M1-7
24	5 Dec 2001	Space Shuttle (STS-108)	UF-1	Expedition Four Crew/MPLM

* A = US Flight R = Russian Flight P = Supply by Progress
S = Exchange of Soyuz UF = Utilisation Flight

Appendix 2

Future Assembly Schedule

(Updated 13 December 2001)

Date	Assembly flight	Launch vehicle	Element(s)
2002 (21 March)	8A	US Orbiter STS-110	• Central Truss Segment (ITS S0) • Mobile Transporter (MT)
2002 (2 May)	UF-2	US Orbiter STS-111	• Multi-Purpose Logistics Module (MPLM) with payload racks • Mobile Base System (MBS) • Crew Rotation
2002 (1 Aug)	9A	US Orbiter STS-112	• First right-side truss segment (ITS S1) with radiators • Crew & Equipment Translation Aid (CETA) Cart A
2002 (6 Sep)	11A	US Orbiter STS-113	• First left-side truss segment (ITS P1) • Crew & Equipment Translation Aid (CETA) Cart B
2003 (16 Jan)	ULF1	US Orbiter STS-114	• Utilisation and Logistics Flight • Spares Pallet (ESP-2) • Crew Rotation
2003	12A	US Orbiter	• Second left-side truss segment (ITS P3/P4) • Solar array and batteries
2003	12A.1	US Orbiter	• Third left-side truss segment (ITS P5) • Logistics and supplies

2003	13A	US Orbiter	• Second right-side truss segment (ITS S3/S4) • Solar array set and batteries (Photovoltaic Module)
2003	13A.1	US Orbiter	• Logistics and supplies
TBD*	3R	Russian Proton	• Universal Docking Module (UDM)
TBD	5R	Russian Soyuz	• Docking Compartment 2 (DC-2)
2003	15A	US Orbiter	• Solar arrays and batteries (Photovoltaic Module S6)
2004	10A	US Orbiter	• US Node 2
2004	9A.1	US Orbiter	• Russian provided Science Power Platform (SPP) with four solar arrays
2004	1J/A	US Orbiter	• Japanese Experiment Module Experiment Logistics Module (JEM ELM PS) • Science Power Platform (SPP) solar arrays with truss
TBD	ATV		• European Automated Transfer Vehicle
2004	1J	US Orbiter	• Kibo Japanese Experiment Module (JEM) • Japanese Remote Manipulator System (JEM RMS)
2005	UF-3	US Orbiter	• Multi-Purpose Logistics Module (MPLM) • Express Pallet
2005	N/A	US Orbiter	• X-38 Flight Demo
2005	1E	US Orbiter	• European Laboratory – Columbus Module
2005	UF-4	US Orbiter	• Express Pallet • Spacelab Pallet carrying "Canada Hand" (Special • Purpose Dexterous Manipulator) • Alpha Magnetic Spectrometer (AMS)

2005	2J/A	US Orbiter	• Japanese Experiment Module Exposed Facility (JEM EF) • Solar array batteries • Cupola
2005	UF-5	US Orbiter	• Multi-Purpose Logistics Module (MPLM) • Express Pallet
TBD	9R	Russian Proton	• Docking and Stowage Module (DSM)
TBD	14A	US Orbiter	• Science Power Platform (SPP) solar arrays • Zvezda Micrometeoroid and Orbital Debris (MMOD) Shields • Port and Starboard MT/CETA Rails
TBD	UF-6	US Orbiter	• Multi-Purpose Logistics Module (MPLM) • Batteries
TBD	20A	US Orbiter	• US Node 3
TBD	8R	Russian Soyuz	• Research Module 1
TBD	16A	US Orbiter	• Habitation Module
TBD	17A	US Orbiter	• Multi-Purpose Logistics Module (MPLM) • Destiny racks • Common Berthing Adapter (CBA – interface between CRV and Node 3)
TBD	18A	US Orbiter	• Crew Return Vehicle (CRV)
TBD	19A	US Orbiter	• Multi-Purpose Logistics Module (MPLM)
TBD	10R	Russian Soyuz	• Research Module 2
TBD	UF-7	US Orbiter	• Centrifuge Accommodation Module (CAM)

* TBD = to be decided.

Notes: All dates and times are subject to change. Additional Progress, Soyuz, H-II Transfer Vehicle and Automated Transfer Vehicle flights for crew transport, logistics and resupply are not listed.

Appendix 3

ISS Crew Biographies

EXPEDITION ONE

The Expedition One crew was launched to the new, uninhabited International Space Station on 31 October 2000 by a Soyuz rocket from the Baikonur launch site in Kazakhstan, and successfully docked with the station on 2 November. The crew was scheduled to spend approximately 4 months aboard the station. Eventually, Shepherd, Gidzenko and Krikalev spent 136 days aboard the space station. They returned to the Kennedy Space Center, Florida, on 20 March 2001 on board the Space Shuttle Discovery, after spending 141 days in space. During the Expedition One mission, the complex was visited by three Space Shuttle missions and the space station more than doubled in size and power with the installation of two giant solar arrays and the Destiny laboratory module.

William M. Shepherd, Mission Commander (Captain, US Navy)
NASA astronaut

Personal background: Born 26 July 1949 in Oak Ridge, Tennessee, but considers Babylon, New York, his hometown. Married to Beth Stringham of Houston, Texas. He enjoys sailing, swimming, and working in his garage. His mother, Barbara Shepherd, resides in Bethesda, Maryland. His father, George R. Shepherd, is deceased.

Education: Graduated from Arcadia High School, Scottsdale, Arizona, in 1967; received a bachelor of science degree in aerospace engineering from the US Naval Academy in 1971, and the degrees of ocean engineer and master of

The Expedition One crew. Left to right: Sergei Krikalev, commander William Shepherd and Yuri Gidzenko. (NASA photo JSC2000-06243)

science in mechanical engineering from the Massachusetts Institute of Technology in 1978.

Organisations: American Institute of Aeronautics and Astronautics (AIAA).

Honours: Recipient of NASA's "Steve Thorne" Aviation Award.

Experience: Shepherd, a 1971 graduate of the US Naval Academy and a Navy Seal, dreamed of becoming a pilot, but did not meet the eyesight requirements. He became a Navy diver instead, applied for the astronaut programme in 1980 and was accepted in 1984. Shepherd graduated from the US Naval Academy in 1971, and served with the Navy's Underwater Demolition Team 11, SEAL Teams 1 and 2, and Special Boat Unit 20.

Selected by NASA in May 1984. He made three flights as a mission specialist on STS-27 (2–6 December 1988), STS-41 (6–10 October 1990) and STS-52 (22 October–1 November 1992), and logged 440 hours in space. From March 1993 to January 1996, Shepherd was assigned to the Space Station Programme and served in various management positions. After selection for the Expedition One crew, he trained extensively in the US and Russia.

Yuri Pavlovich Gidzenko, Soyuz Commander (Lt Colonel, Russian Air Force)
Russian cosmonaut

Personal background: Born 26 March 1962 in the village of Elanets, Elanetsky district, Nikolayev region. Resides in Star City. Married to Olga Vladimirovna (Shapovalova), born 21 December 1961. They have two sons. Gidzenko enjoys team sports, swimming, tennis and football. Gidzenko's parents, Pavel Vasilievich and Galina Mikhailovna, live in the settlement of Berezovka, Odessa region.

Education: Graduated from Kharkov Military Aviation College of Pilots in 1983. Graduated from Moscow State University of land surveying and cartography in 1994.

Honours: Awarded three Armed Forces medals.

Experience: After graduation from Aviation College in 1983, served as a pilot and senior pilot in the Air Force. Third class military pilot. December 1987 to June 1989, took the general space training course. From September 1989, he continued training with the group of test-cosmonauts. Took part in 145 parachute jumps as instructor of paradrop training. From March to October 1994 he trained for the 17th main Mir mission and Euromir-94 flight as commander of the back-up crew. From 3 September 1995 to 29 February 1996, Gidzenko was the commander of the Euromir-95 mission on the Russian Space Station Mir. During his period on Mir, he took part in two EVAs lasting a total of 3 hours 35 minutes.

Sergei Konstantinovich Krikalev, Flight Engineer
Russian cosmonaut

Personal background: Born 27 August 1958 in Leningrad, Russia (later renamed St Petersburg). Married to Elena Terekhina of Samara, Russia. They have one daughter. He enjoys swimming, skiing, bicycle riding, aerobatic flying, and amateur radio operations, particularly from space. His parents, Konstantin and Nadia, live in Leningrad. Her parents, Faina and Yuri, live in Samara, Russia.

Education: Graduated from high school in 1975, received a mechanical engineering degree from the Leningrad Mechanical Institute (now called St Petersburg Technical University) in 1981.

Honours: A member of the Russian and Soviet national aerobatic flying teams – Champion of Moscow in 1983, Champion of the Soviet Union in 1986. For his space flight experience, he was awarded the title of Hero of the Soviet Union, the Order of Lenin, the French Legion of Honour, and the new title of Hero of Russia.

He was also awarded the NASA Space Flight Medal in 1994 and 1998.

Experience: After graduation in 1981, he joined NPO Energia, the Russian industrial organisation responsible for manned space flight activities. He tested space flight equipment, developed space operations methods, and participated in ground control operations. When the Salyut 7 space station failed in 1985, he worked on the rescue mission team, developing procedures for docking with the uncontrolled station and repairing the station's onboard system.

Krikalev was selected as a cosmonaut in 1985, completed his basic training in 1986, and, for a time, was assigned to the Buran shuttle programme. In early 1988, he began training for his first long-duration flight aboard the Mir space station. This training included preparations for at least six space walks, installation of a new module, the first test of the new Manned Manoeuvring Unit (MMU), and the second joint Soviet–French science mission. Soyuz TM-7 was launched 26 November 1988, with Krikalev as flight engineer, commander Alexander Volkov, and French astronaut Jean-Loup Chrétien. The previous crew (Vladimir Titov, Musa Manarov and Valeri Polyakov) remained on Mir for another 25 days, marking the longest period a six-person crew had been in orbit. Krikalev, Polyakov and Volkov continued to conduct experiments aboard Mir. Because arrival of the next crew had been delayed, they prepared Mir for a period of unmanned operations before returning to Earth on 27 April 1989.

In April 1990, Krikalev was a member of the back-up crew for the eighth long-duration Mir mission. In December 1990, he began training for the ninth Mir mission. Soyuz TM-12 launched on 19 May 1991 with Krikalev as flight engineer, commander Anatoli Artsebarski, and British astronaut Helen Sharman. Sharman returned to Earth with the previous crew after one week, while Krikalev and Artsebarski remained on Mir. During the summer, they conducted six EVAs to perform a variety of experiments and station maintenance.

In July 1991, Krikalev agreed to stay on Mir as flight engineer for the next crew, which was scheduled to arrive in October because the next two planned flights had been reduced to one. Both he and Austrian astronaut Franz Viehbok returned with Artsebarski on 10 October 1991. Commander Alexander Volkov remained on board with Krikalev. After the crew replacement in October, Volkov and Krikalev continued their experiments on Mir and conducted another EVA before returning to Earth on 25 March 1992. During this time, the Soviet Union was disbanded and Krikalev came home as a Russian citizen.

In October 1992, NASA announced that an experienced cosmonaut would fly aboard a future Space Shuttle mission. Krikalev was one of two candidates named by the Russian Space Agency. In April 1993, he was assigned as primary mission specialist and became the first Russian to fly on the Space Shuttle. STS-60 was launched on 3 February 1994 for an 8-day mission involving experiments on Spacehab-2 and the first flight of the Wake Shield Facility. Krikalev conducted many of the Remote Manipulator System operations during the flight. STS-60 landed at Kennedy Space Center, Florida, on 11 February 1994.

Krikalev returned to duty in Russia following his American experience on STS-60, but in September 1993, he was the back-up for Vladimir Titov on STS-63. He also worked at the Johnson Space Center in Houston as CapCom in mission control, supporting the joint US–Russian missions STS-63, STS-71, STS-74 and STS-76.

After his selection for the first long-term ISS crew, Krikalev flew on STS-88 (4–15 December 1998), the first International Space Station assembly mission. During this 12-day mission, the Unity module was mated with the Zarya module.

During his first four space flights, Krikalev logged more than 1 year, 3 months and 19 days in space, including seven EVAs.

EXPEDITION TWO

The second long-term crew (two American astronauts and one Russian cosmonaut) was launched aboard Shuttle Discovery on 8 March 2001 and they arrived at the International Space Station two days later. The crew installed and conducted tests on the Canadian-made Space Station Robotic arm (SSRMS), using it for the first time on orbit to berth the US Joint Airlock Quest to the station. They also unloaded the first Italian-made Multi-Purpose Logistics Module, performed internal and external maintenance tasks, and carried out medical and science experiments. Usachev, Helms and Voss spent 167 days in space, including 163 days on board the space station. All three crew members previously flew to the space station on the Shuttle Atlantis during STS-101 in May 2000.

Yuri Vladimirovich Usachev, Mission Commander
Russian cosmonaut

Personal background: Born 9 October 1957 in Donetsk, Rostov in Don Region, Russia. Married to Vera Sergeevna Usacheva (nee Nazarova) of Kalingrad, Moscow region. They have one daughter, Zhenya. His mother, Anna Grigorevna Usacheva, resides in Donetsk. His father is deceased. He has a brother, 5 years older, and a twin sister, 5 minutes older. He enjoys photography and video production. Resides in Star City.

Education: Graduated from Donetsk Public Schools in 1975. Received an engineering diploma from Moscow Aviation Institute in 1985.

The Expedition Two crew. Left to right: James Voss, commander Yuri Usachev, and Susan Helms. (NASA photo ISS02-S-002)

Honours: Awarded the Hero of the Russian Federation and the Pilot/Cosmonaut medals after his first space flight in 1994. After his second flight, in 1996, he was awarded the Order of Service to the Country, Level III. He was also named a Chevalier in the French Legion of Honour. Other awards include the NASA medal for Public Service, and a NASA Space Flight Medal.

Experience: Upon graduation from the Moscow Aviation Institute, he went to work for NPO Energia, participating in groups working with EVA training, future construction in space, public relations, and ergonomics. In 1989, he became a cosmonaut candidate. From 1989 to 1992, he underwent a course of general space training at the Cosmonaut Training Centre. He was a member of the back-up crew for the Mir 13, 14, and 19 missions. From 8 January to 9 July 1994, he served as flight engineer on the Mir space station. From 21 February to 2 September 1996, he again served as flight engineer on the Mir 21 crew. During this mission, he and Yuri Onufrienko were joined by NASA's Shannon Lucid. Their stay was extended more than a month longer than expected by a delay in building a rocket to deliver their replacements. Most recently, he served on the crew of STS-101, the third Shuttle mission devoted to

International Space Station construction. Prior to Expedition Two, commander Usachev had spent a total of 375 days in space and performed six spacewalks during two missions to Mir.

James S. Voss, Flight Engineer (Colonel, US Army, Retd)
NASA astronaut

Personal background: Born 3 March 1949 in Cordova, Alabama, but considers Opelika, Alabama, to be his hometown. Married to the former Suzan Curry of Birmingham, Alabama. They have one daughter. Enjoys woodworking, skiing, softball, racquetball, scuba diving, and flying an aeroplane he built himself. As an undergraduate, he participated on the Auburn University Wrestling Team.

Education: Graduated from Opelika High School, Opelika, Alabama. Received a bachelor of science degree in aerospace engineering from Auburn University in 1972, and a master of science degree in aerospace engineering sciences from the University of Colorado in 1974.

Honours: NASA Outstanding Leadership Award (1996); NASA Exceptional Service Medal (1994); NASA Space Flight Medals (1992, 1993, 1995); Defense Meritorious Service Medal (1993); Defense Superior Service Medal (1992); Outstanding Student Award, USN Test Pilot School (1983); William P. Clements Jr Award for Excellence in Education as the outstanding Professor at the US Military Academy (1982); Meritorious Service Medal (1982); NASA Summer Faculty Research Fellowship (1980); Commandant's List - Infantry Officer Advanced Course (1979); Army Commendation Medal (1978); Honour Graduate and Leadership Award - Ranger School (1975); Distinguished Graduate - Infantry Officer Basic Course (1974).

Experience: After graduation from Auburn and commissioning as a 2nd Lieutenant, Voss went to the University of Colorado to obtain his masters degree under the Army Graduate Fellowship Programme. After attending the Infantry Basic Course, Airborne and Ranger schools, he served with the 2nd Battalion 48th Infantry in Germany as a platoon leader, intelligence staff officer, and company commander. On returning to the United States, he attended the Infantry Officer Advanced Course, then taught for three years in the Department of Mechanics at the US Military Academy. After attending the US Naval Test Pilot School and the Armed Forces Staff College, Voss was assigned to the US Army Aviation Engineering Flight Activity as a Flight Test Engineer/Research and Development Coordinator. He was involved in several major flight test projects before being detailed to NASA's Johnson Space Center in November 1984.

As a Vehicle Integration Test Engineer, he supported Shuttle and payload testing at the Kennedy Space Center for Shuttle flights 51-D, 51-F, 61-C and 51-L. He participated in the STS-51-L accident investigation, and supported the resulting reviews dedicated to returning the Space Shuttle safely to flight. Selected as an astronaut candidate by NASA in June 1987, Voss completed a one-year training and evaluation programme in August 1988, which qualified him for assignment as a Shuttle mission specialist. He has worked as a flight crew representative in the area of Shuttle safety, as a CapCom, and as the Astronaut Office Training Officer. Prior to Expedition Two, he flew on four Shuttle missions.

In 1991, STS-44 deployed a Defense Support Programme satellite. The crew also conducted two Military Man in Space experiments, three radiation monitoring experiments, and numerous medical tests. The mission ended after 110 orbits with a landing on the lakebed at Edwards Air Force Base, California.

STS-53 deployed the classified Department of Defence payload DOD-1 and also performed several Military Man in Space and NASA experiments. After completing 115 orbits, Discovery landed at Edwards Air Force Base on 9 December 1992.

On STS-69 the crew successfully deployed and retrieved a SPARTAN satellite and the Wake Shield Facility. Voss conducted an EVA lasting 6 hours 46 minutes to test space suit modifications and evaluate procedures and tools to be used in construction of the ISS. Endeavour landed at the Kennedy Space Center on 18 September 1995.

Voss served as the back-up crew member for two missions to the Russian Space Station Mir. During this time he lived and trained for two years at the Gagarin Cosmonaut Training Centre in Star City, Russia. He then flew on STS-101, the first Shuttle mission to repair and provision the International Space Station, in May 2000.

Susan J. Helms, Flight Engineer (Colonel, US Air Force)
NASA astronaut

Personal background: Born 26 February 1958 in Charlotte, North Carolina, but considers Portland, Oregon, to be her hometown. She enjoys piano and other musical activities, jogging, travelling, reading, computers, and cooking. Plays keyboard for MAX-Q, a rock-n-roll band. Her parents, Lt Col. (Ret., USAF) Pat and Dori Helms, reside in Albuquerque, New Mexico.

Education: Graduated from Parkrose Senior High School, Portland, Oregon, in 1976. Received a bachelor of science degree in aeronautical engineering from the US Air Force Academy in 1980, and a master of science degree in aeronautics/astronautics from Stanford University in 1985.

Organisations: Women Military Aviators; US Air Force Academy Association of Graduates; Stanford Alumni Association; Association of Space Explorers, Sea/Space Symposium, Chi Omega Alumni.

Honours: Recipient of the Distinguished Superior Service Medal, the Defense Meritorious Service Medal, the Air Force Meritorious Service Medal, the Air Force Commendation Medal, NASA Space Flight Medals, and the NASA Outstanding Leadership Medal. Named a Distinguished Graduate of the USAF Test Pilot School, and recipient of the R.L. Jones Award for Outstanding Flight Test Engineer, Class 88A. In 1990, she received the Aerospace Engineering Test Establishment Commanding Officer's Commendation, a special award unique to the Canadian Forces. Named the Air Force Armament Laboratory Junior Engineer of the Year in 1983.

Experience: Helms graduated from the US Air Force Academy in 1980. She received her commission and was assigned to Eglin Air Force Base, Florida, as an F-16 weapons separation engineer with the Air Force Armament Laboratory. In 1982, she became the lead engineer for F-15 weapons separation. In 1984, she was selected to attend graduate school. She received her degree from Stanford University in 1985 and was assigned as an assistant professor of aeronautics at the US Air Force Academy. In 1987, she attended the Air Force Test Pilot School at Edwards Air Force Base, California. After completing one year of training as a flight test engineer, Helms was assigned as a USAF Exchange Officer to the Aerospace Engineering Test Establishment, Canadian Forces Base, Cold Lake, Alberta, Canada, where she worked as a flight test engineer and project officer on the CF-18 aircraft. She was managing the development of a CF-18 Flight Control System Simulation for the Canadian Forces when selected for the astronaut programme. As a flight test engineer, Helms has flown in 30 different types of US and Canadian military aircraft.

Selected by NASA in January 1990, Helms became an astronaut in July 1991. She flew on four Shuttle missions and logged over 1,096 hours in space before participating in ISS Expedition Two. During STS-54 (13–19 January 1993) the crew deployed a Tracking and Data Relay Satellite (TDRS-F) and carried out an EVA to assess ISS assembly techniques.

On STS-64 (9–20 September 1994) Helms was the flight engineer for orbiter operations and the primary remote arm operator. The major objective of this flight was to validate the Lidar in Space Technology Experiment by gathering data about the Earth's troposphere and stratosphere. Helms helped to deploy and retrieve the SPARTAN-201 free-flying satellite.

On STS-78 (20 June to 7 July 1996) Helms was the payload commander and flight engineer on the first mission to combine microgravity studies and a life science investigations. The Life and Microgravity Spacelab mission served as a model for future studies on board the International Space Station. Mission duration was 16 days, 21 hours, 48 minutes – the longest Space Shuttle mission to date.

STS-101 (19–29 May 2000) was dedicated to the delivery and repair of critical hardware for the International Space Station. Her prime responsibilities during this mission were to perform critical repairs to extend the life of Zarya. She also had prime responsibility for the onboard computer network and served as the mission specialist for rendezvous with the ISS.

Helms was the space station's first long-term female resident. During her stay on board, she operated the Canadarm 2 and was responsible for berthing the Quest airlock to the station.

EXPEDITION THREE

The third long-term crew (one American astronaut and two Russian cosmonauts) was launched to the ISS on 10 August 2001 on board Shuttle

The Expedition Three Crew. Left to right: Mikhail Tyurin, commander Frank Culbertson and Vladimir Dezhurov. (NASA photo ISS003-S-002)

Discovery (STS-105). They arrived at the International Space Station two days later and returned to Earth on board Shuttle Endeavour (STS-108) on 17 December, having completed 117 days as the primary station crew and spent 125 days aboard the station overall. Their total space time for the mission was 129 days. During the mission they received the Pirs docking module and the Andromède Soyuz taxi mission.

Frank L. Culbertson, Jr, Mission Commander (Captain, US Navy, Retd)
NASA astronaut

Personal background: Born 15 May 1949 in Charleston, South Carolina, but considers Holly Hill to be his hometown. Married in June 1987, to the former Rebecca Ellen Dora of Vincennes, Indiana. Five children. He enjoys flying, cycling, squash, running, golf, camping, photography, music and water sports. Member of varsity rowing and wrestling teams at USNA. His parents, Dr and Mrs Frank Culbertson, Sr, reside in Holly Hill, South Carolina. Her mother, Mrs Avanelle Vincent Dora, resides in Vincennes, Indiana. Her father, Robert E. Dora, is deceased.

Education: Graduated from Holly Hill High School, Holly Hill, South Carolina, in 1967. He received a bachelor of science degree in aerospace engineering from the US Naval Academy in 1971.

Organisations: Member of the American Institute of Aeronautics and Astronautics, the Association of Naval Aviators, Aircraft Owners & Pilots Association, the US Naval Institute, the Aviation Boatswains Mate's Association, and the Association of Space Explorers.

Honours: Awarded the Navy Flying Cross, the Defense Superior Service Medal, the NASA Outstanding Leadership Medal, NASA Space Flight Medals, Navy Commendation Medal, Air Force Commendation Medal, Navy Unit Commendation, three Meritorious Unit Commendations, the Armed Forces Expeditionary Medal, the Humanitarian Services Medal, and various other unit and service awards. Distinguished graduate, US Naval Test Pilot School. Awarded Honorary Doctor of Science Degree, College of Charleston, 1994.

Experience: Culbertson graduated from Annapolis in 1971 and served aboard the USS *Fox* in the Gulf of Tonkin for six months before reporting for flight training in Pensacola, Florida. After designation as a Naval Aviator at Beeville, Texas, in May 1973, he received training as an F-4 Phantom pilot at Miramar, California. From March 1974 to May 1976, he was assigned to the USS *Midway*, permanently based in Yokosuka, Japan. He was then assigned as an exchange pilot with the USAF at Luke Air Force Base, Arizona, where he served as Weapons and Tactics Instructor flying F-4C's with the 426th TFTS until September 1978. Culbertson then served as the Catapult and Arresting Gear Officer for the USS *John F. Kennedy* until May 1981 when he was selected to attend the US Naval Test Pilot School, Patuxent River, Maryland. Following graduation with distinction in June 1982, he was assigned to the Carrier Systems Branch of the Strike Aircraft Test Directorate where he served as Programme Manager for all F-4 testing and as a test pilot for automatic carrier landing system tests in the F-4S, and carrier suitability in the F-4S and the OV-10A. He was engaged in fleet replacement training in the F-14A Tomcat at Oceana, Virginia, from January 1984 until his selection for the astronaut candidate programme. He has logged over 5,000 hours flying time in 40 different types of aircraft, and 350 carrier landings.

Culbertson completed basic astronaut training in June 1985. Subsequent technical assignments included: member of the team that redesigned and tested the Shuttle nose wheel steering, tyres, and brakes, to provide more safety margin during landing rollout; member of the launch support team at Kennedy Space Center for Shuttle flights 61-A, 61-B, 61-C, and 51-L; participated in the preparations for the proposed launch at Vandenberg Air Force Base, California, in 1986; worked at the NASA Headquarters Action Center in Washington, DC, assisting with the Challenger accident investigations conducted by NASA, the Presidential Commission, and Congress; lead astronaut at the Shuttle Avionics Integration Laboratory, involved in the checkout and verification of the computer software and hardware interfaces for STS-26 and subsequent flights; lead of the First Emergency Egress Team, which conducts periodic tests of improvements to the Shuttle ground egress systems; member of the Astronaut Office Safety Branch; lead spacecraft communicator (CapCom) in the Mission Control Center for seven missions (STS-27, 29, 30, 28, 34, 33, and 32) following his first flight, he served as the Deputy Chief of the Flight Crew Operations in the Space Station Support Office as well as the lead astronaut for Space Station Safety. While assigned to STS-51, he was a member of the team evaluating the proposed mission to dock with the Russian Space Station Mir. He was Chief of the Astronaut Office Mission Support Branch, Chief of the Johnson Space Center Russian Projects Office and Programme Manager for the ISS Phase 1 Shuttle–Mir programme.

During two Shuttle flights, Culbertson logged over 344 hours in space. He was the pilot on STS-38 a five-day mission in 1990, during which the crew conducted Department of Defense operations. In 1993, he was the crew commander on STS-51, a 10-day mission during which the crew deployed the Advanced Communications Technology Satellite and the Shuttle Pallet Satellite, ORFEUS/SPAS.

Culbertson replaced Ken Bowersox as commander of Expedition Three.

Vladimir Nikolaevich Dezhurov, Soyuz Commander/Mission Pilot (Lt Colonel, Russian Air Force)
Russian cosmonaut

Personal background: Born 30 June 1962 in Yavas settlement, Zubovo-Polyansk district, Mordovia, Russia. He resides in Star City. Married to Elena Valentinovna Dezhurova (nee Suprina). Two daughters. Nikolai Serafimovich Dezhurov, father, and Anna Vasilevna Dezhurova, mother, reside in Yavas settlement, Zubovo-Polyansk district, Mordovia, Russia.

Education: Graduated from the SI Gritsevits Kharkov Higher Military Aviation School in 1983 with a pilot-engineer's diploma.

Honours: Awarded three Armed Forces medals.

Experience: After graduating from the aviation military school in 1983, he served as a pilot and senior pilot in the Air Force. In 1987, he was assigned to the Cosmonaut Training Centre. From December 1987 to June 1989, he underwent a course of general space training. Since September 1989, he has continued training as a member of the group of test cosmonauts. Since 1991, he has been a correspondence student at the Yuri Gagarin Air Force Academy. In March 1994, Dezhurov began flight training as commander of the prime crew of the Mir 18 mission. The crew, which included NASA astronaut Norman Thagard – the first American to fly on Mir – launched from Baikonur Cosmodrome on 14 March 1995 aboard Soyuz TM-21. Following a 115-day flight, the mission concluded with a landing at the Kennedy Space Center, Florida, aboard Space Shuttle Atlantis on 7 July 1995. He and Gennadi Strekalov carried out five EVAs lasting more than 19 hours in total, including several related to the docking of the Spektr module. Dezhurov was assigned as back-up for the first ISS mission before taking part in the third ISS expedition.

Mikhail Tyurin, Flight Engineer
Russian Space Agency cosmonaut

Personal background: Born 2 March 1960 in Kolomna, Russia (about 60 miles from Moscow) where his parents still reside. Tyurin lives in Korolev, a small city outside of Moscow. He is married to Tatiana Anatoleyvna. They have a daughter, Alexandra, born in 1982. He enjoys sailing.

Education: He graduated from the Moscow Aviation Institute in 1984 with a degree in engineering and a specialisation in creating mathematical models related to mechanical flight. He is currently doing graduate work in his field of research.

Experience: After graduating from the Aviation Institute he began working at the Energia Corporation as an engineer. His main specialisations have been dynamics, ballistics, and software development. His personal scientific research is connected with the psychological aspects of cosmonauts' training for the manual control of spacecraft motion. In 1993 he was selected to begin cosmonaut training, and since 1998 he has trained as a flight engineer for the third ISS mission and as a back-up crew member for the first ISS mission.

EXPEDITION FOUR

The fourth crew to live on the International Space Station was launched aboard Shuttle Endeavour (STS-108) on 5 December 2001. The crew of three (two American astronauts and one Russian cosmonaut) was scheduled to return aboard STS-111 in May 2002. Their mission was to perform flight tests of the station hardware, conduct internal and external maintenance tasks, and develop the scientific capability of the station.

The Expedition Four Crew. Left to right: Daniel Bursch, commander Yuri Onufrienko and Carl Walz. (NASA photo ISS004-S-002)

Yuri Ivanovich Onufrienko, Mission Commander (Colonel, Russian Air Force)
Ukrainian cosmonaut

Personal background: Born 6 February 1961 in Ryasnoe, Zolochev district, Kharkov region, Ukraine. Married to Valentina Mikhailovna Onufrienko (nee Ryabovol). Two sons and one daughter. His parents are deceased. He has two older brothers. He enjoys tennis, cooking, fishing, chess, and flying. Resides in Star City.

Education: Graduated from the V.M. Komarov Eisk Higher Military Aviation School for Pilots in 1982 with a pilot-engineer's diploma. Graduated from Moscow State University in 1994 with a degree in cartography.

Honours: Hero of Russia. Awarded two Armed Forces medals. He was also named a Chevalier in the French Legion of Honour.

Experience: Upon graduation from aviation school, he served as a pilot and then senior pilot in the Air Force. In 1989, he became a cosmonaut candidate at the Cosmonaut Training Centre. From September 1989 to January 1991, he underwent a course of general space training. Starting April 1991, he underwent training as a member of a group of test cosmonauts. Starting March 1994, he began flight training as the commander of the stand-by crew of the Mir 18 expedition. From 21 February to 2 September 1996, he served as commander on Mir 21. One month after he and Yuri Usachev arrived, they were joined by NASA's Shannon Lucid. During Mir 21 he performed numerous research experiments and participated in six EVAs. He and Usachev were also briefly joined by French cosmonaut Claudie André-Deshays (also known as Claudie Haigneré). Before Expedition Four, he had logged 193 days in space. He has also flown over 800 hours in military aircraft, including the L-29, SU-7, SU-17 (M1-4), and L-39.

Daniel W. Bursch, Flight Engineer (Captain, US Navy)
NASA astronaut

Personal background: Born 25 July 1957 in Bristol, Pennsylvania, but considers Vestal, New York, to be his hometown. Married to the former Roni J. Patterson of Modesto, California. Four children. He enjoys tennis, softball, windsurfing, skiing, and woodworking. His father, Dudley Bursch, resides in Stuart, Florida. His mother, Betsy Bursch, is deceased. Roni's mother, Gayle Hutcheson, resides in Modesto, California. Her father, Jack Patterson, resides in Truckee, California.

Education: Graduated from Vestal Senior High School, Vestal, New York, in

1975; received a bachelor of science degree in physics from the United States Naval Academy in 1979, and a master of science degree in engineering science from the Naval Postgraduate School in 1991.

Organisations: Member of the US Naval Academy Alumni Association.

Honours: Awarded the Defense Superior Service Medal, NASA Space Flight Medals, the Navy Commendation Medal and the Navy Achievement Medal. Distinguished graduate, US Naval Academy and US Naval Test Pilot School.

Experience: Bursch graduated from the US Naval Academy in 1979, and was designated a naval flight officer in April 1980 at Pensacola, Florida. After initial training as an A-6E Intruder bombardier/navigator, he reported to Attack Squadron 34 in January 1981, and deployed to the Mediterranean aboard the USS *John F. Kennedy*, and to the North Atlantic and Indian Ocean aboard the USS *America*. He attended the US Naval Test Pilot School in January 1984. Upon graduation in December he worked as a project test flight officer flying the A-6 Intruder until August 1984, when he returned to the US Naval Test Pilot School as a flight instructor.

In April 1987, Bursch was assigned to the Commander, Cruiser-Destroyer Group 1, as Strike Operations Officer, making deployments to the Indian Ocean aboard the USS *Long Beach* and the USS *Midway*. Redesignated an Aeronautical Engineering Duty officer, he attended the Naval Postgraduate School in Monterey, California, from July 1989 until his selection to the astronaut programme. He has clocked over 2,900 flight hours in more than 35 different aircraft.

Selected by NASA in January 1990, Bursch became an astronaut in July 1991. He was initially assigned to the Astronaut Office Operations Development Branch, working on controls and displays for the Space Shuttle and Space Station. He was also a spacecraft communicator (CapCom) in mission control and Chief of Astronaut Appearances.

Bursch served as a mission specialist with Carl Walz on STS-51 in September 1993. During the 10-day mission, the crew deployed the Advanced Communications Technology Satellite (ACTS), and the Shuttle Pallet Satellite (SPAS) with NASA and German scientific experiments on board. Following a spacewalk by two crew members to evaluate Hubble Space Telescope repair tools, Bursch recovered the SPAS using the Remote Manipulator System.

STS-68, carrying Space Radar Lab 2, launched on 30 September 1994. Three advanced radars and a carbon-monoxide pollution sensor in Endeavour's cargo bay studied Earth's surface and atmosphere.

STS-77 launched on 19 May 1996. In the fourth Spacehab module were 12 separate materials processing, fluid physics and biotechnology experiments. STS-77 completed a record four rendezvous with the SPARTAN 207/Inflatable

Antenna Experiment and the Passive Aerodynamically-stabilised Magnetically-damped Satellite/Satellite Test Unit.

Bursch logged over 746 hours in space during his three Shuttle flights.

Carl E. Walz, Flight Engineer (Colonel, US Air Force)
NASA astronaut

Personal background: Born 6 September 1955 in Cleveland, Ohio. Married to the former Pamela J. Glady of Lyndhurst, Ohio. They have two children. He enjoys piano and vocal music, sports, and is lead singer for MAX-Q, an astronaut rock-n-roll band.

Education: Graduated from Charles F. Brush High School, Lyndhurst, Ohio, in 1973; received a bachelor of science degree in physics from Kent State University, Ohio, in 1977, and a master of science in solid state physics from John Carroll University, Ohio, in 1979.

Organisations: American Legion, KSU Alumni Association.

Honours: Graduated Summa Cum Laude from Kent State University. Awarded the Defense Superior Service Medal, the USAF Meritorious Service Medal with one Oak Leaf Cluster, the Defense Meritorious Service Medal with one Oak Leaf, the USAF Commendation Medal, and the USAF Achievement Medal with one Oak Leaf Cluster. Distinguished Graduate from the USAF Test Pilot School, Class 83A. Inducted into the Ohio Veterans Hall of Fame. Awarded three NASA Space Flight Medals, NASA Exceptional Service Medal. Distinguished Alumnus Award, Kent State University, 1997.

Experience: From 1979 to 1982, Walz was responsible for analysis of radioactive samples from the Atomic Energy Detection System at the 1155th Technical Operations Squadron, McClellan Air Force Base, California. The subsequent year he studied to be a Flight Test Engineer at the USAF Test Pilot School, Edwards Air Force Base, California. From January 1984 to June 1987, Walz served as a Flight Test Engineer to the F-16 Combined Test Force at Edwards Air Force Base, where he worked on a variety of F-16C airframe avionics and armament development programmes. From July 1987 to June 1990, he served as a Flight Test Manager at Detachment 3, Air Force Flight Test Center.

Selected by NASA in January 1990, Walz is a veteran of three Shuttle flights. During the STS-51 mission in September 1993, the crew deployed the Advanced Communications Technology Satellite (ACTS), and the Shuttle Pallet Satellite (SPAS) with NASA and German scientific experiments. Walz served as a mission specialist alongside Bursch and participated in a 7-hour space walk to evaluate tools for the Hubble Space Telescope servicing mission.

In July 1994 he was flight engineer on STS-65, a 15-day flight of the second International Microgravity Laboratory (IML-2) Spacelab module. The crew conducted more than 80 experiments focusing on materials and life sciences research in microgravity. The mission completed 236 orbits of the Earth and set a new flight duration record for the Shuttle programme.

He was then a mission specialist on STS-79 in September 1997. During the 10-day mission, the Shuttle Atlantis docked with the Russian Mir station, delivered food, water, US scientific experiments and Russian equipment, and exchanged NASA long-duration crew members. The Atlantis–Mir complex set a record for docked mass in space.

He has logged over 833 hours on the Shuttle.

EXPEDITION FIVE

The crew will comprise Russian cosmonauts Valeri Korzun, the mission commander, cosmonaut Sergei Treschov and astronaut Peggy Whitson (PhD). This will be the first space flight for both Treschev and Whitson. This crew was assigned as the back-up team for Expedition Three. Astronaut Scott J. Kelly (Lt Cmdr USN), who flew on STS-103 in 1999, will serve as Whitson's back-up. Dmitri Kondratiev will back up Korzun and Treschev. Their mission is scheduled to begin in May 2002 and end five months later.

Digital photos of Valeri Korzun, Sergei Treschev and Peggy Whitson.

Valeri Grigorievich Korzun, Mission Commander (Colonel, Russian Air Force)

Russian cosmonaut

Personal background: Born 5 March 1953 in Krasny Sulin. He has a son,

Nikita. His father is Korzun Grigori Andreevich, and his mother, Korzun Maria Arsentievna.

Education: Graduated from Kachin Military Aviation College in 1974; Commander Department of the Gagarin Air Force Academy in 1987.

Honours: Awarded six Air Force Medals.

Experience: After graduation from the Kachin Military Academy in 1974, he served as a pilot, senior pilot, and then commander of a squadron in the Russian Air Force. In March 1987, he was selected as a cosmonaut for training at the Gagarin Cosmonaut Training Centre after graduating from the Gagarin Military Air Force Academy. Between December 1987 and June 1989, he took a course of general space training. Korzun was certified as a test cosmonaut in September 1989 and for the next three years he trained as part of the test-cosmonauts group. He was assigned as back-up for Soyuz TM-13 in 1991, and from October 1992 to March 1994, he took a training course as a Soyuz TM commander.

From March 1994 to June 1995, he trained as a group member for flight on Mir and was deputy director of the 27KC crew training complex flight as supervisor of communication with the crew. June 1995 until August 1996, he completed training as a flight engineer for the Mir 22/NASA-3 and French "Cassiopeia" programmes.

Between 17 August 1996 and 2 March 1997, he completed a 197-day flight on Mir. The mission included joint operations with a succession of NASA astronauts Shannon Lucid, John Blaha and Jerry Linenger, French astronaut Claudie André-Deshays and German astronaut Reinhold Ewald. Korzun performed two spacewalks totalling 12 hours and 33 minutes.

Korzun is a first-class military pilot. He has logged 1,473 hours, and has flown four types of aircraft. He is an instructor of parachute training, and has completed 377 parachute jumps. He has been the Chief Cosmonaut since 1998.

Sergei Yevgenyevich Treschev, Flight Engineer

Russian cosmonaut

Personal background: Born 18 August 1958 in Volynsky District, Lipetsk Region (Russia). Married to Elvira Victorovna Trescheva. There are two sons in the family, Dmitry and Alexy. His father is Yevgeny Georgievich Treschev, and his mother is Nina Davydovna Trescheva. His hobbies include soccer, volleyball, ice hockey, hiking, tennis, music, photography and video.

Education: Graduated from the Moscow Energy Institute in 1982.

Experience: From 1982 to 1984, he served as a group leader in an Air Force regiment. From 1984 to 1986, he worked as a foreman and as an engineer at the RSC Energia. His responsibilities included the analysis and planning of cosmonaut activities aboard the orbital station and their in-flight technical training. He also developed technical documentation and was involved in setting up cosmonaut flight training with the Yuri Gagarin Cosmonaut Training Centre. He supported crew training aboard Mir in order to maintain their skills in performing certain descent and emergency escape operations. He also participated in testing of the ground-based complex (transport vehicle/Mir core module/Kvant 2 module) to optimise its life-support system.

In March 1992, he was enrolled in the RSC Energia cosmonaut group of civilian engineers. From 1992 to 1994, he completed his basic training course, then from 1994 to 1996 he underwent a course of advance training as a test cosmonaut. He served as a member of the back-up crew for the Soyuz TM-27 mission to Mir in January 1998. From June 1999 to July 2000 Treschev trained as a flight engineer for the Soyuz-TM back-up ISS contingency crew.

Peggy A. Whitson (PhD), Flight Engineer
NASA astronaut

Personal Background: Born 9 February 1960 in Mt Ayr, Iowa. Married to Clarence F. Sams, PhD. She enjoys windsurfing, biking, basketball and water skiing.

Education: Graduated from Mt Ayr Community High School, Mt Ayr, Iowa, in 1978; received a bachelor of science degree in biology/chemistry from Iowa Wesleyan College in 1981, and a doctorate in biochemistry from Rice University in 1985.

Organisations: American Society for Biochemistry and Molecular Biology.

Honours: Two patents approved (1997, 1998); Group Achievement Award for Shuttle–Mir Programme (1996); American Astronautical Society Randolph Lovelace II Award (1995); NASA Tech Brief Award (1995); NASA Space Act Board Award (1995, 1998); NASA Silver Snoopy Award (1995); NASA Exceptional Service Medal (1995); NASA Space Act Award for Patent Application; NASA Certificate of Commendation (1994); Submission of Patent Disclosure for "Method and Apparatus for the Collection, Storage, and Real Time Analysis of Blood and Other Bodily Fluids" (1993); Selected for Space Station Redesign Team (March–June 1993); NASA Sustained Superior Performance Award (1990); Krug International Merit Award (1989); NASA–JSC National Research Council Resident Research Associate (1986–88); Robert A. Welch Postdoctoral Fellowship (1985–86); Robert A. Welch Predoctoral

Fellowship (1982–85), Summa Cum Laude from Iowa Wesleyan College (1981); President's Honor Roll (1978–81); Orange van Calhoun Scholarship (1980); State of Iowa Scholar (1979); Academic Excellence Award (1978).

Experience: From 1981 to 1985, Whitson conducted her graduate work in biochemistry at Rice University, Houston, Texas, as a Robert A. Welch Predoctoral Fellow. Following completion of her graduate work she continued at Rice University as a Robert A. Welch Postdoctoral Fellow until October 1986. Following this position, she began her studies at NASA Johnson Space Center, Houston, Texas, as a National Research Council Resident Research Associate. From April 1988 until September 1989, Whitson served as the Supervisor for the Biochemistry Research Group at KRUG International, a medical sciences contractor at Johnson Space Center. From 1991 to 1997 Whitson was also an Adjunct Assistant Professor in the Department of Internal Medicine and Department of Human Biological Chemistry and Genetics at University of Texas Medical Branch, Galveston, Texas. In 1997, she became Adjunct Assistant Professor at Rice University in the Maybee Laboratory for Biochemical and Genetic Engineering.

From 1989 to 1993, Whitson worked as a Research Biochemist in the Biomedical Operations and Research Branch at NASA–JSC. In 1990, she was also Research Adviser for the National Research Council Resident Research Associate. From 1991 to 1993, she served as Technical Monitor of the Biochemistry Research Laboratories in the Biomedical Operations and Research Branch. From 1991 to 1992 she was the Payload Element Developer for a Bone Cell Research Experiment aboard STS-47, and was a member of the US–USSR Joint Working Group in Space Medicine and Biology. In 1992, she was named Project Scientist for the Shuttle–Mir Programme and served in this capacity until the conclusion of the Phase 1A Programme in 1995. From 1993 to 1996 Whitson was also Deputy Division Chief of the Medical Sciences Division at NASA–JSC. From 1995 to 1996 she served as Co-Chair of the US–Russian Mission Science Working Group.

In April 1996, she was selected as an astronaut candidate and started training in August 1996. Upon completing two years of training and evaluation, she was assigned technical duties in the Astronaut Office Operations Planning Branch and served as the lead for the Crew Test Support Team in Russia from 1998 to 1999. Whitson trained as a back-up crew member for Expedition Three.

Digital photos of Kenneth Bowersox, Nikolai Budarin and Donald Thomas.

EXPEDITION SIX

Expedition Six is scheduled to launch in September 2002 for a 4-month stay on board the ISS. This crew was assigned as the back-up team for Expedition Four. Carlos Noriega (Lt Col, USMC) who flew on STS-84 in 1997 and STS-97 in 2000, is the back-up commander to Bowersox. Donald Pettit (PhD), a space rookie, will back up Thomas. Cosmonaut Salizhan Sharipov is the back-up for Budarin.

Kenneth D. Bowersox, Mission Commander (Captain, USN)

NASA astronaut

Personal background: Born 14 November 1956 in Portsmouth, Virginia, but considers Bedford, Indiana, to be his hometown.

Education: Graduated from Bedford High School, Bedford, Indiana, in 1974; received a bachelor of science degree in aerospace engineering from the United States Naval Academy in 1978, and a master of science degree in mechanical engineering from Columbia University in 1979.

Experience: Bowersox received his commission in the United States Navy in 1978 and was designated a Naval Aviator in 1981. He was then assigned to Attack Squadron 22, aboard the USS *Enterprise*, where he served as a Fleet A-7E pilot, logging over 300 carrier arrested landings. Following graduation from the United States Air Force Test Pilot School at Edwards Air Force Base, California, in 1985, he moved to the Naval Weapon Center at China Lake, California, where he spent the next year and a half as a test pilot flying A-7E and F/A-18 aircraft until his selection to the astronaut programme in June 1987.

Bowersox completed a one-year training and evaluation programme in

August 1988. Since then, he has held a variety of assignments including: flight software testing in the Shuttle Avionics Integration Laboratory; Technical Assistant to the Director of Flight Crew Operations; Astronaut Office representative for Orbiter landing and rollout issues; Chief of the Astronaut Office Safety Branch; Chairman of the Spaceflight Safety Panel; during several Shuttle missions he served as a spacecraft communicator (CapCom) in the Houston Mission Control Center. He also served as back-up to the first International Space Station crew.

Bowersox has logged over 50 days in space during four Shuttle flights. STS-50 (25 June–9 July 1992) was the first flight of the United States Microgravity Laboratory and the first Extended Duration Orbiter flight. In a two-week mission, the crew aboard Columbia conducted a wide variety of materials processing and fluid physics experiments.

STS-61 (2–13 December 1993) serviced and repaired the Hubble Space Telescope. During the 11-day flight, the HST was captured and restored to full capacity through a record five spacewalks by four astronauts.

STS-73 (20 October–5 November 1995) was the second flight of the United States Microgravity Laboratory. The mission focused on numerous microgravity scientific experiments housed in the pressurised Spacelab module.

STS-82 (11–21 February 1997) was the second Hubble Space Telescope maintenance mission. The crew retrieved and secured the HST in Discovery's payload bay. In five spacewalks, two teams installed two new instruments and replaced insulation. HST was then boosted to a higher orbit and redeployed.

Bowersox also trained as a back-up crew member to Bill Shepherd on the Expedition One crew. He was originally assigned as commander of Expedition Three, but was replaced by Frank Culbertson.

Donald A. Thomas (PhD), Flight Engineer

NASA astronaut

Personal background: Born 6 May 1955 in Cleveland, Ohio. Married to the former Simone Lehmann of Göppingen, Germany. They have one son. He enjoys swimming, biking, camping, flying. His mother, Mrs Irene M. Thomas, resides in Bloomington, Indiana. Her parents, Margrit and Gerhard Lehmann, reside in Göppingen, Germany.

Education: Graduated from Cleveland Heights High School, Cleveland Heights, Ohio, in 1973; received a bachelor of science degree in physics from Case Western Reserve University in 1977, and a master of science degree and a doctorate in materials science from Cornell University in 1980 and 1982, respectively. His dissertation involved evaluating the effect of crystalline defects and sample purity on the superconducting properties of niobium.

Organisations: Tau Beta Pi; Association of Space Explorers.

Honours: Graduated with Honours from Case Western Reserve University in 1977. Received NASA Sustained Superior Performance Award, 1989. Also four NASA Group Achievement Awards, four NASA Space Flight Medals, two NASA Exceptional Service Medals, and the NASA Distinguished Service Medal.

Experience: After graduation from Cornell University in 1982, Thomas joined AT&T Bell Laboratories in Princeton, New Jersey, working as a Senior Member of the Technical Staff. His responsibilities there included the development of advanced materials and processes for high-density interconnections of semiconductor devices. He was also an adjunct professor in the Physics Department at Trenton State College in New Jersey. He holds two patents and has authored several technical papers. He left AT&T in 1987 to work for Lockheed Engineering and Sciences Company in Houston, Texas, where his responsibilities involved reviewing materials used in Space Shuttle payloads. In 1988 he joined NASA's Johnson Space Center as a materials engineer. His work involved lifetime projections of advanced composite materials for use on Space Station Freedom. He was also a Principal Investigator for the Microgravity Disturbances Experiment, a middeck crystal growth experiment which flew on STS-32 in January 1990. This experiment investigated the effects of Orbiter and crew-induced disturbances on the growth of crystals in space.

He is a private pilot with over 250 hours in single engine land aircraft and gliders, and over 800 hours flying as mission specialist in NASA T-38 jet aircraft.

Selected by NASA in January 1990, Thomas became an astronaut in July 1991. He has served in the Safety, Operations Development, and Payloads Branches of the Astronaut Office. He was CapCom (spacecraft communicator) for Shuttle missions 47, 52 and 53. From July 1999 to June 2000 he was Director of Operations for NASA at the Gagarin Cosmonaut Training Centre in Star City, Russia.

A veteran of four space flights, he has logged over 1,040 hours in space. He was a mission specialist on STS-65 (8–23 July 1994), the second International Microgravity Laboratory (IML-2). This set a new flight duration record for the Space Shuttle programme. During the 15-day flight the crew conducted more than 80 experiments focusing on materials and life sciences research in microgravity.

During STS-70 (13–22 July 1995), Thomas was responsible for the deployment of the sixth and final Tracking and Data Relay Satellite from the Space Shuttle.

The STS-83 Microgravity Science Laboratory (MSL-1) Spacelab mission (4–8 April 1997) was cut short because of problems with one of the Shuttle's three fuel cells. STS-94 (1–17 July 1997) was a re-flight of the MSL-1 mission, and focused on materials and combustion science research in microgravity.

Nikolai Mikhailovich Budarin, Flight Engineer
Russian cosmonaut

Personal background: Born 29 April 1953 in Kirya, Chuvashia (Russia). Married to Marina Lvovna Budarina (nee Sidorenko). Two sons, Dmitri and Vladislav. His hobbies include fishing, skiing, picking mushrooms. His father, Mikhail Romanovich Budarin, died in 1984. His mother, Alexandra Mikhailovna Budarina, died in 1986.

Education: Graduated from the S. Ordzhonikidze Moscow Aviation Institute in 1979 with a mechanical engineering diploma.

Special honours: Awarded the titles of Hero of Russia, and a Pilot-Cosmonaut of the Russian Federation.

Experience: Since 1976, Budarin has occupied the positions of engineer and leading engineer at RSC Energia. In February 1989 he was enrolled in the Energia cosmonaut detachment as a candidate test cosmonaut. From September 1989 to January 1991, he underwent a complete basic space training course at the Gagarin Cosmonaut Training Centre and passed a State examination to qualify as a test cosmonaut. From February 1991 to December 1993, he took an advanced training course for the Soyuz-TM transport vehicle and the Mir station.

From 27 June to 11 September 1995, Budarin was flight engineer of the 19th long-term Mir expedition. The crew was launched by the Space Shuttle and landed in Soyuz TM-21 in a flight that overlapped with that of NASA astronaut Norman Thagard. Budarin and Anatoli Solovyov participated in three EVAs lasting more than 14 hours.

From 28 January to 25 August 1998, he was the flight engineer on the 25th long-term expedition to Mir. Also on board were Talgat Musabayev and the final NASA–Mir astronaut, Andrew Thomas. Budarin participated in an internal spacewalk to attempt a repair to a leaking airlock, and then undertook five EVAs to replace an attitude control system. The total time spent outside in these spacewalks was 30 hours 32 minutes.

Digital photos of Yuri Malenchenko, Sergei Moschenko and Edward Lu.

EXPEDITION SEVEN

Expedition Seven is scheduled to begin in January 2003 and end in summer 2003. Alexander Poleshchuk was originally named as Flight Engineer, but he was replaced by Sergei Moschenko, the first representative of the Khrunichev Space Centre to visit the ISS. This crew was assigned as the back-up team for Expedition Five. Astronaut Paul Richards will serve as a back-up to Edward Lu. Sergei Krikalev and Sergei Volkov are the back-ups for Malenchenko and Moschenko.

Yuri Ivanovich Malenchenko, Mission Commander (Colonel, Russian Air Force)

Russian cosmonaut

Personal background: Born 22 December 1961 in Svetlovodsk, Kirovograd Region, Ukraine. He has one son, Dmitri. His mother, Nina Ivanovna, father, Ivan Karpovich, and brother, Sergei Ivanovich live in Svetlovodsk.

Education: Graduated from Svetlovodsk public schools. In 1983, he graduated from S.I. Gritsevets Kharkov Higher Military Aviation School with a pilot-engineer's diploma. In 1993, he graduated from the Zhukovsky Air Force Engineering Academy.

Honours: Awarded the Hero of the Russian Federation medal, and the National Hero of Kazakhstan medal.

Experience: After graduating from Military Aviation School, he served as pilot, senior pilot, and multi-ship flight lead. In 1987, he was assigned to the

Cosmonaut Training Centre. From December 1987 to June 1989 he underwent a course of general space training, then continued training as a member of a group of test cosmonauts. He was the commander of the back-up crew for Mir 15. From 1 July to 4 November 1994, he served as commander of Mir-16, which included a joint flight with European Space Agency astronaut Ulf Merbold and world endurance record holder Valeri Polyakov. During this flight he controlled the first manual docking of a Progress craft and participated in two lengthy spacewalks.

On 8–20 September 2000, he served on the crew of STS-106, a Shuttle mission to prepare the ISS for the arrival of the first permanent crew. Five astronauts and two cosmonauts delivered several tonnes of supplies and installed batteries, power converters, a toilet and a treadmill on the Space Station. Malenchenko and Edward Lu performed a 6 hour 14 minute spacewalk to connect power, data and communications cables to the newly arrived Zvezda Service Module and the station. After his second space flight, Malenchenko had logged over 137 days in space, including three EVAs totalling over 18 hours.

Sergei Moschenko, Flight Engineer

Russian cosmonaut

Personal background: Born 12 January 1954 on Krekshino state farm near Moscow.

Experience: Engineer with the Khrunichev Scientific Industrial Centre. Selected as a member of the civilian engineer cosmonaut group on 27 June 1997. In training since January 1998.

Edward Tsang Lu (PhD), Flight Engineer

NASA astronaut

Personal Background: Born 1 July 1963, in Springfield, Massachusetts. Considers Honolulu, Hawaii, and Webster, New York, to be his hometowns. Unmarried. He enjoys aerobatic flying, coaching wrestling, piano, tennis, surfing, skiing and travel. His parents, Charlie and Snowlily Lu, reside in Fremont, California.

Education: Graduated from R.L. Thomas High School, Webster, New York, in 1980. Bachelor of science degree in electrical engineering from Cornell University, 1984. Doctorate in applied physics from Stanford University, 1989.

Organisations: American Astronomical Society, Aircraft Owners and Pilots Association, Experimental Aircraft Association.

Special honours: Cornell University Presidential Scholar, Hughes Aircraft Company Masters Fellow.

Experience: Since obtaining his PhD, Lu has been researching in the fields of solar physics and astrophysics. He was a visiting scientist at the High Altitude Observatory in Boulder, Colorado, from 1989 until 1992, the final year holding a joint appointment with the Joint Institute for Laboratory Astrophysics at the University of Colorado. From 1992 until 1995, he was a postdoctoral fellow at the Institute for Astronomy in Honolulu, Hawaii. Lu has developed a number of new theories that help to understand the underlying physics of solar flares. He has published articles on many topics, including solar flares, cosmology, solar oscillations, statistical mechanics, and plasma physics. He holds a commercial pilot certificate with instrument and multi-engine ratings, and has over 1,000 hours of flying time.

He was selected by NASA in December 1994 and reported to the Johnson Space Center in March 1995. After completing a year of training and evaluation, he qualified for assignment as a mission specialist. Among technical assignments held since then Lu has worked in the astronaut office computer support branch, and has served as lead astronaut for Space Station training issues.

Lu has logged over 504 hours in space. He first flew as a mission specialist on STS-84 (15–24 May 1997), the sixth Shuttle mission to rendezvous and dock with the Russian Space Station Mir. He was then a mission specialist and payload commander on STS-106 (8–20 September 2000). During the 12-day mission, the crew delivered more than 3 tonnes (6,600 lb) of supplies and installed batteries, power converters, life-support and exercise equipment to the International Space Station. Lu and Yuri Malenchenko performed a 6 hour and 14 minute spacewalk in order to connect power, data and communications cables to the newly arrived Zvezda service module.

EXPEDITION EIGHT

Expedition Eight is scheduled to take place during the spring and summer of 2003. Leroy Chiao (PhD) is training as back-up commander for Foale, and John Phillips (PhD) will be McArthur's back-up. Chiao is a space veteran with three missions, STS-65, STS-72 and STS-92, and Phillips has flown on STS-100, which launched in April 2001. Back-up for Tokarev is cosmonaut Mikhail Kornienko.

Digital photos of Michael Foale, William McArthur and Valeri Tokarev.

C. Michael Foale (PhD), Mission Commander
NASA astronaut

Personal background: Born 6 January 1957 in Louth, England, but considers Cambridge, England, to be his hometown. Married to the former Rhonda R. Butler of Louisville, Kentucky. They have two children. He enjoys many outdoor activities, particularly wind surfing. Private flying, soaring, and project scuba diving have been his other major sporting interests. He also enjoys exploring theoretical physics and writing children's software on a personal computer. His parents, Colin and Mary Foale, reside in Cambridge, England. Her parents, Reed and Dorothy Butler, reside in Louisville, Kentucky.

Education: Graduated from Kings School, Canterbury, in 1975. He attended the University of Cambridge, Queens' College, receiving a bachelor of arts degree in physics, natural sciences tripos, with first-class honours, in 1978. While at Queens' College, he completed his doctorate in laboratory astrophysics at Cambridge University in 1982.

Experience: As a postgraduate at Cambridge University, Foale participated in the organisation and execution of scientific scuba diving projects. Pursuing a career in the US space programme, he moved to Houston, Texas, to work on Space Shuttle navigation problems at McDonnell Douglas Aircraft Corporation. In June 1983, he joined NASA's Johnson Space Center in the payload operations area of the Mission Operations Directorate. In his capacity as payload officer in the Mission Control Center, he was responsible for payload operations on Shuttle missions STS-51G, 51-I, 61-B and 61-C.

He was selected as an astronaut candidate in June 1987. Before his first flight he flew the Shuttle Avionics Integration Laboratory simulator to provide verification and testing of the Shuttle flight software, and later developed crew rescue and integrated operations for International Space Station Alpha. Foale has served as Deputy Chief of the Mission Development Branch in the Astronaut Office, and Head of the Astronaut Office Science Support Group. In preparation for a long-duration flight on the Russian Space Station Mir, Foale trained at the Yuri Gagarin Cosmonaut Training Centre, Star City, Russia. He also served as Chief of the Astronaut Office Expedition Corps.

A veteran of five space flights, Foale has logged over 168 days in space including three space walks totalling 18 hours and 49 minutes. STS-45 (24 March to 2 April 1992) was the first of the ATLAS series of missions to study the atmosphere and its interaction with the Sun. STS-56 (9–17 April 1993) carried ATLAS-2 and the Spartan retrievable satellite which made observations of the solar corona. STS-63 (2–11 February 1995) was the first rendezvous with the Russian Space Station Mir. During the flight he made a 4 hour 39 minute spacewalk, evaluating the effects of extremely cold conditions on his spacesuit, as well as moving the Spartan satellite as part of a mass handling experiment.

Foale then spent 4½ months on board Mir. He launched with the crew of STS-84 on board Space Shuttle Atlantis on 15 May 1997 and joined the crew on Mir two days later. Foale spent the next 134 days conducting various science experiments and helping the crew resolve and repair numerous malfunctioning systems. On 6 September 1997 he and Commander Anatoli Solovyov conducted a 6-hour EVA to inspect damage to the Spektr module caused by the June 25 collision with a Progress resupply ship. Foale returned on 6 October 1997 with the crew of STS-86 on board Shuttle Atlantis.

His fifth flight, STS-103 (19–27 December 1999) was an 8-day mission to install new instruments and upgraded systems on the Hubble Space Telescope. During an 8 hour 10 minute EVA, Foale and Nicollier replaced the telescope's main computer and fine guidance sensor. He was subsequently appointed Assistant Director (Technical) at Johnson Space Center.

William Surles "Bill" McArthur, Jr, Flight Engineer (Colonel, US Army)
NASA astronaut

Personal background: Born 26 July 1951 in Laurinburg, North Carolina. His hometown is Wakulla, North Carolina. Married to the former Cynthia Kathryn Lovin of Red Springs, North Carolina. They have two daughters. He enjoys basketball, running, and working with personal computers. Bill's stepfather, Mr Weldon C. Avant, resides in Red Springs. His parents, Brigadier General William S. McArthur and Mrs Edith P. Avant, are deceased. Cynthia's mother, Mrs A.K. Lovin, resides in Red Springs, North Carolina.

Education: Graduated from Red Springs High School, Red Springs, North

Carolina, in 1969; received a bachelor of science degree in applied science and engineering from the United States Military Academy, West Point, New York, in 1973, and a master of science degree in aerospace engineering from the Georgia Institute of Technology in 1983.

Organisations: Member of the American Institute of Aeronautics and Astronautics, the Army Aviation Association of America, the Association of the United States Army, the United States Military Academy Association of Graduates, the West Point Society of Greater Houston, MENSA, Phi Kappa Phi, and the Association of Space Explorers.

Honours: Recipient of the Defense Superior Service Medal, the Defense Meritorious Service Medal, the Meritorious Service Medal (First Oak Leaf Cluster), the Army Commendation Medal, the NASA Space Flight Medal, and the NASA Exceptional Service Medal. Distinguished Graduate of the US Army Aviation School. Honorary Doctor of Science degree from the University of North Carolina at Pembroke. Recipient of the Order of the Long Leaf Pine, North Carolina's highest civilian award. Member of the Georgia Tech Academy of Distinguished Engineering Alumni. 1996 American Astronautical Society Flight Achievement Award. Recipient of the Ellis Island Medal of Honor.

Experience: McArthur graduated from West Point in June 1973 and was commissioned as a Second Lieutenant in the US Army. Following a tour with the 82nd Airborne Division at Fort Bragg, North Carolina, he entered the US Army Aviation School in 1975. He was the top graduate of his flight class and was designated an Army aviator in June 1976. He subsequently served as an aeroscout team leader and brigade aviation section commander with the 2nd Infantry Division in the Republic of Korea. In 1978 he was assigned to the 24th Combat Aviation Battalion in Savannah, Georgia, where he served as a company commander, platoon leader, and operations officer. After completing studies at Georgia Tech, he was assigned to the Department of Mechanics at West Point as an assistant professor. In June 1987, he graduated from the US Naval Test Pilot School and was designated an experimental test pilot. Other military schools completed include the Army Parachutist Course, the Jumpmaster Course, and the Command and General Staff Officers' Course. As a Master Army Aviator, he has logged over 4,000 flight hours in 37 different aircraft.

McArthur was assigned to NASA's Johnson Space Center in August 1987 as a Space Shuttle vehicle integration test engineer. Duties involved engineering liaison for launch and landing operations of the Space Shuttle. He was actively involved in the integrated test of the flight control system for each Orbiter for its return to flight and was a member of the Emergency Escape and Rescue Working Group.

Selected by NASA in January 1990, McArthur became an astronaut in July 1991. Since then, McArthur has held various assignments within the Astronaut Office particularly relating to the solid rocket booster and new solid rocket motors. Most recently, he served as Chief of the Astronaut Office Flight Support Branch, supervising astronaut support of the Mission Control Center, prelaunch Space Shuttle processing, and launch and landing operations. A veteran of three space flights, McArthur has logged 35 days 2 hours 25 minutes in space.

STS-58 (18 October–1 November 1993) involved various biomedical experiments on the crew and 48 rats, as well as engineering tests on board the Orbiter and 20 Extended Duration Orbiter Medical Project experiments.

STS-74 (12–20 November 1995) was the second Space Shuttle mission to rendezvous and dock with the Russian Space Station Mir. The crew successfully attached a permanent docking module to Mir and transferred one and a half tonnes of supplies.

During STS-92 (11–24 October 2000) the seven-member crew attached the Z1 Truss and Pressurised Mating Adapter 3 to the International Space Station using Discovery's robotic arm and performed four spacewalks to configure these elements. This prepared the station for its first resident crew. McArthur's two EVAs totalled 13 hours and 16 minutes.

Valeri Ivanovich Tokarev, Flight Engineer (Colonel, Russian Air Force)
Russian cosmonaut

Personal background: Born 29 October 1952 at the town of Kap-Yar, Astrakhan Region. Resides at Star City, Moscow Region. His wife, Irina Nikolaevna (Tokareva), was born 25 February 1955. They have two children: a daughter, Olya, and a son, Ivan. His mother, Lidiya Nikolaevna (Tokareva), lives in the city of Rostov, Yaroslavl Region. His father, Ivan Pavlovich, died in an auto accident in 1972. Valery Tokarev enjoys nature, automobiles, aeroplanes, and game sports.

Education: Master's degree in State Administration from the National Economy Academy affiliated with the Russian Federation Government in Moscow.

Experience: In 1973, Tokarev graduated from the Stavropol Higher Military School for Fighter Pilots. In 1982, he graduated from the Test Pilot Training Centre with honours. He later graduated from the Yuri Gagarin Air Force Academy in the town of Monino, Moscow Region, the National Economy Academy affiliated with the Russian Federation Government in Moscow. He is a first-class Air Force pilot, and a first-class test pilot. Tokarev has gained proficiency and flight experience with 44 types of aeroplanes and helicopters, and has participated in tests of fourth-generation carrier-based aircraft and vertical/short take-off and landing jets.

In 1987, Tokarev was selected to join the cosmonaut corps to fly the Buran space shuttle. Since 1994, he has served as commander of a group of cosmonauts specialising in aerospace systems and, since 1997, as a test cosmonaut for the Yuri Gagarin Cosmonaut Training Centre.

His first space flight was on STS-96 (27 May–6 June 1999). During the 10-day mission the crew delivered 4 tonnes of logistics and supplies to the International Space Station in preparation for the arrival of the Expedition One crew.

Selected Web Sites

National Contributions

NASA human spaceflight (ISS and Shuttle)
http://www.spaceflight.nasa.gov/

NASA's Human Exploration and Development of Space (HEDS) Enterprise
http://www.hq.nasa.gov/osf/

European Space Agency
http://www.estec.esa.nl/spaceflight/

Russian Aerospace Agency
http://www.rosaviakosmos.ru/

NASDA
http://www.nasda.go.jp/

Canadian Space Agency
http://www.space.gc.ca/

Italian Space Agency
http:/www.asi.it/

German Space Agency
http://www.raumstation.dlr.de/

French Space Agency (CNES)
http://www.cnes.fr/

Brazilian Space Agency (INPE)
http://www.agespacial.gov.br/

Marshall Space Flight Center
http://www.msfc.nasa.gov/news

Johnson Space Center
http:/www.jsc.nasa.gov/

Kennedy Space Center
http:/www.ksc.nasa.gov/

Shuttle press kits
http://www.shuttlepresskit.com/

Science on the ISS

US Science Experiments
http://scipoc.msfc.nasa.gov/

Scientific Research
http://spaceresearch.nasa.gov

US National Space Biomedical Research Institute
http://www.nsbri.org/

Space Station Experiments
http://scipoc.nsfc.nasa.gov/factchron.html

Japanese Experiment Module
http:/jem.tksc.nasda.go.jp/index_e.html

NASA Spinoffs
http:/www.thespaceplace.com/nasa/spinoffs.html

Life on the ISS

Living in Space
http://science.nasa.gov/

Astronaut and Cosmonaut Biographies
http://www.jsc.nasa.gov/bios/

European Astronaut Centre
http:/www.estec.esa.int/spaceflight/astronaut

Orbital debris
http://www.orbitaldebris.jsc.nasa.gov/

Industrial Contractors

Boeing
http:/www.boeing.com/defense-space/space

Lockheed Martin
http://lmms.external.lmco.com/

Astrium
http://www.astrium-space.com/

Alenia Spazio
http://www.alespazio.it/

Energia
http://www.Energia.ru/english/index.html

Commercialisation of Space

NASA Commercial Space Centers
http://spd.nasa.gov/csc.html

NASA Commercial Development of Space
http://commercial.nasa.gov/

MirCorp
http://www.mirstation.com/

ISS General

ISS crew code of conduct
http://www.spaceref.com/news/viewpr.html?pid=3418

Location and Visibility of the ISS

Marshall's "Liftoff to Space Exploration" web site
http://liftoff.msfc.nasa.gov/

Johnson Space Center's Skywatch web site
http://spaceflight.nasa.gov/realdata/sightings/

Heavens Above
http://www.heavens-above.com/

General Space Sites

Space.com
http://www.space.com/

Spaceref.com
http://www.spaceref.com/

CNN
http://cnn.com/TECH/space/

BBC
http://news.bbc.co.uk/hi/english/sci/tech/

Selected Reading List

Human Spaceflight – General

Bond, Peter, *Heroes in Space, from Gagarin to Challenger* (Blackwell, 1987)
Bond, Peter, *Reaching for the Stars* (Cassell, 1996)
Harvey, Brian, *Russia in Space: The Failed Frontier?* (Springer–Praxis, 2001)
Harvey, Brian, *The New Russian Space Programme* (Wiley–Praxis, 1996)
Shayler, David, *Disasters and Accidents in Manned Spaceflight* (Springer–Praxis, 2000)

Space Stations

Bizony, Piers, *Island in the Sky: Building the International Space Station* (Aurum, 1996)
Burrough, Bryan, *Dragonfly: NASA and the Crisis Aboard Mir* (Fourth Estate, 1998)
Foale, Colin, *Waystation to the Stars: The Story of Mir, Michael and Me* (Headline, 1999)
Harland, David, *The Mir Space Station* (Wiley–Praxis, 1997)
Linenger, Jerry, *Off the Planet* (McGraw-Hill, 2000)
Messerschmid, Ernst and Bertrand, Reinhold, *Space Stations: Systems and Utilisation* (Springer-Verlag 1999)
NASA, *International Space Station Familiarization Manual* (NASA–JSC, 1998)
Neri, R., *Manned Space Stations: Their Construction, Operation and Potential Applications* (ESA, 1990)
Portree, David, *Mir Hardware Heritage* (NASA–JSC, 1994)
Shayler, David, *Skylab* (Springer–Praxis, 2001)

Life and Science in Space

Bond, Peter, *Zero G: Life and Survival in Space* (Cassell, 1999)
ESA, *The International Space Station: European Users Guide* (ESA, 2001)
Freeman, Marsha, *Challenges of Human Space Exploration* (Springer–Praxis, 2000)
NASA, *Improving Life on Earth and in Space* (NASA, 1998)
NASA, *International Space Station User's Guide* (NASA, 2000)

Others

Aviation Week & Space Technology special issue (8 December, 1997)
Beardsley, Tim, *Science in the Sky* (*Scientific American*, June 1996)
CNES, *CNES Magazine* (quarterly magazine)
ESA, *On Station Newsletters* (December 1999 onwards)
ESA, *ESA Bulletin* (quarterly journal)
NASDA, *NASDA Report* (monthly newsletter)
Popular Science, special issue (May 1998)
Report of the Cost Assessment and Validation Task Force on the International Space Station (21 April 1998)
Report by the ISS Management and Cost Evaluation Task Force (November 2001)
Spaceflight (British Interplanetary Society monthly magazine)
Van Allen, J., *Space Science, Space Technology and the Space Station* (*Scientific American*, January 1986)

Index

2001 A Space Odyssey, 23, 27
Abbey, George, 15
Active Rack Isolation System, 276, 287, 308, 311, 316
AERCam Sprint satellite, 17
Afanasyev, Viktor, 57, 61, 95, 96, 219, 232, 313
Aldrin, Edwin ˆBuzz¤, 29
Alexandrov, Alexander , 49, 55
Almaz space station, 31, 32, 41, 43, 48
Alpha space station, 120, 121, 122, 123
Alpha Magnetic Spectrometer, 284, 336
Ames Research Center, 18
Anderson, Walt, 97
André-Deshays, Claudie, 76, 147, 219–20, 231–3, 268, 313, 314
Androgynous Peripheral Attach System (APAS), 158
Andromède mission, 219–20, 231–3, 287, 312
animals, 56, 90, 95, 306, 312,
Apollo Applications Programme, 36
Apollo–Soyuz Test Programme, 64, 65, 67, 104, 134
Ariane 5 rocket, 135, 143–5
Armstrong, Neil, 29
Artyukhin, Yuri, 41
Artsebarski, Anatoli, 58, 59
Ashby, Jeffrey, 199
ASI (*see* Italian Space Agency)
Astrium, 18, 19
astrophysics, 9, 46, 54, 56, 284–5, 293, 296
Atkov, Oleg, 49
Atlantic Monthly magazine, 21
ATV (*see* Automated Transfer Vehicle)
Automated Landing Flight Experiment (ALFLEX) 153
Automated Transfer Vehicle, 11, 135, 144, 150–2, 163, 256, 325
Avdeyev, Sergei, 59, 71–2, 93, 96, 97

Baikonur cosmodrome, 32, 43, 96, 125, 126, 129, 133–4, 141,166, 172, 176, 182, 221
Balandin, Alexander, 56, 57
Bantle, Jeffrey, 163
Barry, Dan, 173, 213, 214
Bartoe, John-David, 275
Baturin, Yuri, 92, 93, 202, 204, 298, 299
Beggs, James, 104, 105, 107
Bella, Ivan, 95, 96
Berezovoi, Anatoli, 48

biomedical research, 34, 65, 69, 87, 90, 218, 264
biotechnology, 205, 206, 275, 282, 287, 290, 293, 295, 297, 305, 310, 316
Blaha, John, 77–8
Bloomfield, Michael, 186
Boeing, 121, 275
Bonestell, Chesley, 27
Bowersox, Kenneth, 364–5
Boxer, Congresswoman Barbara, 114
Brazil, 3, 11, 100, 127–8, 289, 332
Brick Moon, 21, 22
Brinkley, Randy, 91–2
Bristol-Myers Squibb, 277, 302
Brown, Congressman George, 124–5, 127
Budarin, Nikolai, 68, 70, 88, 89, 176, 364, 367
Buhlman, Edelgard, 329
Buran shuttle, 56, 118
Burbank, Dan, 179
Bursch, Dan, 356, 357–9
Bush, President George, 65, 113, 115, 116
Bush, President George W., 320, 321
Bus 1 module, 127

Cabana, Robert, 168, 169, 170
Canada, 2, 8, 100, 105, 107, 108, 122, 125, 203, 301, 308, 312, 318, 329
Canadarm 2, 5, 6, 9, 138, 199, 200–2, 204, 206, 207, 219, 220, 289, 319, 339
Cape Canaveral, 37, 131–2, 173, 194
Carr, Gerald, 39–40
Carter, President Jimmy, 65, 104, 116
Carter, Layne, 256
Centrifuge, 4, 10, 14, 283, 288, 295, 321, 328, 332
Centrifuge Accommodation Module (*see* Centrifuge)
Chelomei, Vladimir, 31, 41
Chernomyrdin, Prime Minister Viktor, 66, 119
Chiao, Leroy, 181
China, 334–6
Chrétien, Jean-Loup, 48, 55
Clifford, Michael, 74
Clinton, President William, 66, 116, 117, 121, 122, 319, 320
CNES, 135, 219, 232
Cockrell, Kenneth, 191
Cold War, 66, 278, 281
Collier's magazine, 25, 26, 27
Columbus module, 4, 9, 108, 117, 122, 125, 288, 290, 291, 293, 321
commercial activities, 75, 297–302, 317
Congress, 14, 112, 113, 114, 115, 117, 119, 121, 126–7, 273, 320, 330
Conrad, Charles^Pete¤, 38
Cosmos 557, 36, 41
Cosmos 1267, 48
Cosmos 1443, 49
Cosmos 1669, 51
Cosmos 1686, 51, 52
Crew and Equipment Translation Aid, 319
Crew Return Vehicle, 14, 154–6, 320, 321, 324, 325, 327
Culbertson, Frank, 85, 90, 91, 212, 215, 216, 217, 218, 219, 220, 223, 225, 266, 352, 353–4
cupola, 10
Curbeam, Robert, 192
Currie, Nancy, 168
Curry, John, 183

Daimler-Chrysler Aerospace (*see* Astrium)
debris, 269–70
Destiny module, 4, 9, 157, 160, 163, 189, 190–4, 195, 196, 197, 198, 199, 200, 201, 205, 210, 212, 215, 216, 223, 240, 255, 261, 264, 268, 288, 289, 291, 306, 307, 311, 312, 318
Dezhurov, Vladimir, 69, 212, 213, 215, 218, 222, 224, 225, 312, 352, 355
Disney, Walt, 27
Dittemore, Ron, 181, 199, 212
Dobrovolsky, Georgi, 34

Docking Compartment 1 (*see* Pirs module)
docking, 38, 44, 60, 72, 157–60, 217
Dreamtime Holdings Inc., 220, 300, 311, 314
Dryden Flight Research Center, 155, 325
Duma, 94
Dumbacher, Dan, 161
Dunbar, Bonnie, 126
Dzhanibekov, Vladimir, 44, 50

E-Nose, 241–2
Earth observation, 10, 34, 39, 41, 43, 46, 68, 199, 207, 285, 291, 297, 304, 305, 309, 311, 314, 317
Edeen, Marybeth, 256
Eisenhower, President Dwight, 28
ELDO (European Launcher Development Organisation), 101
Elektron, 54, 81, 82, 184, 218, 219, 239
Energia, NPO/RKK/RSC, 90, 93, 95, 97, 98, 123, 148,150, 299, 322, 323, 324, 332
Energia rocket, 118, 176
Enhanced Gaseous Nitrogen Dewar, 287, 309, 316
Engstrom, Fredrick, 114
Enterprise module, 322, 323, 324
ergometer, 43, 68, 176, 198, 262
ESA (*see* European Space Agency)
ESRO (European Space Research Organisation), 101
Euromir missions, 61–2,
European Automated Transfer Vehicle 143
European Robotic Arm, 6, 8, 9, 296
European Space Agency, 4, 8, 19, 100, 103, 117, 119, 122, 124, 125, 135, 150, 155, 219, 285, 291, 293, 301, 322, 324, 329, 332, 336
Ewald, Reinhold, 80, 81
exercise, 69, 262–4
Expedition One, 142, 162, 182–5, 195, 196, 197, 234, 236, 238, 239, 296, 304–6, 339, 343–7
Expedition Two, 194–212, 233, 240, 259, 267, 302, 306–10, 317, 339, 347–52
Expedition Three, 85, 212–25, 234, 257, 266, 268, 310–14, 339, 352–6
Expedition Four, 165, 222, 223, 225–6, 302, 314–17, 339, 356–60
Expedition Five, 318, 360–4
Expedition Six, 318, 364–8
Expedition Seven, 368–70
Expedition Eight, 370–5
Explorer 1 satellite, 28, 29, 131
Express pallet, 11, 289,
EXPRESS rack, 287, 288, 308, 310–11, 316
extravehicular activity (*see* spacewalks)
Eyharts, Leopold, 88, 89

Faris, Mohammed, 55
Ferring, Mark, 201
Feustel-Buechl, Jorg, 325, 329, 332
FGB (*see* Zarya module)
FGB-2 module, 323, 324
fire, 35, 80–1, 283
Flade, Klaus-Dietrich, 59, 60
Flight control, 162–3
flight directors 196
Foale, Michael, 82–7, 230, 371–2
food, 39, 40, 69, 204, 250–4, 298, 299
Forrester, Patrick, 212, 213, 214
France, 93, 107, 135
Freedom space station, 66, 108, 112–13, 114, 115, 116, 117, 118, 119, 121, 122, 125
Fregat rocket stage, 19
French Space Agency (*see* CNES)
Frutkin, Arnold, 103

Gagarin, Yuri, 28–9, 30, 31, 133
Garneau, Marc, 186
Gemini spacecraft, 36, 37, 193
Germany, 3, 18, 24–5, 45, 107, 305, 312
Gernhardt, Michael, 209, 211
Gibbons, John, 123
Gibson, Robert, 68, 69

Gidzenko, Yuri, 71, 160, 166, 183, 184, 189, 194, 195, 230, 257, 268, 271, 305, 345
Glenn, John, 227
Glenn Research Center, 287
Goddard, Robert, 23
Goddard Spaceflight Center, 110
Godwin, Linda, 74, 223, 224
Goldin, Daniel, 65, 97, 115–16, 117, 121, 124, 125, 126, 190, 198, 203, 319, 321, 326, 329
Goodman, Jerry, 261
Gorbachev, President Mikhail, 58, 59, 60, 65
Gore, Vice President Al, 66, 117, 119
Gorie, Dominic, 223
Graf, Jochen, 302
gravity well, 274
Grechko, Georgi, 43, 44, 46
Griner, Carolyn, 278
ground control centres, 99, 123, 162, 172, 200, 221, 224, 234, 286, 312
Grumman, 121
Grunsfeld, John, 79
Gubarev, Alexei, 43, 46
Guidoni, Umberto, 199, 202

H-II Orbiting Plane - Experimental (*see* HOPE-X spaceplane)
H-II rocket, 126
H-IIA rocket 136, 145–6, 152
H-II Transfer Vehicle, 11, 136, 146, 152–3, 161, 256
Habitation module, 14, 15, 114, 241, 320, 321, 327
Hadfield, Chris, 72, 199, 200, 202
Haigneré, Claudie (*see* André-Deshays, Claudie)
Haigneré, Jean-Pierre, 95, 96, 231
Hale, Edward Everett, 21, 22
Hale, Wayne, 224
Hawes, Michael, 320, 321,193
Helms, Susan, 174, 176, 194, 195, 196, 197, 200, 201, 202, 203, 206, 209, 210, 214, 261, 348, 350–2
Hermes shuttle, 108, 125, 281
High Manoeuvrable Experimental Space Vehicle (*see* HIMES)
HIMES, 153
Hobaugh, Charles, 209
Hodge, John, 105
Holloway, Tommy, 176
HOPE-X spaceplane 153–4
Horowitz, Scott, 253
Houston, 11, 36, 106, 109, 162, 202, 214, 229, 230, 232, 233, 236, 246
HTV (*see* H-II Transfer Vehicle)
Human Research Facility, 196, 198, 205, 262, 289–90, 306, 316
Hungary, 47, 308
Huntsville, 11, 25, 28, 29, 109, 303
HYFLEX, 153
hygiene, 254
Hypersonic Flight Experiment (*see* HYFLEX)

Igla rendezvous system, 53
IMAX, 73, 168, 175, 199, 300
Inflatable Re-entry and Descent Technology, 19–20
Inspektor satellite, 18, 88
Interim Control Module, 127, 319
International Geophysical Year, 28
International Space Station, 88, 90, 92, 166
  Air, 2, 4, 209, 211, 238–41, 242, 258
  Assembly, 340–2
Attitude control, 3, 6
Communications, 6, 13, 195, 198, 199–200, 201, 296, 308
  Computers, 4, 9, 200, 201–2, 205, 210, 212
  Cost, 275, 278, 279, 301–2, 319, 320–1, 325, 326, 327, 329, 331, 332
  Orbit, 2, 123, 225
  Partners
  Power supply, 6–8, 200, 215
  Re-entry, 336–7
  Robotic arm (*see* Canadarm 2)
  Size, 1, 2, 227, 318

Temperature, 270
TV, 196, 198, 203, 208, 213, 216, 219, 224, 300, 322
International Standard Payload Rack, 288, 291, 292
Iskra satellite, 48
ISS Management and Cost Evaluation Task Force, 326–8
Italy, 3, 107, 312, 324
Italian Space Agency, 321
Ivanchenkov, Alexander, 45
Ivins, Marsha, 192

Japan, 2, 100, 105, 107, 108, 117, 119, 122, 124, 126, 135–6, 146, 180–2, 198, 301, 304, 312
Japanese Experiment Module, 4, 6, 9, 288, 290, 294–5, 321, 332
Japanese Space Agency (*see* NASDA)
Jernigan, Tamara, 160, 173
Jett, Brent, 186
Johnson Space Center, 15, 36, 109, 110, 121, 162, 224, 229, 232, 264, 275, 315, 320
Joint Airlock (*see* Quest airlock)
Jones, Thomas, 192, 194
Jupiter-C rocket, 25

Kaleri, Alexander, 59, 76, 78, 79, 80, 81, 97, 98
Kavandi, Janet, 209
Kazakhstan, 32, 59, 60, 96, 126, 133–4, 204
KC-135 aircraft, 18, 229
Kelly, Mark, 223, 225
Kennedy, President John, 29, 30, 132
Kennedy Space Center, 131, 132, 139, 161
Kerwin, Joe, 38
Khrunichev enterprise, 140, 161, 165, 172, 174, 177, 323
Kibo (*see* Japanese Experiment Module)
Kiriyenko, Prime Minister, 92
Kiselov, Anatoli, 161, 177
Kizim, Leonid, 49, 51, 52
Kohoutek, comet, 40
Kohrs, Richard, 114, 117
Komarov, Vladimir 146
Kondakova, Elena, 61, 62
Koptev, Yuri, 65, 80, 88, 92, 94, 95, 116, 117, 124, 166, 296
Korolev (*see* ground control centres)
Korolev, Sergei, 30, 31, 32, 141
Korzun, Valeri, 76, 78, 79, 80, 81, 86, 360–1
Kourou 144
Kovalyonok, Vladimir, 45
Kozeev, Konstantin, 219, 232
Kraft, Chris, 163
Krasnov, Alexei, 298
Krikalev, Sergei, 55, 58, 59, 60, 61, 67, 165, 167, 170, 183, 184,189, 194, 230, 260, 271, 345–7
Kristall module, 56, 57, 60, 61, 66, 68, 72, 73, 77, 86, 166
Krueger, Senator Bob, 116
Kubasov, Valeri, 34,
Kurs rendezvous system, 58, 78, 148–9, 160, 190, 205, 220
Kvant 1 module, 53, 54, 56, 58, 79, 80, 82
Kvant 2 module, 56, 57, 58, 68, 70, 86, 88, 89, 166

Laika, 28
Lampson, Congressman Nick, 328–9
Larson, Kirsten, 320
Laveikin, Alexander, 53
Lawrence, Wendy, 87
Lazutkin, Alexander, 80, 82, 84, 85
Lebedev, Valentin, 47, 48–9
Lego, 298, 299
leisure, 52, 265–7
Leonardo (*see* Multi-Purpose Logistics Module)
Lewis Research Center, 110
life sciences, 294, 308, 311, 315, 316–17
life support systems, 254–6
Lindsey, Steve, 17, 208, 209, 211
Linenger, Jerry, 78–82, 243
Llewellyn, Peter, 95

Lonchakov, Yuri, 199, 266
Lopez-Alegria, Michael, 181, 237
Lounge, John, 322
Low, George, 64, 103
Lu, Edward, 179, 368, 369–70
Luch satellite, 13
Lucian, 21
Lucid, Shannon, 74–8
Lyakhov, Vladimir, 45, 49

Makarov, Oleg, 44
Malenchenko, Yuri, 61, 178, 179, 368–9
Manakov, Gennadi, 57, 78
Manarov, Musa, 55, 58
Manber, Jeffrey, 123
Manned Orbiting Laboratory, 36, 37
Mark, Hans, 104
Mars, 15, 62, 274, 281, 283
Marshall Space Flight Center, 29, 36, 109, 110, 161, 168, 205, 206, 229, 255, 258, 275, 278,
303, 306, 310, 315
Martin, David, 264
Massachusetts Institute of Technology, 284, 304–5
materials processing, 40, 46, 56, 283–4, 290–1, 292, 294–5, 311, 316
McArthur, William, 181, 371, 372–4
McDonnell Douglas, 101, 116, 121
medical care,
Melroy, Pamela, 181
Merbold, Ulf, 61–2, 103
Mercury programme, 28, 29, 163
meteors, 15, 60, 61, 71, 95, 269, 312
microbes, 257–9, 284, 293
Micros satellite, 18
Mini Station 1, 332–4
Mir space station, 6, 42, 51, 52–63, 65–99, 117, 118, 164,166, 174, 204, 205, 227, 231, 243,
259, 260, 271, 275, 281, 321, 337
Mir 2 space station, 8, 118–19
MirCorp, 97, 98, 332–4
Mishin, Vasili, 32, 35
Mobile Base System, 6, 318
Mobile Servicing System, 5, 8, 17
Mobile Transporter
Mohmand, Abdul, 55
Montreal, 12
Moon, 21, 27, 29, 30, 31, 274, 283
Morukov, Boris, 179
Moschenko, Sergei, 369
Moscow, 12, 126, 184
MPLM (*see* Multi-Purpose Logistics Module)
Multi-Purpose Logistics Module, 11,139, 194, 196, 199, 200, 202, 212, 214, 223, 225, 250,
256, 257, 314, 339
Mulville, Daniel, 330
Muratore, John, 325
Musabayev, Talgat, 61, 88, 89, 91, 202, 204, 205, 298, 299

N-1 rocket, 31
NACA, 28
NASA, 28, 70, 94, 100, 101, 102, 107, 108, 109, 117, 126, 202, 259, 300, 301, 320, 321, 324,
325, 326, 328, 329–31
NASDA (Japanese Space Agency), 219, 307, 312, 317
National Advisory Committee for Aeronautics (*see* NACA)
National Aeronautics and Space Administration (*see* NASA)
Naval Research Laboratory, 319
Newman, James, 168, 171
Newton, Sir Isaac, 21, 22
New York, 216, 222
Nikolayev, Andrian, 32
Nixon, President Richard, 36, 100, 101, 104
Node 1 (*see* Unity module)
Node 2, 4, 9, 10, 321
Node 3, 14, 241, 256, 327
noise, 259–61
Noordung, Herman (*see* Potocnik, Herman)
Noriega, Carlos, 186, 188

OMS (*see* orbital manoeuvring system)
Orbital manoeuvring system 138–9
Oberpfaffenhofen, 12
Oberth, Hermann, 23
Ochoa, Ellen, 173
O'Connor, Bryan, 119–20
OKB-1, 30, 31, 32, 41
OKB-52, 31, 32, 48
O'Keefe, Sean, 14, 328
Onufrienko, Yuri, 74, 76, 223, 224, 225, 226, 356, 357
Orbital Re-entry Experiment (*see* OREX)
OREX, 153
Ostroumov, Boris, 174

P6 Solar Array Assembly, 188
Padalka, Gennadi, 93, 95, 176
Paine, Tom, 100
Parazynski, Scott, 199, 200
Patsayev, Viktor, 34–5
Payette, Julie, 173
Payload Operations Center, 206, 288, 302–4, 312, 315
Personal satellite assistants, 17–18
Physics of Colloids in Space experiment, 309
Pickering, James, 29
Pirs module, 3, 6, 8, 148, 157–8, 206, 217–18, 219, 220, 246, 324, 339
Pizza Hut, 176, 298
plants, 34, 207, 308, 309, 315
Pogue, William, 39
Poleshchuk, Alexander,
Poliakov, Valeri, 55, 61, 62–3
Popovich, Pavel, 41
Popular Mechanics magazine, 298, 299
Porter, Ron, 275, 287
Potocnik, Herman, 23, 27
power supply, 84, 110
pressure suits, 56, 77, 79, 87, 208, 243–4, 245
Pressurised Mating Adapter, 169, 181, 190, 195–6, 199
Primakov, Prime Minister, 95
Priroda module, 66, 67, 74–5, 76, 78, 85, 86, 87, 96, 180
Progress spacecraft, 43, 44, 45, 47, 49, 50, 51
Progress M spacecraft, 57, 58, 59, 60, 61, 74, 75, 76, 78, 79, 81, 82, 83, 84, 87, 89, 90, 93, 94, 95, 96, 98, 148–150, 165, 189, 191, 198, 199, 215, 221, 339
Progress M1 spacecraft, 11, 97, 98, 99, 143, 206, 207, 211, 221, 222, 305, 308, 318, 319, 336, 339
Progress M2 spacecraft, 118
Pronin, Nikolai, 97
protein crystals, 87, 205, 212, 281–2, 290, 291, 293, 295, 304, 308, 309, 310, 311, 314, 315, 316
Proton rocket, 3, 31, 35, 129, 133, 139–40, 174, 339
psychology, 70, 207, 211, 286

Quest airlock, 5, 13, 18, 207, 208, 209, 210, 211, 222, 237, 339

R-7 rocket, 133, 141
radiation, 70, 205, 270–2, 306–8, 312, 316
RadioShack, 203, 298, 299
Raduga satellite, 165
Raffaello (*see* Multi-Purpose Logistics Module)
Reagan, President Ronald, 104, 106, 107, 108, 112, 182, 273
Rees, Sir Martin, 280–1
Reilly, James, 209, 210, 211, 247
Reiter, Thomas, 71–2, 74
Remek, Vladimir, 45, 46
Richards, Paul, 196
robots, 9, 15–18, 274, 279
rockets, 2, 22, 23–5, 26, 28, 31, 129–30, 134, 225
Rockwell International, 101, 121
Roemer, Congressman Tim, 121, 127, 321
Rohrabacher, Congressman Dana, 320, 336

Roman, Monsi, 258
Romanenko, Yuri, 44, 46, 53
Rominger, Kent, 159
Ross, Jerry, 168, 171
Rothenberg, Joseph, 321
Russia, 2, 3, 8, 14, 19, 59, 64, 100, 103, 122, 126, 133, 219, 259, 296–7, 301, 302, 304, 305, 317
Russian Aviation and Space Agency, 65, 74, 78, 92, 117, 202, 219, 231, 298, 322, 323, 329, 332
Ryan, Amy, 241–2
Ryumin, Valeri, 45, 47, 61, 90, 91
SAFER manoeuvring unit, 74, 247, 248
Salyut 1 space station, 32–5, 42, 146
Salyut 2 space station, 35–6, 41
Salyut 3 space station, 41, 42
Salyut 4 space station, 41–3
Salyut 5 space station, 43
Salyut 6 space station, 42, 44–8
Salyut 7 space station, 48–52
Salyut Design Bureau 140
Saturn V rocket, 36, 37, 39, 101, 129, 132
Savinykh, Viktor, 50, 51
Savitskaya, Svetlana, 48
Schwartzenberg, Roger-Gérard, 298
Science Power Platform, 8, 124, 296, 323, 324
scientific experiments, 67, 86, 87, 205, 207, 213, 215, 218, 220, 274, 276–8, 279–80, 281–97, 299, 300, 304–17, 331–2
Semenov, Yuri, 93, 94, 95, 117, 176, 178
Sensenbrenner, Congressman James, 125, 127
Serebrov, Alexander, 55, 56, 60, 92
Service Module (*see* Zvezda module)
Sevestyanov, Vitali, 32, 96
Sharman, Helen, 58
Shenzhou spacecraft, 334–5
Shepard, Alan, 39
Shepherd, William, 74, 183, 184, 186, 190, 194, 195, 196, 230, 238, 267, 271, 343–5
Shuttle-Mir, 61–3, 65–92, 230
Shuttleworth, Mark, 297
Siber spacecraft,
SkyCorp, 300
Skylab space station, 35, 36–41, 45, 63, 100, 129, 257, 303
sleep, 41, 52, 69, 197, 268
solar arrays, 6–8, 38, 44, 49, 50, 55, 58, 68, 69, 70, 73, 75, 82, 84, 88, 94, 98, 111, 187, 221, 224
Solovyov, Anatoli, 56, 57, 59, 68, 70, 85, 88, 89, 165
Solovyov, Vladimir, 49, 51, 52, 85
Soyuz rocket, 61, 94, 129, 130, 141–3, 206, 339
Soyuz spacecraft, 31, 32, 35, 41, 65, 165, 178
Soyuz T spacecraft, 47
Soyuz TM spacecraft, 11, 52, 66, 81, 84, 85, 120, 134, 146–8, 230, 308, 318, 325, 328, 332, 333, 336, 339
Soyuz 9, 32
Soyuz 10, 33–4
Soyuz 11, 34–5, 41, 47, 146
Soyuz 18, 43
Soyuz 19, 43
Soyuz 24, 43
Soyuz 25, 44
Soyuz 26, 44
Soyuz 27, 44
Soyuz 32, 46, 47
Soyuz 33, 46
Soyuz 34, 46–7
Soyuz T-1, 47
Soyuz T-2, 47
Soyuz T-8, 49
Soyuz T-10, 49
Soyuz T-15, 51
Soyuz TM-3, 54, 55
Soyuz TM-5, 55
Soyuz TM-6, 55
Soyuz TM-9, 56
Soyuz TM-16, 57, 60
Soyuz TM-17, 61
Soyuz TM-22, 71
Soyuz TM-23, 74

Soyuz TM-25, 80
Soyuz TM-27, 88
Soyuz TM-29, 95
Soyuz TM-31, 194, 199
Soyuz TM-32, 219, 220
Soyuz TM-33, 12, 219, 231
Spacehab, 11, 74, 172, 322, 323, 339
Spacelab, 69, 101, 102, 103, 107, 139, 199, 202
Space Media, 299
Space Shuttle, 2, 3, 6, 11, 13, 17, 18–19, 40, 61, 64, 65–7, 70, 86, 100, 101, 102, 104, 105, 123, 130, 132, 136–9, 202, 231, 274, 287, 288, 303, 327, 328, 336
Atlantis, 68, 70, 71, 73, 74, 77, 78, 86, 136, 176, 179, 180, 190, 191, 192, 194, 206, 207, 211, 226, 318
Challenger, 19, 111,
Columbia, 18, 102, 136
Discovery, 6, 61, 62, 67, 90, 136, 159, 173, 182, 189, 195, 196, 197, 206, 212, 214, 242, 310
Endeavour, 88, 93, 136, 167, 169, 170, 186, 189, 199, 201, 203, 223, 224, 225, 236, 314, 318
space stations, 22, 23, 24, 26, 27, 31, 100, 101, 103, 104, 105, 106, 107, 108, 109, 110,
spacewalks, 2, 12, 13, 15, 18, 38, 39, 44, 47, 48, 49–50, 51, 53–4, 55, 57, 58, 59, 60, 68, 71, 72, 74, 75, 82, 85, 87, 88, 89, 96, 187, 192, 194, 195, 196,199, 200, 206, 208, 209, 210, 211, 213, 214, 218–19, 220, 222, 223, 243, 245–6, 247, 248–9, 318
Special Purpose Dexterous Manipulator, 9
Spektr module, 66, 67, 68, 69, 70, 71, 74, 75, 76, 78, 80, 83, 84, 85, 86, 87, 89, 98
Spring, Sherwood, 112
Sputnik, 28, 88, 94
SRB (*see* solid fuel rocket boosters)
Solid fuel rocket boosters 137, 146
Star City, 67, 126, 229, 230, 231, 232, 233
Starshine, 173
Steklov, Vladimir, 97
Strekalov, Gennadi, 57, 68
StelSys, 300
Stephenson, Art, 303
Strela crane, 6, 60, 71, 175, 219, 220
STS-61B, 112
STS-79, 77
STS-81, 79
STS-86, 247
STS-88, 339
STS-91, 284
STS-92, 157, 339
STS-96, 158, 339
STS-97, 239, 339
STS-98, 287, 339
STS-100, 199
STS-101, 339
STS-102, 339
STS-104, 207, 208
STS-105, 206, 212, 214, 253, 309, 310, 339
STS-106, 262, 305, 339
STS-108, 223, 224, 314, 339
STS-110, 315
STS-122, 324
Sturckow, Rick, 169

Tanegashima, 135–6, 146
Tani, Dan, 223, 224
Tanner, Joseph, 186, 188
Technological Experiment Facility
Telerobotically Operated Rendezvous System, 60, 61, 81, 82–3, 149, 157,160, 166, 169, 184, 189, 206
telescopes, 46, 56, 285, 293
Thagard, Norman, 62, 63, 64, 67–70, 77, 126
Thomas, Andrew, 88, 89, 90, 196
Thomas, Donald, 364, 365–7
Ting, Samuel, 284
Tito, Dennis, 98, 147, 202, 203, 204, 297
Titov, Vladimir, 55, 61, 67, 87

TKS (*see* Transport Logistics Spacecraft)
toilet, 39, 183, 197, 208, 256–7
Tokarev, Valeri, 173, 371, 374–5
TORU (*see* Telerobotically Operated Rendezvous System)
Toulouse, 12
Tracking and Data Relay Satellites, 324
training, 12, 79, 228–33
Transport Logistics Spacecraft, 43, 49
Tracking and Data Relay Satellite, 13, 265
Transhab module, 14–15
treadmill, 46, 83, 205, 207, 215, 224, 262, 263, 304
Treschev, Sergei, 360, 361–2
Truly, Richard, 113, 114
truss, 6, 181, 224, 289, 318, 319, 339
Tsibliev, Vasili, 60, 80, 81, 82, 85
Tsiolkovsky, Konstantin, 21–23
Tsukuba, 12, 163
Tyurin, Mikhail, 158, 212, 215, 220, 224, 312, 352, 355–6

Ukraine, 11, 126, 324
Unity module, 3, 94, 159, 162, 171, 172, 173, 174, 176, 181, 185, 186, 193, 195, 199, 200, 202, 207, 208, 223, 237, 242, 304, 339
Uri, John, 275, 310, 312, 315
US Air Force, 104, 317
US National Academy of Sciences, 331
US National Commission on Space, 274
US National Science Foundation, 278
US Propulsion Module, 127, 320, 336
Usachev, Yuri, 61, 74, 76, 174, 194, 195, 197, 203, 206, 209, 210, 212, 213, 214, 240, 246, 248, 255, 259, 263, 266, 267, 298, 299, 347–9
Ustinov, Dmitri, 43

V2 rocket, 24–5, 26, 131
Van Allen, James, 29, 279
Vanguard rocket, 28
Vasyutin, Vladimir, 51
Vest committee, 117
Viktorenko, Alexander, 55, 56, 59, 62
Vinogradov, Pavel, 85, 88
Volkov, Alexander, 51, 55, 59, 60
Volkov, Vladislav, 34–5
von Braun, Wernher, 24–8, 29, 36
Voss, James, 174, 175, 194, 195, 196, 198, 200, 203, 206, 209, 214, 246, 260, 298, 349–50
Vostok rocket, 31
Vozdukh air-purification system, 188, 192, 198, 240

WAC Corporal rocket stage, 131
Wakata, Koichi, 180, 181
Walz, Carl, 223, 356, 359–60
water, 69, 77, 216, 250, 254–6, 258
Weinberger, Caspar, 105
Weitz, Paul, 38
Weldon, Congressman Dave, 97
Wetherbee, James, 194
White, Ed, 193
White Sands, 26
Whitson, Peggy, 360, 362–3
Wilcutt, Terence, 77
Williams, Jeffrey, 175
Wisoff, Jeff, 181
Wolf, David, 87, 88
Woodard, Lybrease, 310

X-38 spaceplane, 155–6, 325
X-ray astronomy, 39, 54, 285, 286

Yeltsin, President Boris, 65, 116, 122, 126, 204

Zaletin, Sergei, 97, 98
Zarya space station (*see* Salyut 1)
Zarya module, 3, 6, 8, 94, 119, 124, 127, 133, 139, 141, 148, 151, 157, 159–160, 162, 165, 167, 168, 169, 171, 172,173, 177, 179, 182, 184, 189, 193, 194, 197, 206, 219, 232, 237, 242, 259, 266, 269, 322, 323, 324, 339
Zarya spacecraft, 118
Zholobov, Vitali, 43

Zvezda module, 3, 4, 8, 98, 127, 139–140, 148–9, 157, 160, 174, 177, 179, 183, 185, 191, 198, 206, 208, 215, 217, 218, 219, 220, 221, 222, 225, 232, 240, 246, 253, 254, 259, 260, 262, 263, 267, 268, 269, 270, 296, 297, 305, 311, 312, 323, 336, 339